MINITAB Manual

for use with

Elementary Statistics
A Step by Step Approach

Fifth Edition

Allan G. Bluman
Community College of Allegheny County

Prepared by
Gerry Moultine
Northwood University

Boston Burr Ridge, IL Dubuque, IA Madison, WI New York San Francisco St. Louis
Bangkok Bogotá Caracas Kuala Lumpur Lisbon London Madrid Mexico City
Milan Montreal New Delhi Santiago Seoul Singapore Sydney Taipei Toronto

The McGraw-Hill Companies

MINITAB Manual for use with
ELEMENTARY STATISTICS: A STEP BY STEP APPROACH, FIFTH EDITION
ALLAN G. BLUMAN

Published by McGraw-Hill Higher Education, an imprint of The McGraw-Hill Companies, Inc., 1221 Avenue of the Americas, New York, NY 10020.

This book is printed on recycled, acid-free paper containing 10% postconsumer waste.

2 3 4 5 6 7 8 9 0 QPD/QPD 0 9 8 7 6 5 4

ISBN 0-07-254908-4

www.mhhe.com

MINITAB Manual for Elementary Statistics

The goal of this manual is to provide students with instructions that parallel the sequence of topics in a typical elementary statistics course. Examples are from the textbook *Elementary Statistics: A Step by Step Approach*, *5th Edition* by Allan G. Bluman, published by McGraw-Hill (2004).

The instructions are given for MINITAB® Release 14 (October, 2003). The computer system used is Windows XP®. Some menus or dialog boxes will appear differently if a different operating system is used.

The MINITAB software is not included with this manual. Check with your McGraw-Hill representative for information about purchasing the text bundled with the Student Edition of MINITAB®. The new Student Edition for Windows is based on MINITAB Release 14.

This manual does not teach the subject of statistics. The objective is to learn how to use the software. It is assumed that the student has access to MINITAB® Release 14 in a school computer lab or the new student edition of MINITAB.

1--A demo version of MINITAB Release 14 can be downloaded from the Minitab website at (http://www.minitab.com). Use is permitted for a limited time.

2--Students may rent the use of MINITAB software. See www.e-academy.com/minitab

There are no problem sets in this manual. The student should use this manual along with the textbook and the textbook exercises. The software can and should be used to help with many of the textbook exercises. At the end of each chapter there will be a list of suggested exercises in the textbook and a place for notes. Caution! The data sets used in this text have been chosen to aid in the instruction of statistics. The user should make no assumption regarding the validity of the results.

File Locations

Data files that are installed with the MINITAB program will be in the default directory C:\Program Files\MINITAB 14. There are several subdirectories. One of these directories is Studnt12 and another is Studnt14. Verify the location on your system then write the path for these files in the space below.

Studnt12 = ______________________________

Studnt14 = ______________________________

You will be asked to save files. In the manual a temporary folder on the hard drive will be used. Instructions will show this directory as Projects. Some files may be large if they contain graphic images.

They may not fit on a floppy disk. Ask your instructor where you should save your files then write the path for these files in the space below.

Projects = ______________________________

Ask your instructor about the location of data files used in exercises, examples, etc. in the textbook, *Elementary Statistics: A Step by Step Approach 5/e*. The files may be located on a network server or some directory on the hard drive. They are included on the CD-ROM that came with your new textbook.

The MTW extension indicates the file contains data in a MINITAB worksheet. If the extension is MTP, the file is in a portable format that can be opened by release 13 and earlier versions of MINITAB. Instructions show this directory as **BlumanMTP**. Write the path for the textbook data files in the space below:

BlumanMTP = ______________________________

File Names for Textbook Data

The files are named according to the following convention. The initial letter denotes the type of problem.

E = Example	MT = MINITAB Examples
CT = Critical Thinking	EC = Extending the Concepts
I = Introduction	RP = Review Exercises
P = Exercise	TI83 = TI-83 Calculator Exercises
Q = Quiz	XL = Excel Examples

The remainder of the file name denotes chapter, section and problem number. For example, P-C02-S04-18.MTW refers to Chapter 2 Section 4 Exercise 18. Filename E-C02-S04-05.MTW refers to the data for Chapter 2 Section 4 Example 5.
Release 14 will not read projects and worksheets created in Release 14 or earlier. However, it will read worksheets in portable format.

Completing an Assignment

When completing an assignment, you will want to *print* what you need to print and *save* files that you may need later. Finally, you will either *quit* the program or begin a new project. Your instructor will tell you which of these options to do for class assignments.

Print

1. Print the Session Window.
   Be sure to type your name and identification info at the top.
2. Print Graph windows.
3. Print Worksheet(s).
   Use the worksheet description to identify the file and describe the data.
4. Create and Print a report.

Save

5. Save the Worksheet(s).
6. Save the Graph Windows.

7. Create and Save a report.
8. Save the Project. Use project description to describe the assignment and your identity.

Quit

9. Exit MINITAB or start a new project.

Help menus, StatGuide and Tutorials are all provided with the installation of MINITAB. Once you become comfortable with the program, use these resources to "Help" yourself learn about additional commands that may not be included in this elementary introduction.

Minitab Inc.
Quality Plaza
1829 Pine Hall Road
State College, Pa 16801-3008 USA
Tel: 814.238.3280
Fax: 814.238.4383
E-mail: info@minitab.com

ACKNOWLEDGEMENTS

My thanks go to all of the individuals who have helped with the preparation of this manuscript.

Thank you, Randy Welch and others at McGraw-Hill, for your persistent attention to detail and your patience. Thanks to Larry and the rest of my family for indulging me as I explored the new release of MINITAB and created these instructions. Thank you, Cyrilla and Annie, for your suggestions and corrections. Thank you, Minitab, Inc. for generous permission to include MINITAB output and for the great software you have created. Release 14 is a significant upgrade!

And last but not least, thank-you readers for validating all of our hard work when you bought this manual. I hope you are pleased with the result.

G. M.

Table of CONTENTS

Chapter 1 Introduction to MINITAB

In this manual, do the instructions in numbered steps. Other information or explanations are not numbered. Read and make note of the information.

1-1 *Start MINITAB*

Step 1. Use the program list in the start menu of Windows XP.

a. Click on **Start>All Programs>MINITAB14.**

b. Double click the Program icon in the list.

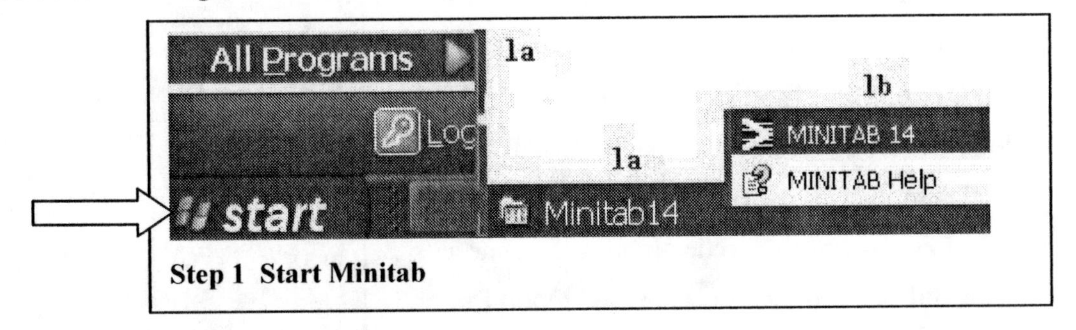

Step 1 Start Minitab

After a few seconds, the program will open. There are three windows named Session, Worksheet 1*** and Project Manager.

Step 2. Click the Project Manager icon to bring it to the front.

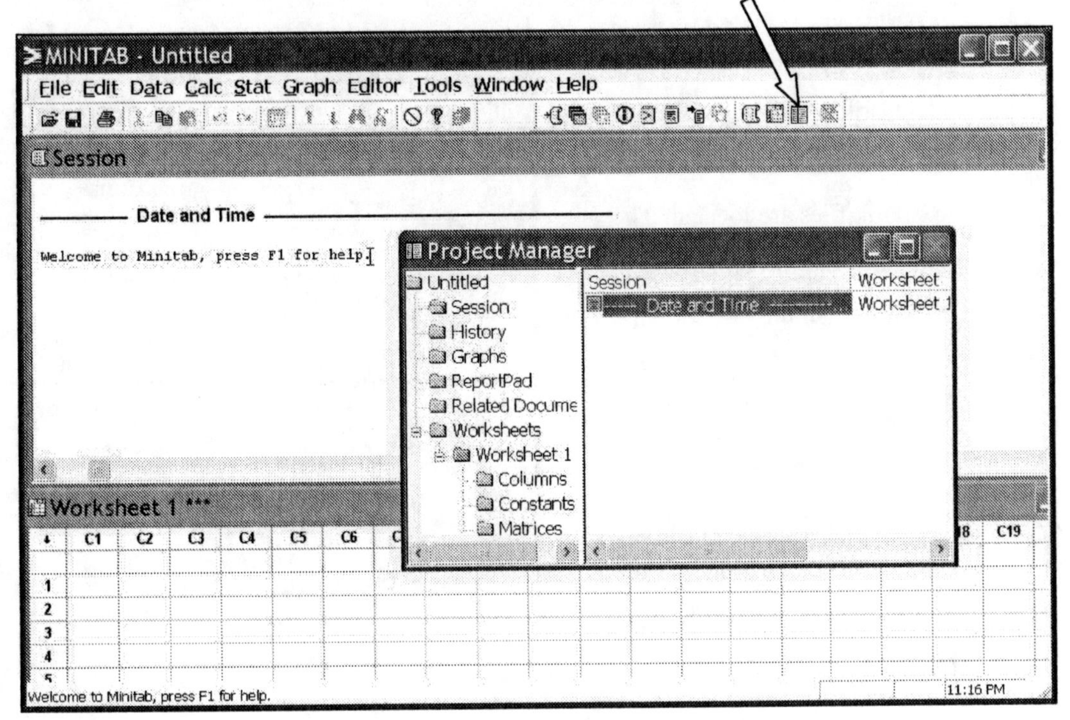

Open the Databank File

The data is shown in Appendix D of the textbook. Here is how to copy the data from a file into a worksheet.

Step 3. Click **File>Open Worksheet**.

The Open Worksheet dialog box will be displayed.

A sequence of menu instructions will be shown this way. Menus and menu selections that you will be using in the software dialog boxes are bold-faced and in set in a different typeface. For example,

File>Open Worksheet indicates you should select the File menu then select Open Worksheet. Make the selection by clicking on them with the mouse.

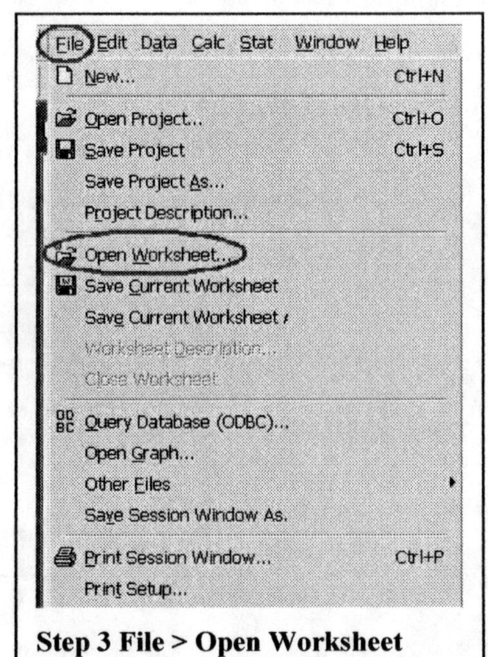

Step 3 File > Open Worksheet

NOTE: This menu selection is NOT the same as clicking the File icon. If the dialog box says Open Project instead of Open Worksheet, click **[Cancel]** then start again selecting from the menu.

[ ] surround a button in a dialog box, **[Cancel]**

< > surround a key on the keyboard, <Tab>

You must check three items in this dialog box.

a. The Look in: dialog box should show the directory where the file textbook files are located. The name of your directory may be different than the one shown. Make sure the Files of type: shows the correct type. Files of data used in the textbook are saved in a portable format and the type should say Minitab Portable [*.mtp]. They are numbered by type, chapter number, section number, then the example or problem number. **E-C02-S04-12.MTP** is the MINITAB worksheet containing the data for Example 12 shown in the textbook in Chapter 2 Section 4 on page 74.

b. Double click the file name in the list box, Databank.MTP.

A dialog box *may* open to inform you that a copy of this file is about to be added to the project. Click on the check box if you do not want to see this warning again. The data will be copied into a worksheet. If a data file was already open, it will not be deleted or replaced. A second worksheet will open with the contents of the new file. Part of the worksheet is shown.

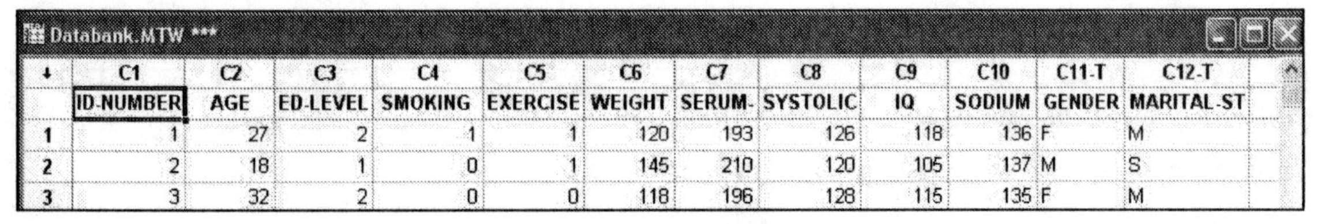

Databank.MTW ***

↓	C1	C2	C3	C4	C5	C6	C7	C8	C9	C10	C11-T	C12-T
	ID-NUMBER	AGE	ED-LEVEL	SMOKING	EXERCISE	WEIGHT	SERUM-	SYSTOLIC	IQ	SODIUM	GENDER	MARITAL-ST
1	1	27	2	1	1	120	193	126	118	136	F	M
2	2	18	1	0	1	145	210	120	105	137	M	S
3	3	32	2	0	0	118	196	128	115	135	F	M

There is a row for each person's data and a column for each variable. The first person is a 27-year-old married female who smokes less than a pack a day and has a college degree. C12-T Marital Status has a T appended to the label to indicate it contains text characters not numbers. This data is qualitative!

Summarize Qualitative Data

Step 4. To summarize qualitative data in a table select **Stat>Tables>Tally Individual Variables... .**

a. The triangle on the right side of the menu indicates another menu will be next.

b. The three dots, or ellipses, after the menu command indicate a dialog box will be next.

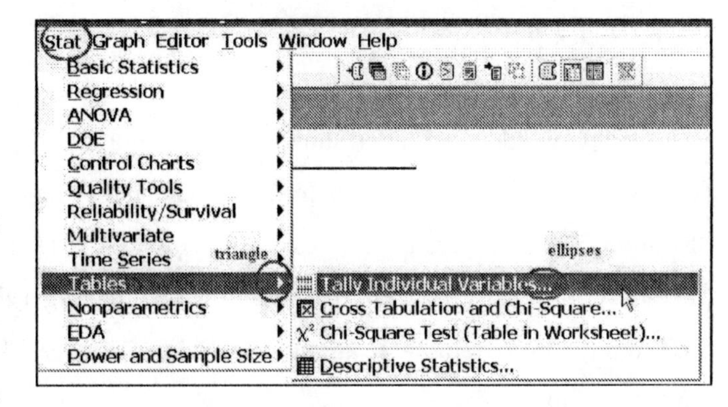

In the dialog box you must select one or more variables and indicate the statistics you would like to view.

c. The I-beam cursor should be blinking in the dialog box Variables:. You may need to click inside before the list of columns is visible.

Double click C12 MARITAL-STATUS in the list of variables.

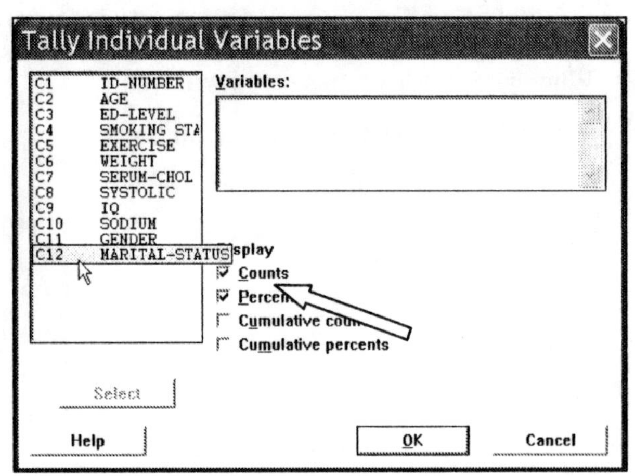

d. Check the boxes for Counts and Percents.

e. Click **[OK]**.

The table will be displayed in the Session window. Fifty-three out of the 100 in the sample are married. Sixteen percent are divorced.

Tally for Discrete Variables: MARITAL-STATUS

MARITAL-STATUS	Count	Percent
D	16	16.00
M	53	53.00
S	22	22.00
W	9	9.00
N=	100	

Summarize Quantitative Data

Step 5. Select **Stat>Basic Statistics>Graphical Summary…**

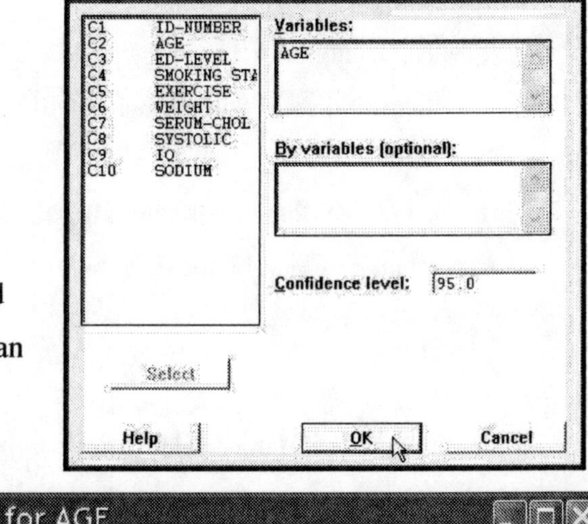

a. Double click **C2 Age** in the variable list to select it for analysis. Age is a continuous quantitative variable.

b. Click **[OK]**.

A graph window will open showing several graphs and a table with statistical measures such as the mean and standard deviation. The mean age is 38.4 years. The youngest is 18 and the oldest is 74 years old. Half of the sample were younger than 36. From the histogram, there are more individuals under 40 than above 60.

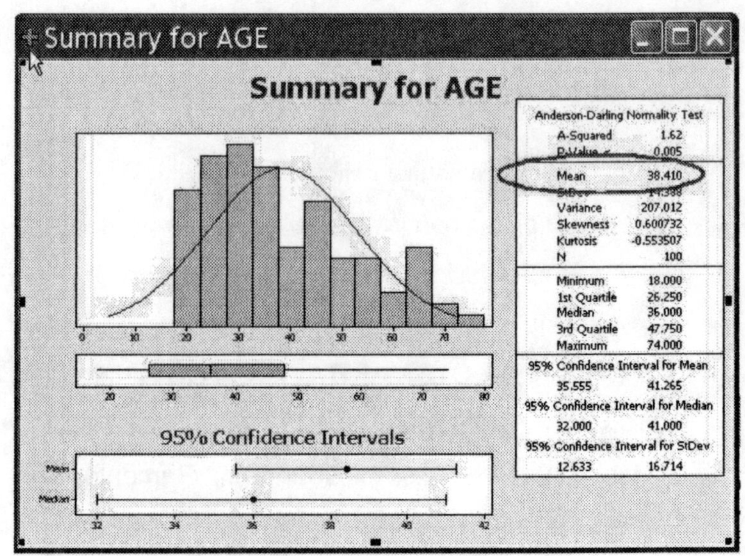

When you finish a project, you will want to print, save files and quit the program. Save the files you may need again. If you were not finished, you can save the work you have done so far in a project then open the file later to complete the task. Print windows or reports as directed by your instructor. The instructions here will show you how to print the graph and session windows then save the project.

Finish the Project

Print the Graph Window

Step 6. To print the graph, click the Project Manager icon on the toolbar.

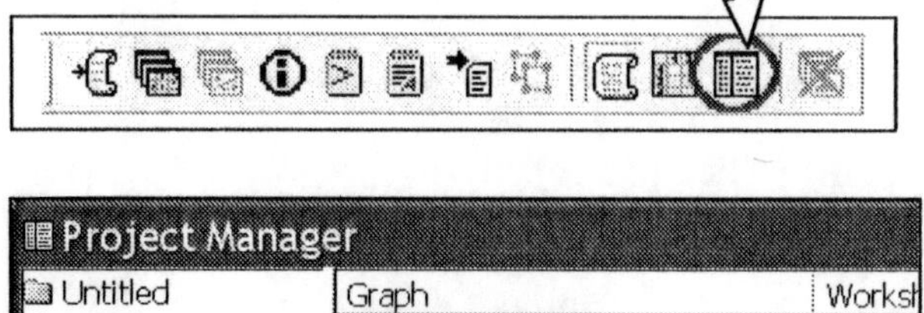

a. Click the Graphs folder.

b. Right click on Summary for AGE.

c. Click Print... .

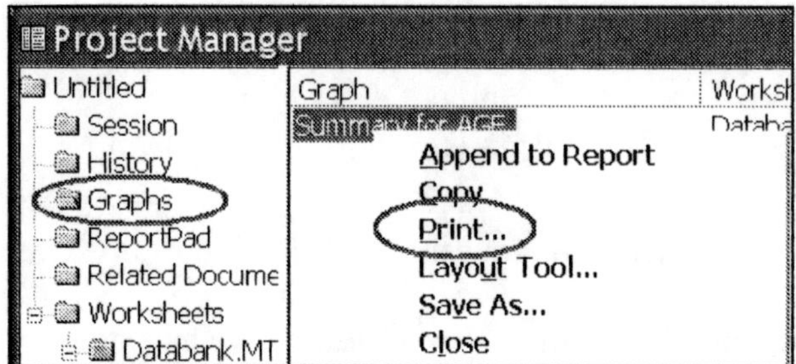

Print the Session Window

Step 7. To print the Session window, right click the Session folder in the Project Manager window, then click Print (not shown).

The two windows will print on separate pages.

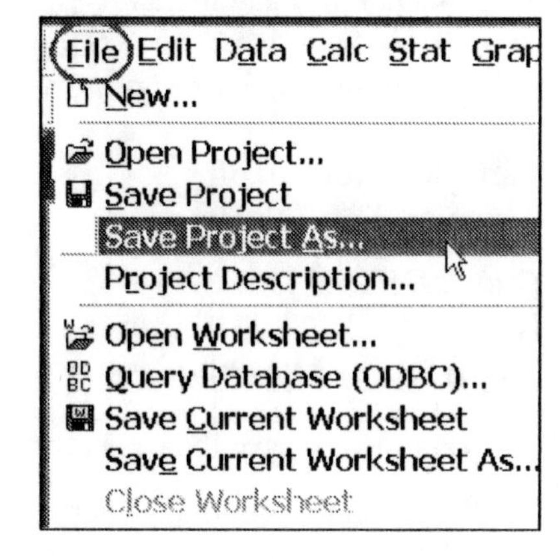

Save the Project

Step 8. Select **File >Save Project As**.

a. Navigate to your Projects folder.
   Use the drop-down arrow in the Save in: text box to view the directory tree.

b. Type in the name of the file, **Intro01.**

c. Click on **[Description]**.

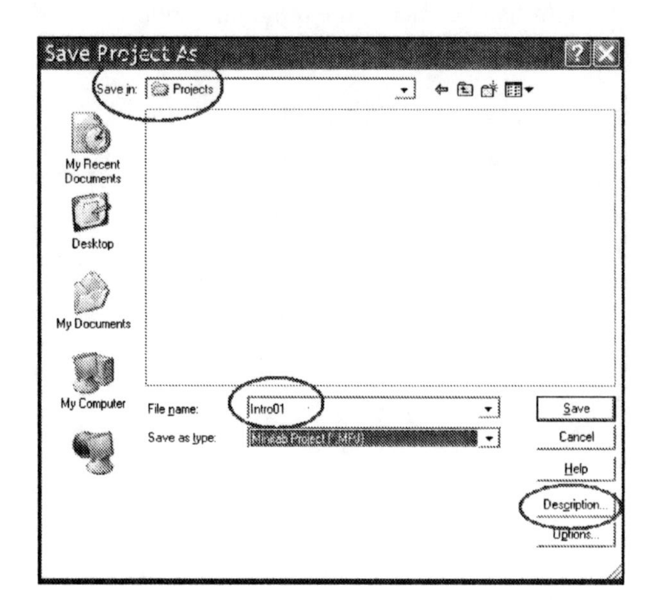

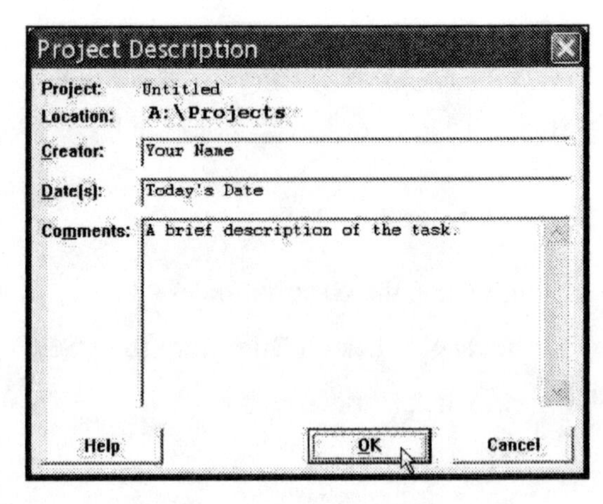

1. Type in **your name** as the creator.
   Press <Tab> and enter **today's date.**
2. Press <Tab>, then type in a brief description of the task in the Comments: dialog box.
3. Click **[OK]**.

d. Click **[Save]**.

The Data window, the Session window, the Graphical Summary, and all program settings are stored in one file, a project file. Projects have an extension of MPJ. This file will appear as Intro01.mpj. The MPJ extension is automatically added to the name, Intro01. The two worksheets, the session window results, and program settings are saved in this project file. When a project file is opened, the program will start where you left off.

Quit MINITAB

Step 9. Select **File>Exit.** If any changes have been made since the project was saved, you would be prompted to save the project. If no changes have been made, the program will close and return you to the Windows desktop. Alternately, you may quit the program by clicking the Close icon in the upper right-hand corner of the program window.

Step 10. Start MINITAB again (step 1) then

a. Click the Session window to make it the active window.

b. Select File on the menu bar.

c. Continue to the next section.

1-2 Using Menus

The File Menu

This section is informative. Read through this information.
Until a project is saved, the default project name is **MINITAB.MPJ**.

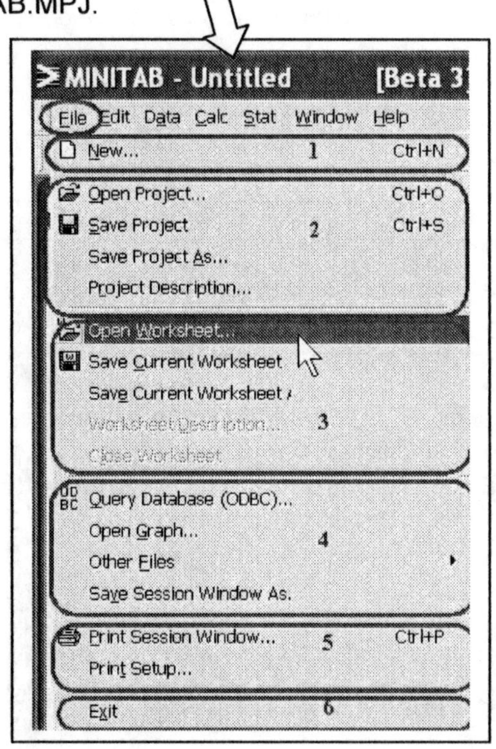

The file menu is divided into six sections.

1) The **New...** menu item allows us to add a new worksheet to the project or to start a new MINITAB project.
2) The second section contains options for projects.
3) Next are items for worksheets.
4) The fourth section manages graphs and other file functions.
5) The **Save** and **Print** options change depending on which window is the active window.
6) Last is **Exit.** Selecting this item quits the MINITAB program and returns you to the Windows desktop.

The symbols at the end of the menu items indicate what will happen next if that item is selected.

1) The three dots, or ellipses, indicate a dialog box will be next.
   Example: **Open Project...**
2) A small triangle at the end of the line indicates another menu is next.
   Example: **Other Files** ▸
3) No symbol means that command will be carried out as soon as you click the menu item.
   Example: **Save Project**

There are four ways to select menu items.

1) Click on the item in the menu.
2) Use the arrow keys to move the cursor so it highlights the desired item and then press <Enter>.
3) Use keyboard shortcuts.
   a) Pressing <Alt> + the underlined letter of the item will select that menu item. For example,
   Example: <Alt> + F will open the file menu.
   b) <Ctrl> + Letter are shortcuts even if a menu is not visible. These shortcuts are displayed on the right side of the menu.
   Example: <Ctrl> + P will print the active window.
4) Some menu commands have an icon on a toolbar. On the left side of the standard toolbar you will see the file management tools, open, save and print.

Note: Clicking the Disk icon on is the same as selecting **File>Save Project**.

Clicking the File icon is the same as selecting **File>Open Project**.

Clicking the Printer icon prints the active window.

1-3 Toolbars and the Session Window

Toolbars

Step 1. MINITAB should already be running. If not, start MINITAB. (See instructions, Section 1-1.)

Step 2. Click the Session window icon to make it the active window. The menu bar and toolbars are located at the top of the program window. The Standard toolbar and Project Manager toolbar are shown here.

Step 3. Click **Tools>Toolbars.**

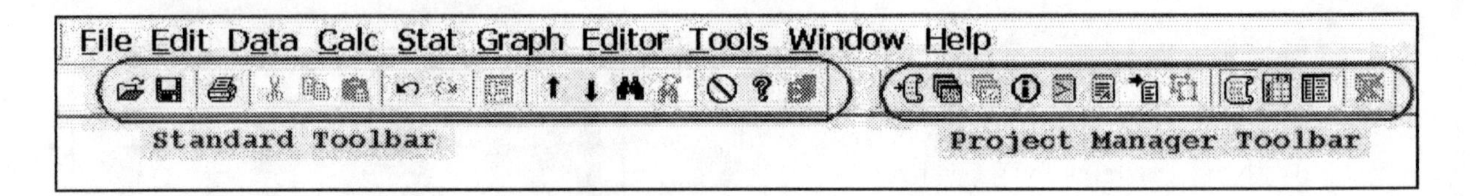

In the menu, Standard and Project Manager are checked. These are switches that toggle the feature on and off. If checked, they are on. If you click, the check is removed and the toolbar is off, no longer available as a toolbar. Do not change this setting for now.

Customize the Session Window

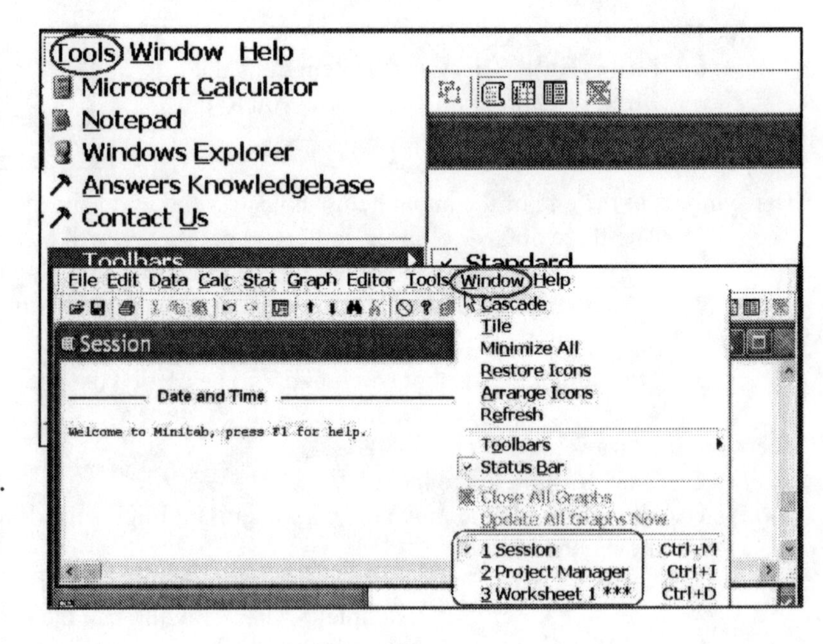

Step 4. On the menu bar, select Window.

At the bottom of this menu the windows of MINITAB are listed.

The Session window should be checked. This is not a toggle. Only one window will have a check mark. Only one can be the active window. Be sure the window you need has a check mark.

Step 5. Check Session.

If you ever "lose" a window, click Window on the menu bar and look for the lost window in the list at the bottom. All open windows will be listed there whether they are visible or not.

The date and time are displayed at the top of the Session window along with a message that welcomes you to MINITAB. We are going to erase the welcome line and customize the session by typing in our user information. Here is how.

Step 6. Drag the mouse over the welcome line then press the <Delete> key on the keyboard.

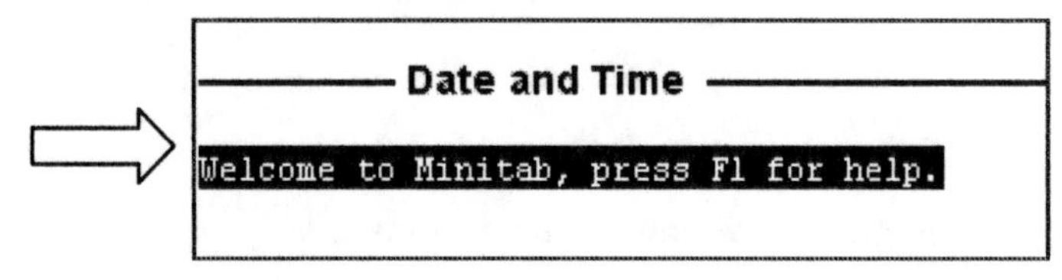

The line should erase with the cursor at the left edge.

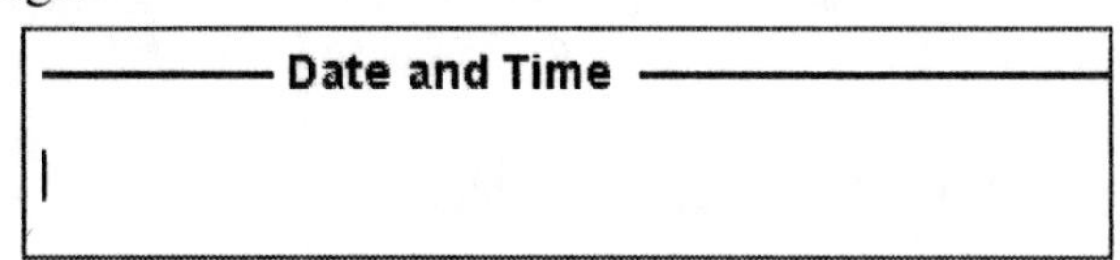

Note: If the line did not erase, a simple setting needs to be changed. If the line erased, skip to Step 8.

Step 7. Select **Editor**.

Make sure the item **Output Editable** is checked. This is a toggle. Clicking it turns the feature on and off. When checked, you may erase or change output in the session window. If unchecked, MINITAB results are protected. Turn it on and off as needed to protect your work from unwanted changes.

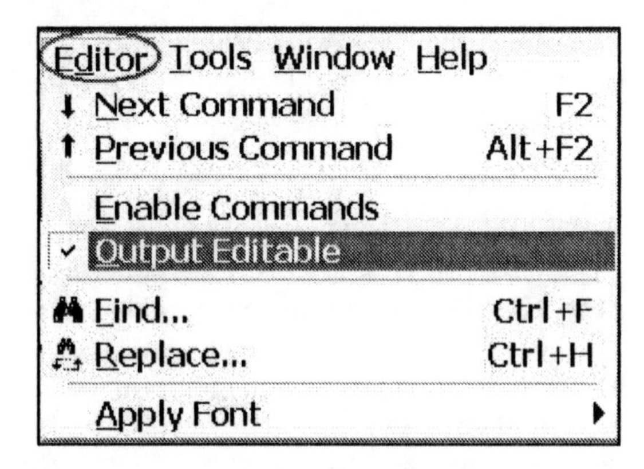

Step 8. Select **Tools>Options**.

a. Click the small plus sign to the left of **Session Window.** The list expands and shows the options you can change.

b. Choose **Comment Font.**
Scroll the list of fonts until you find Arial, then click it. Also click **Italic style** and **Size 14**. The Input/Output Font is Courier New 9. Do not change the I/O font or MINITAB results will not be spaced properly.

c. Click **[OK]**.

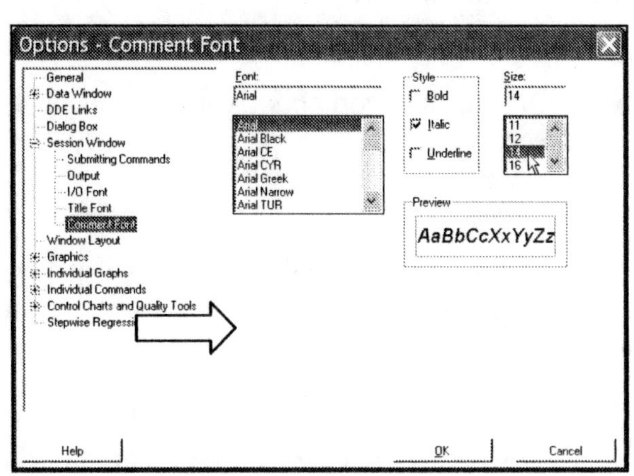

Step 9. Type your name and course information in the Session window at the cursor. It should appear in this new font. Anything you type on a new line in the Session window is considered a comment. When you change the comment font, every comment line is immediately changed to the new font.

The first line with the date is in the Title Font, Arial 14 Italic.

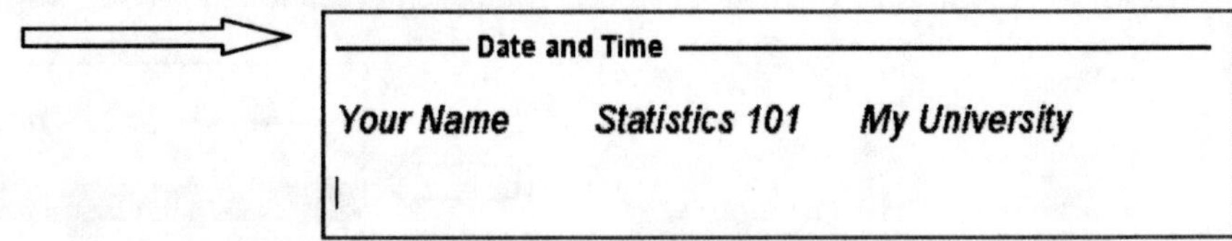

The Session window is ready for work.

Step 10. Click the Disk icon then save the file as **NewParks.MPJ.** Keep going!

The Project Manager Toolbar

The Project Manager toolbar provides a convenient way to switch active windows.
The icons on the far right side of the toolbar allow you to switch MINITAB program windows.

Step 11. Click the Session window icon.
The Session window is now the active window.

Step 12. Click the maximize button on the Session window title bar.

The Session window will enlarge to fill the screen. The close button is dimmed since this window cannot be closed.

Step 13. Click the Worksheet icon on the toolbar. The empty worksheet will now fill the screen.

Note: The icon furthest to the right is the Graph Window icon. Since there are no graphs in this project, it is dimmed. Options that cannot be used are dimmed in this way on menus, toolbars and dialog boxes.

In the next section you will learn how to edit data in a worksheet window. The Session window and Data worksheets are essential for analyzing data. The Project Manager window will allow you to manage your projects. The Project Manager will be introduced as needed, a little at a time!

1-4 Worksheets

A worksheet in MINITAB is used to enter and edit data. Data will either be typed in using the keyboard or copied from a file. All of the data files used in this manual were installed with the MINITAB program or they are included on the CD-ROM that came with your new textbook.

Open the Ballpark Worksheet

Step 1. Click File on the menu bar and then click Open Worksheet.
A dialog box will open. Navigate to the directory, C:\Program Files\MINITAB 14\Data.

If the dialog box title says Open Project instead of Open Worksheet, you either clicked the Open File icon or you clicked **File>Open Project** in error. Press the <ESC> key and select again.

Step 2. Click the icon for Up One Level.

The list will move up one notch in the directory tree.

Step 3. Click the drop-down arrow.

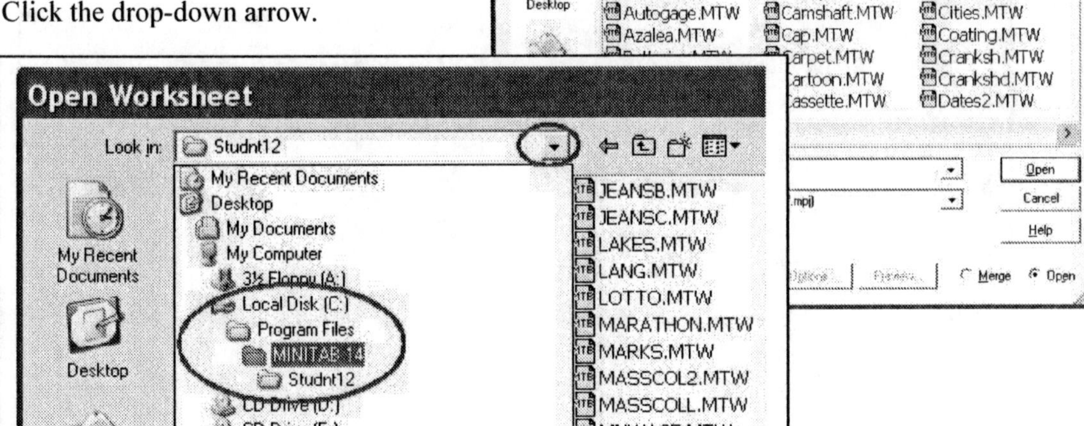

Step 4. Double click on the Studnt12directory icon.
Step 5. The file type should be Minitab [*mtw, mpj].

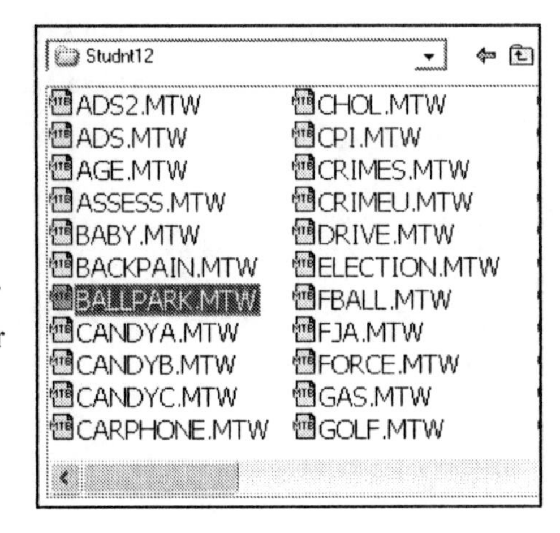

Step 6. Double click the file named BALLPARK.MTW.

This data will be copied into Worksheet 1. The name of the Worksheet window will change to the name of the file. The first twelve rows of this data are shown as they appear in the worksheet.

A popup menu may warn you that the worksheet will be added to the project. If so, click **[OK]**.

Columns are numbered as C1, C2, C3, and so on. These columns do not behave like a spreadsheet. Don't try to enter titles or comments as it will mess up your data. Rows are not in a column of data, they are only numbered for reference on the screen.

Notice the first column has a text designation, a T after the column number. The data is qualitative. The column with the league name is also qualitative. Text data is left justified in the cells. The year the park was built is quantitative. Quantitative data is right justified in the cells.
It is typical for the data of each variable such as League, to be placed on one column of the worksheet while the data for each individual (ballpark) is placed in a row.

BALLPARK.MTW ***

↓	C1-T	C2-T	C3	C4	C5	C6
	Team	League	ParkBlt	Capacity	Attend	Pct
1	Anaheim Angels	American	1966	64593	21819	0.519
2	Baltimore Orioles	American	1992	48262	45816	0.605
3	Boston Red Sox	American	1912	33871	27827	0.481
4	Chicago White Sox	American	1991	44321	23913	0.497
5	Cleveland Indians	American	1994	43863	42559	0.534
6	Detroit Tigers	American	1912	46945	17280	0.488
7	Kansas City Royals	American	1973	40625	19710	0.416
8	Milwaukee Brewers	American	1953	53192	18513	0.484
9	Minnesota Twins	American	1982	48678	17421	0.420
10	New York Yankees	American	1923	57545	33083	0.593
11	Oakland Athletics	American	1966	43662	15965	0.401
12	Seattle Mariners	American	1976	59856	39494	0.556

Step 7. Select **Window > 2. Project Manager.**

In the left-hand pane of the Project Manager you should see Worksheets at the bottom of the list.

The folder for the active worksheet, BALLPARK.MTW, is green.

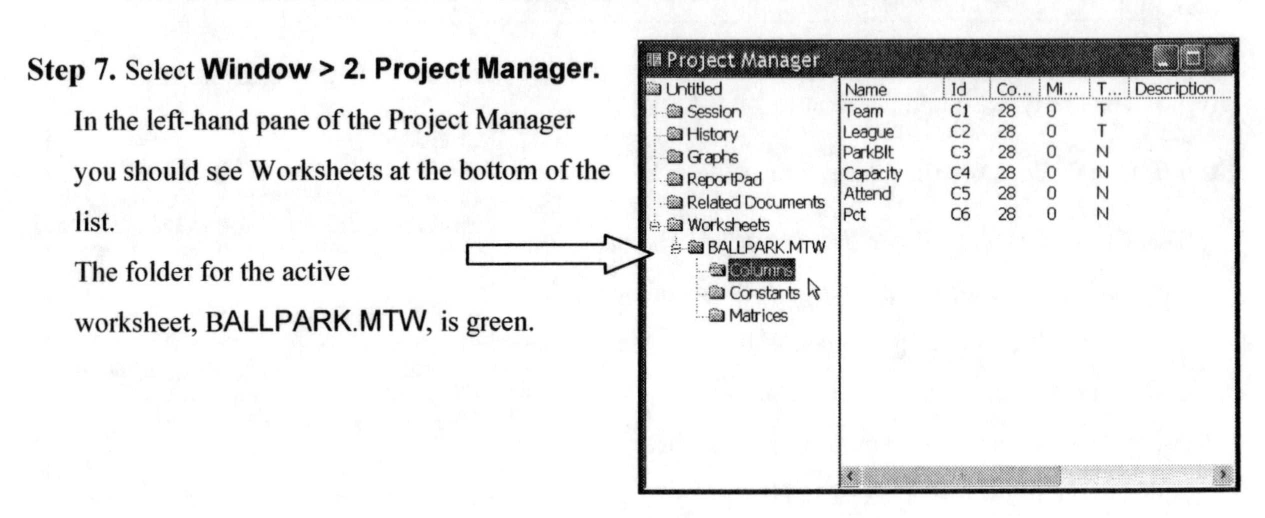

Step 8. Click the **Columns** folder.

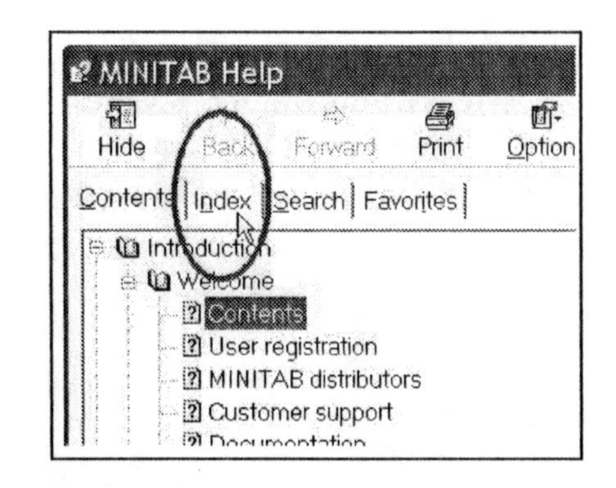

The right-hand pane will show information about all the columns that contain data. For example, C1 Team has 28 rows of text with no missing values.

Open the MINITAB Help File

Step 9. On the keyboard, press <F1>, the <Help> key.

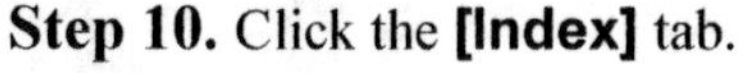

Step 10. Click the **[Index]** tab.

Step 11. Type the first few letters into the dialog box.

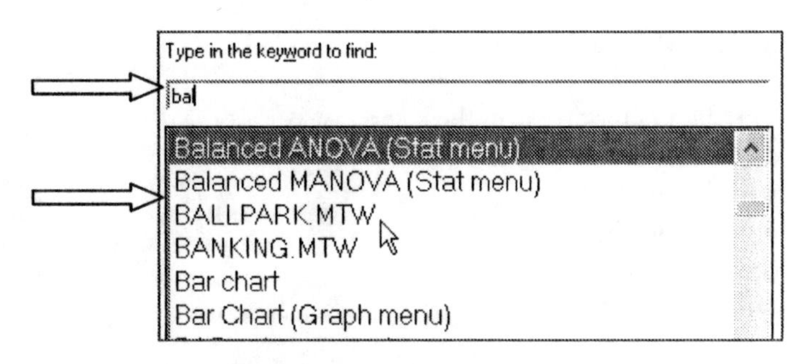

As you type the list will jump to the appropriate location.

Step 12. When visible, double click **BALLPARK.MTW.**

The summary for this worksheet is shown. There are 28 rows of data and six columns.

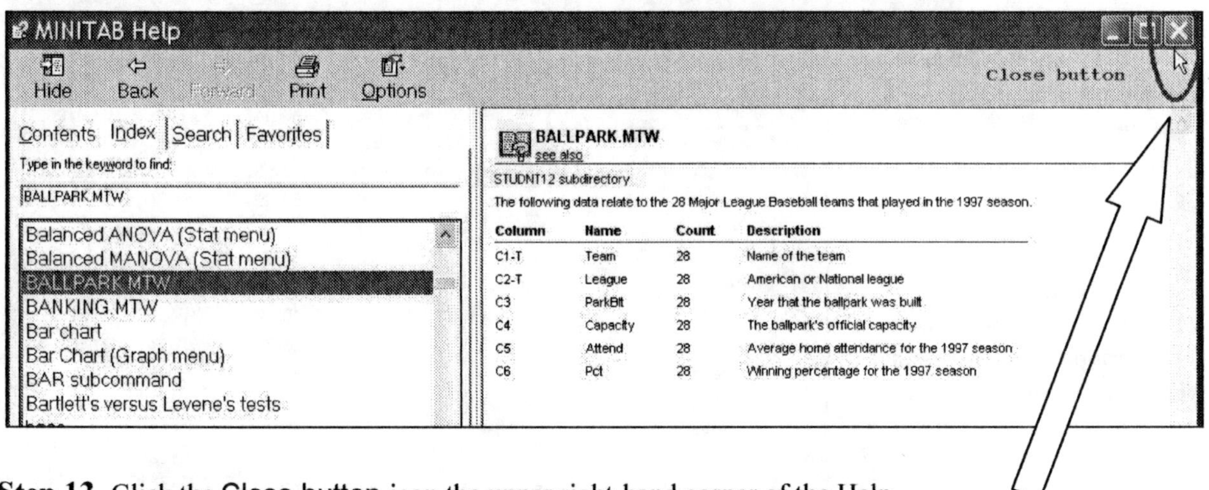

Step 13. Click the **Close button** icon the upper right-hand corner of the Help window.
This data is old and needs to be edited. There have been a lot of changes! That will be our task.

Step 14. Click the **Worksheet** icon in the Project Manager toolbar to switch back to the worksheet.

Save the Project

Step 15. Select **File>Save Project.** If prompted for a file name, type **NewParks**.

Keep going. Do not exit MINITAB.

Notes about Ballpark.mtw

Here is a nifty feature. You must have a scrolling mouse for it to work.
My laptop touch pad does not do this. ☹

Step 16. With the cursor over the worksheet area, hold down <Ctrl> as you scroll the mouse. What happens? Cool............The worksheet zooms in and out.

Step 17. Press <Ctrl> + <Home> to jump the cursor to **Row 1** of **C1**.

The project name and the name of the active worksheet are shown in the MINITAB Program title bar.

The data entry arrow in the upper left corner of the worksheet is a toggle switch. If pointing to the right, when you press the return key the cursor will move to the right.

Step 18. Click it to change the direction of the data entry arrow to down. It will always be either right or down. Click it again. We want the arrow to be pointing right as shown.

The block cursor shows the location of the mouse pointer. It will not be an arrow pointer when it is over the worksheet, only when it is over a menu or toolbar where you can select items.

MINITAB - NewParks.MPJ [BALLPARK.MTW ***]

File Edit Data Calc Stat Graph Editor Tools Window Help

→	C1-T	C2-T	C3	C4	C5	C6	C7
	Team	League	ParkBlt	Capacity	Attend	Pct	
1	Anaheim Angels	American	1966	64593	21819	0.519	
2	Baltimore Orioles	American	1992	48262	45816	0.605	
3	Boston Red Sox	American	1912	33871	27827	0.481	
4	Chicago White Sox	American	1991	44321	23913	0.497	
5	Cleveland Indians	American	1994	43863	42559	0.534	
6	Detroit Tigers	American	1912	46945	17280	0.488	
7	Kansas City Royals	American	1973	40625	19710	0.416	
8	Milwaukee Brewers	American	1953	53192	18513	0.484	
9	Minnesota Twins	American	1982	48678	17421	0.420	
10	New York Yankees	American	1923	57545	33083	0.593	
11	Oakland Athletics	American	1966	43662	15965	0.401	
12	Seattle Mariners	American	1976	59856	39494	0.556	
13	Texas Rangers	American	1994	49166	36815	0.475	
14	Toronto Blue Jays	American	1989	50516	31967	0.469	
15	Atlanta Braves	National	1996	49714	42765	0.623	
16	Chicago Cubs	National	1914	38884	27725	0.420	
17	Cincinnati Reds	National	1970	52952	22322	0.469	
18	Colorado Rockies	National	1995	50200	48006	0.512	
19	Florida Marlins	National	1987	41855	29555	0.568	
20	Houston Astros	National	1965	54370	25269	0.519	
21	Los Angeles Dodgers	National	1962	56000	40974	0.543	
22	Montreal Expos	National	1976	46500	18489	0.481	
23	New York Mets	National	1964	55775	22643	0.543	
24	Philadelphia Phillies	National	1971	62363	19359	0.420	
25	Pittsburgh Pirates	National	1970	47972	20713	0.488	
26	St. Louis Cardinals	National	1966	49676	32819	0.451	
27	San Diego Padres	National	1967	59690	26117	0.469	
28	San Francisco Giants	National	1960	62000	21135	0.556	

Step 19. Click in the cell for the **Atlanta Braves**. The cell will be highlighted.

Step 20. Click the Disk icon or select **File>Save Project**. The project name should still be NewParks.mpj. Do this to save your work as you go. In the event of a system crash, very little will have to be done over.

Keep going!

1-5 Edit a Worksheet

Wait for Step 1, the numbered steps that follow this list of changes, before making any of the changes. You will learn how to do these editing procedures.

Insert a Row

A new team, the Arizona Diamondbacks, joined the National League in 1998, capacity: 48,500.
Insert the Tampa Bay Devil Rays in the list of American League teams, 1998, capacity: 45,200.

Move a Row

The Milwaukee Brewers moved to the National League.

Edit Cells

Seven teams have opened new ball parks. Change the year and capacity.

Delete a Column

The attendance and percentage figures are incorrect. Erase these two columns.

Insert a New Column

Enter the data for the length of center field, the distance from home plate to the fence.

Begin! To begin, add the worksheet editing tools to the menu bar. This is new in Release 14.

Here is how.

Step 1. Select **Tools>Toolbars** and check Worksheet.

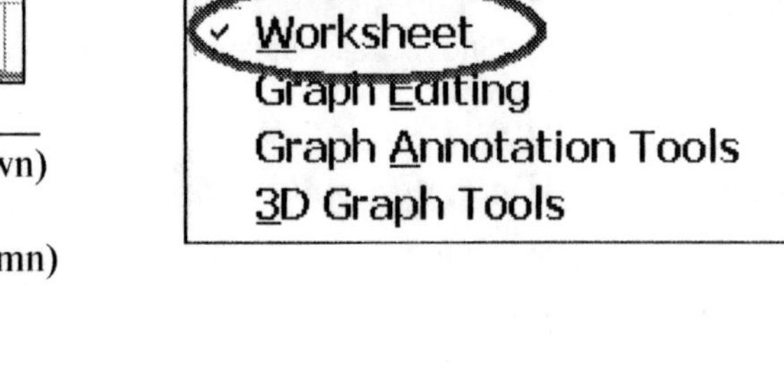

The Worksheet toolbar

Icons from left to right:

Insert a cell	(the rest of the cells move down)
Insert a row of cells	(rows move down.)
Insert a column	(to the left of the current column)
Move a column	

The two dimmed icons are used to edit graphs.

Clear cell contents (same as backspace)

Insert a Row

Step 2. Maximize the worksheet.

Step 3. The data entry arrow should be horizontal.

This arrow toggles from down to right and down again every time you click it.

When you press <Enter>, the cursor will move across the row.

Step 4. Move this cursor to Row 15 in C1, the cell for the Atlanta Braves, then click.

The cell is highlighted with a black outline.

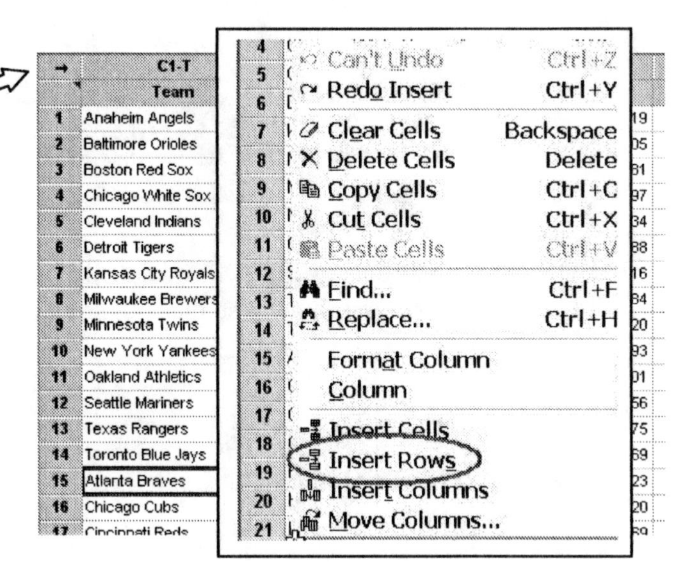

Step 5. Right click then select Insert Rows.

It doesn't matter where the cursor is when you right click as long as a cell in that row is highlighted. This is the same as clicking the Insert Row icon on the Worksheet toolbar. Instantly, rows 15 through 28 will move down and a row of empty cells will be inserted into the worksheet. Empty cells in a numeric column will contain an asterisk symbol (*), the missing value used for numeric data. The text columns will contain a blank cell, the missing value for cells in a text column.

14	Toronto Blue Jays	American	1989	50516	31967	0.469
15			*	*	*	*
16	Atlanta Braves	National	1996	49714	42765	0.623
17	Chicago Cubs	National	1914	38884	27725	0.420

Step 6. Type the information for the Arizona Diamondbacks, one item in each cell:

Arizona Diamondbacks National 1998 49075

Leave the asterisks in the last two cells because there is no data for this team in 1997.

Note: Do not type in the commas for large numbers. Type 49075 not 49**,**075!

Step 7. Repeat steps 4 through 6 to insert a row for the Tampa Bay Devil Rays in the American League between Seattle and Texas. Be sure it will be in alphabetical order. Their ballpark was built in 1990 with a capacity of 44,027.

Move a Row of Data: Cut and Paste

Step 1. Click any cell in the row for Montreal Expos.
Step 2. Click the **Insert row** icon on the Worksheet toolbar.

An empty row opens up and the rest of the rows move down. Notice the asterisk in the empty quantitative columns and the blank spaces in the qualitative columns.

22	Los Angeles Dodgers	National	1962	46500	18489	0.543
23			*	*	*	*
24	Montreal Expos	National	1976	55775	22643	0.481
25	New York Mets	National	1964	62363	19359	0.543

Next you will copy and paste the Milwaukee data.

Step 3. Click the row number **8** so the entire row of Milwaukee data is highlighted.

7	Kansas City Royals	American	1973	40625	19710	0.416
8	Milwaukee Brewers	American	1953	53192	18513	0.484
9	Minnesota Twins	American	1982	48678	17421	0.420

Step 4. Right click and select **Cut Cells** from the menu. The Milwaukee Brewers data will be stored on the clipboard.

Step 5. To paste it into its new location, move the block cursor to the row number, 23. Right click and select **Paste Cells**. Move completed!

23	Milwaukee Brewers	American	1953	53192	18513	0.484
24	Montreal Expos	National	1976	46500	18489	0.481

Edit a Cell

There are changes that need to be made because there are so many new major league baseball parks. Here is how to make a correction one cell at a time. Use the instructions to make each change listed at the end. First, correct Milwaukee's info. Here is how.

Step 1. Click inside the cell for Milwaukee's league. The cell borders should change to wide dark lines with a vertical cursor located somewhere in the cell. In the next step, change from American to National League.

Step 2. To erase the contents of the cell, right click and select **Clear cells** or click the Clear contents icon on the Worksheet toolbar.

Caution:

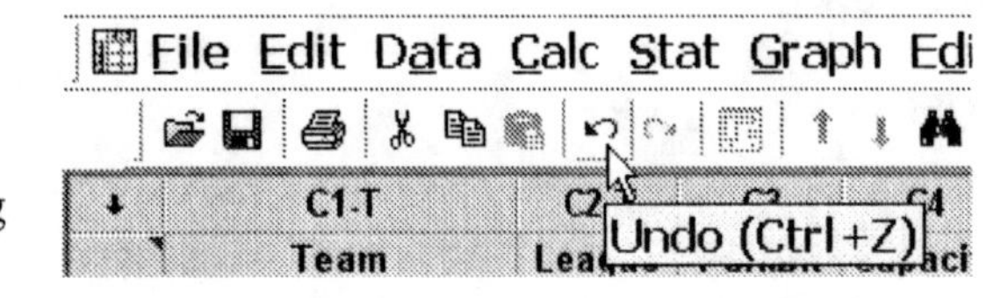

Make sure you clear the cells. Don't delete them or the cell is removed and all data in that column will move up. If you make a mistake, correct it immediately by clicking the Undo icon on the standard toolbar. Only one operation can be undone in MINITAB. Remember, don't type the comma when entering large numbers such as 43,000.

Continue **Step 2**.

a. Type the new league name, **National**, and press <Enter>.

b. In the columns for Year and Capacity type **2001** and **43000**.

Row 23 should now have this data:

Milwaukee Brewers   National   2001   43000 18513 0.484

Repeat steps 1 and 2 to make the following changes:

Team	Pkblt	Capacity
Detroit	2000	40000
Seattle	1999	46621
Cincinnati	2003	42000
Houston	2000	42000
Pittsburgh	2001	38000
San Francisco	2000	40800

Step 3. Click the Disk icon to save changes to the project, NewParks.mpj.

Keep going!

Insert a Column

Step 1. Click any cell in column 5, **C5 Attend**.

Step 2. Right click and then click **Insert Column** from the right-click menu OR select the Insert column icon on the Worksheet toolbar.

Step 3. Name the column **Center**.

Step 4. Type in the lengths of center field. Consult a reference in the library. Search the Internet for the correct data or refer to the completed worksheet on the next page.

Optional: Extra!
Use the Internet to find the names of the fields for each park then insert a column with this information in the worksheet.

Delete a Column

There is so much missing data from the capacity and attendance columns that we are going to erase them.

Step 1. Drag the mouse over the column numbers C6 and C7 so the both columns are highlighted as shown.

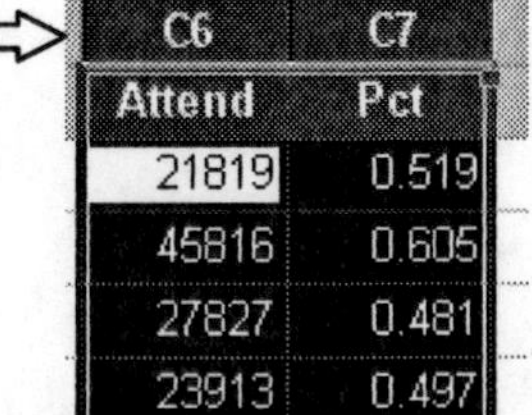

Step 2. Press the <Delete> key. The two columns of data are gone.

The completed worksheet is shown on the next page.

Finish the project.

Print the Worksheet

Step 1. Select **File>Print Worksheet.**

a. In the dialog box enter the project information and your name and date. The dialog box is shown. This is the best place to put the title you will want on your print out.

b. Click **[OK]**.

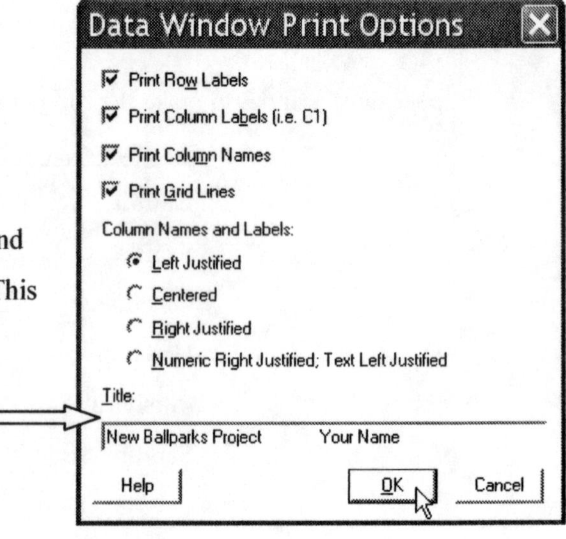

Save the project

Step 2. Select **File>Save Project**. The file will be saved as NewParks.mpj.

Quit MINITAB or Start a New Project

Step 3. Select **File>Exit** to quit or **File>New>Minitab Project** to continue.

New Ballparks Worksheet

	C1	C2	C3	C4	C5
	Team	**League**	**ParkBlt**	**Capacity**	**Center**
1	Anaheim Angels	American	1966	64593	400
2	Baltimore Orioles	American	1992	48262	410
3	Boston Red Sox	American	1912	33871	390
4	Chicago White Sox	American	1991	44321	400
5	Cleveland Indians	American	1994	43863	405
6	Detroit Tigers	American	2000	40000	422
7	Kansas City Royals	American	1973	40625	400
8	Minnesota Twins	American	1982	48678	408
9	New York Yankees	American	1923	57545	408
10	Oakland Athletics	American	1966	43662	400
11	Seattle Mariners	American	2000	40800	405
12	Tampa Bay Devil Rays	American	1990	44027	407
13	Texas Rangers	American	1994	49166	400
14	Toronto Blue Jays	American	1989	50516	400
15	Arizona Diamondbacks	National	1998	49075	407
16	Atlanta Braves	National	1996	49714	401
17	Chicago Cubs	National	1914	38884	400
18	Cincinnati Reds	National	2003	42000	404
19	Colorado Rockies	National	1995	50200	415
20	Florida Marlins	National	1987	41855	434
21	Houston Astros	National	2000	42000	436
22	Los Angeles Dodgers	National	1962	56000	395
23	Milwaukee Brewers	National	2001	43000	400
24	Montreal Expos	National	1976	46500	404
25	New York Mets	National	1964	55775	410
26	Philadelphia Phillies	National	1971	62363	408
27	Pittsburgh Pirates	National	2001	38000	399
28	St. Louis Cardinals	National	1966	49676	402
29	San Diego Padres	National	1967	59690	405
30	San Francisco Giants	National	2000	40800	404

1-6 Data Analysis

Data analysis exercises throughout the textbook use the data in the Databank file described in Appendix D of the textbook on pages 745 through 747. Data for 100 employees were collected. There is a row for each person's information and a column for each variable or survey question. Many of the exercise will require a random sample from this data. Here is how.

Open the Databank File

Step 1. Start MINITABusing the instructions from section 1-1.
Type **your name** and the **date** at the top of the Session window.

Step 2. Click **File>Open Worksheet.**

Step 3. Navigate to the folder containing the textbook files then double click the file name in the list box, Databank.MTP. See more detailed instructions earlier in section 1-1 on page 2.

Step 4. Select **File>Save Project As**. Save the file as DataAnalysis01.MPJ.

It is easy to make MINITAB select a simple random sample from any column.

The sample size of 15 used here is arbitrary. Any sample size could be used.

Select a Random Sample

Step 5. Select **Calc>Random Data>Sample From Columns.**

a. Type in the sample size, **15**.

b. Press <Tab> or click in the dialog box for column(s). The variables list will not be visible until now. Double click C2 Age.

c. Press <Tab> or click in the Store samples in:.

d. Type **Age15,** the name for a new column. Do not use the original column name, AGE, or the original data will be overwritten.

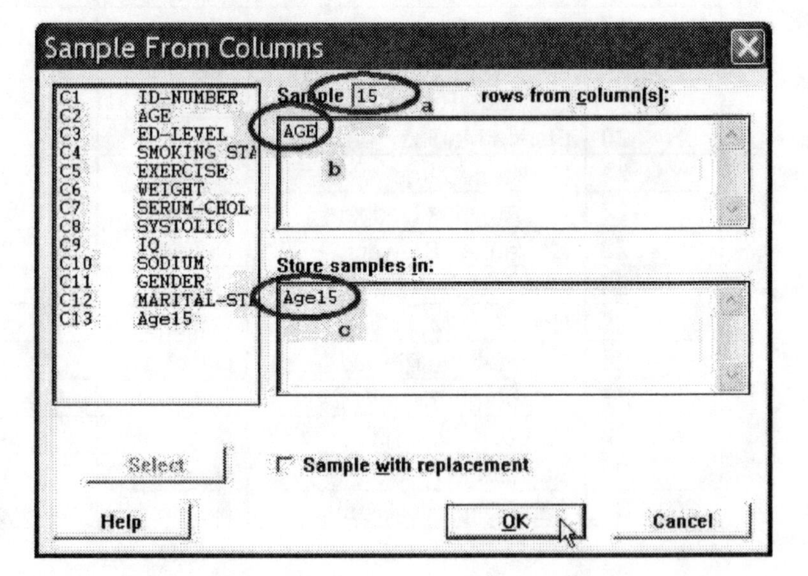

Note: Checking the box for Sample with replacement will allow duplicates. Usually this is not desirable. Don't check it.

e. Click **[OK]**. A simple random sample will appear in the first available column with a name of Age15. It is not likely that your sample will match the example in this example.

f. Select **Data>Display Data** to put the data in the Session window.
1. Double click C13 Age15.
2. Click **[OK]**.

Step 6. Repeat Step 5 to select a random sample of 35 ages, store them in Age35, and then display them in the Session window.

Finish the Project

Step 7. Be sure to type **your name** and the **date** at the top of the Session window.
Step 8. Print the Session Window **File>Print Session Window**
Step 9. Save the Project **File>Save Project** (DataAnalysis01.MPJ)
Step 10. Quit MINITAB **File>Exit**

Data Types

Choosing an appropriate statistical technique depends on correctly classifying the data as quantitative or qualitative. MINITAB uses three types of data: numeric, date, or text. Numeric data and dates are quantitative. Text data is qualitative. In addition to the data type, the level of measurement has an impact on choosing the appropriate technique. For example, Education Level is an ordinal level that has been coded with numbers for ease of entry. It may look like it is quantitative but it is categorical, qualitative. The level of measurement is ordinal.

The following table classifies each variable by type for the Databank file.

Variable Name	Type / MINITAB	Level of measurement	Unit of measure/ abbrev.
Age	quantitative / numeric	ratio	years
Education Level	qualitative / numeric	ordinal, codes for categories 0 no high school 1 high school diploma 2 some college 3 college degree	
Smoking Status	qualitative / numeric	ordinal, codes for categories 0 don't smoke 1 smoke less than 1 pack 2 smoke 1 pack or more	
Exercise	qualitative / numeric	ordinal, codes for categories 0 none 1 light 2 moderate 3 heavy	
Weight	quantitative / numeric	Ratio	pounds / lbs.
Serum Cholesterol	quantitative / numeric	Ratio	milligram percent / mg%
Systolic pressure	quantitative / numeric	Ratio	millimeters of mercury / mmHg
IQ	quantitative / numeric	interval	
Sodium	quantitative / numeric	ratio	Milli equivalents per liter / mEq/1
Gender	qualitative / text	nominal, M or F	
Marital Status	qualitative / text	nominal, letter codes D divorced M married S single W widowed	

Chapter 1 : EndNotes

Chapter 2 Frequency Distributions and Graphs

2-1 Introduction: The Least You Need to Know about Tables and Graphs

The first step when analyzing data is to summarize the data in a frequency distribution and a graph. Once a table is constructed and the graph is made, the spread and concentration of the data can easily be described. The tables can also be used to check data accuracy. Data entry errors may show up in this initial analysis. There are several ways to make frequency distributions from data depending on the data type. Directions for summarizing categorical and continuous data will be given in this introduction. Detailed instructions for other tables and graphs will follow in their appropriate section.

Enter the Data (Qualitative)

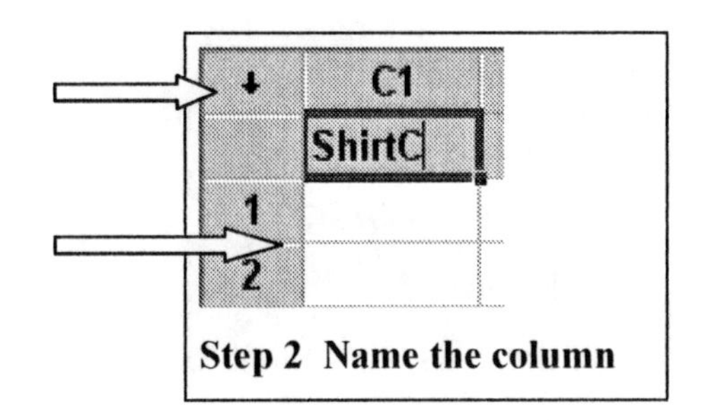

Step 2 Name the column

Step 1. The data entry arrow should be pointing down.
If it is not, click on it and it will change direction.

Step 2. Name the column.

a. Click in the space above row 1 in C1.

b. Type **ShirtColor.**

c. Press <Enter>.

The cursor will be in the in the first row of the worksheet.

ShirtColor									
W	W	BR	Y	BL	BL	W	W	Y	G
W	W	BL	BR	BL	BR	BL	BL	BR	Y
BL	G	W	BL	W	W	BL	W	BL	BR
Y	BL	G	BR	G	BR	W	W	BR	Y
W	BL	Y	W	W	BL	W	BR	G	G

In this example, the data is read across the table. Type all of the data into the first column.

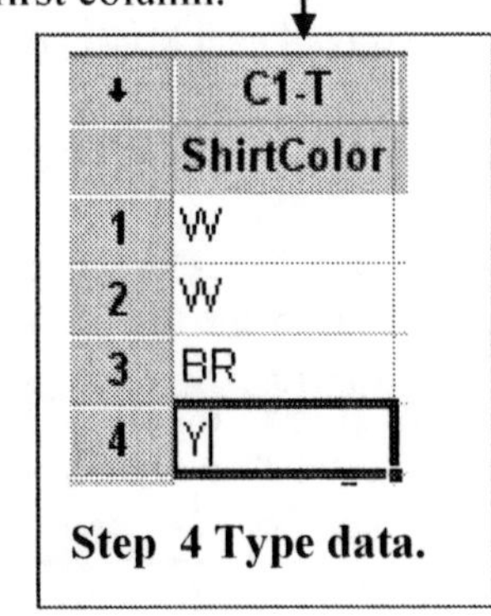

Step 4 Type data.

Step 3. Press the <Caps Lock> key so all data will be capitalized.

Step 4. Type **W**. Press <Enter>.

Note: The data type entered into the first cell determines the data type for the entire column. The column label will change to C1-T indicating the data is text. Text data will be left justified. Continue across the rows of the data table. Type down the column.

a. Type **W**. Press <Enter>.

b. Type **BR**. Press <Enter>.

c. Type **Y**. Press <Enter>. Continue to enter each code for the shirt colors.
All 50 color codes should be in the first column, C1 ShirtColor.

Save the Worksheet

Step 5. Select **File>Save Current Worksheet As... .**

Be sure to use the File menu not the File icon on the menu bar. Check the title on the dialog box to make sure you are not saving a project. Check three items:

a. Save in:

The drop-down list for Save in: will show the My Computer directory tree. Navigate to the directory where you are saving files. Projects

b. Type in the file name:

P-C02-S02-07

Problem 7 in Chapter 2, Section 2.

Step 5 Save the Worksheet

c. Save as type: Minitab.

The program will add the extension MTW when it saves the file.

d. Click on the **[Description...]** button.

In the dialog box type your name, today's date, and a brief description of the data.

Step 6. Click **[Save].**

Step 7. Press <Caps Lock> to turn caps lock off.

Step 8. Click **[OK]** then **[SAVE].**

The file named P-C02-S02-07.MTW will be saved as a MINITAB Release 14 worksheet.

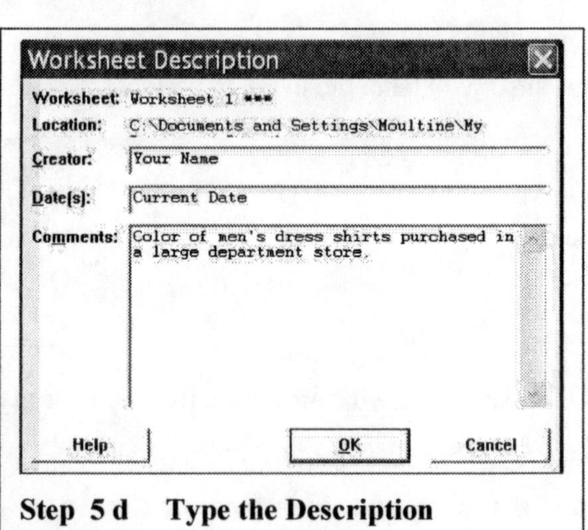

Step 5 d Type the Description

The name of the worksheet in the Program window will change to the name of the file. Since it is the active window the title also shows the three asterisks. The project is untitled.

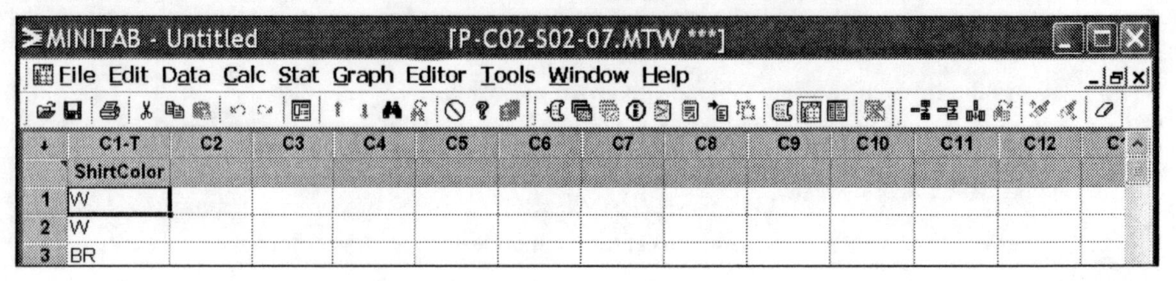

Construct a Categorical Frequency Distribution

Step 1. Select **Stat>Tables>Tally Individual Variables... .**

The cursor should be blinking in the variables dialog box. If not, click inside the box. A list in the left panel will contain the names of columns with suitable data types.

a. Double click C1-T ShirtColor in the list box. Only the label will be displayed in the variables box. The column number is not shown.

b. Check the boxes for Counts and Percents.

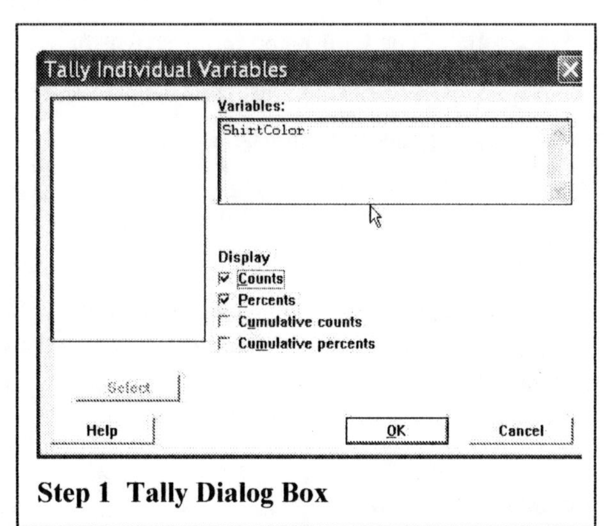

Step 1 Tally Dialog Box

Step 2. Click **[OK].**

The results will be displayed in the Session window. Twenty-six percent of the shirts were blue. Sixteen out of 50 were white.

The colors are listed in alphabetical order, the default order for text variables.

Tally for Discrete Variables: ShirtColor

ShirtColor	Count	Percent
BL	13	26.00
BR	9	18.00
G	6	12.00
W	16	32.00
Y	6	12.00
N=	50	

Construct a Pie Chart

Step 1. Select **Graph>Pie Chart... .**

a. Click in the box for Categorical variables.

b. Double click on C1-T ShirtColor.

Step 2. Click **[Labels...].**

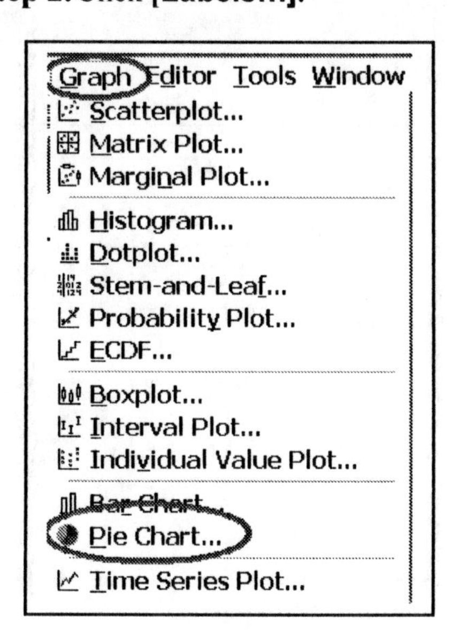

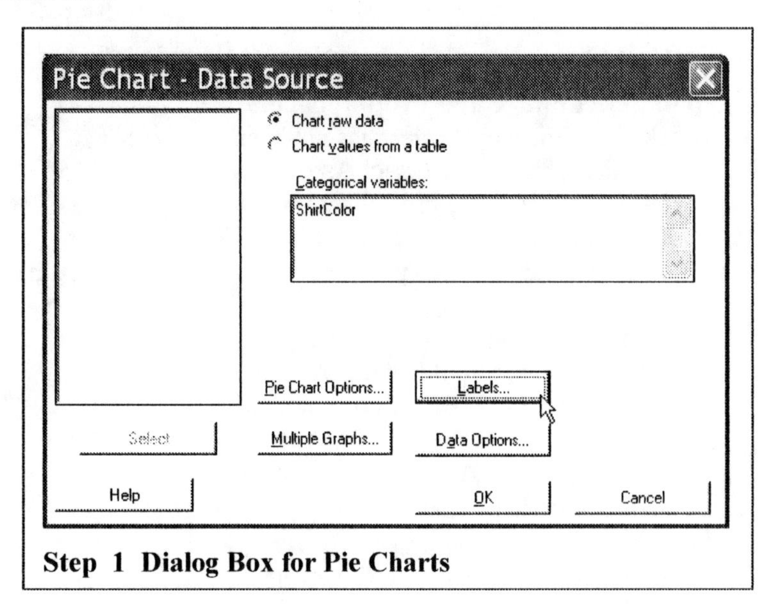

Step 1 Dialog Box for Pie Charts

a. Type in an appropriate title such as **Popular Shirt Colors.**

b. Type your name in the space for Footnote 1.

c. Click **[OK]** twice.

The completed pie chart is shown in a new graph window.

Note: The colors of the pie do not match the colors of the shirts. Computers aren't that clever! We will not change them now.

Pie Chart of Shirt Colors

Print the Graph

Step 3. Right click the graph's control box (upper left corner of the graph window, a green plus sign) then select Print graph. The Graph window will be printed on a page by itself.

Before we continue this introduction, save your work.

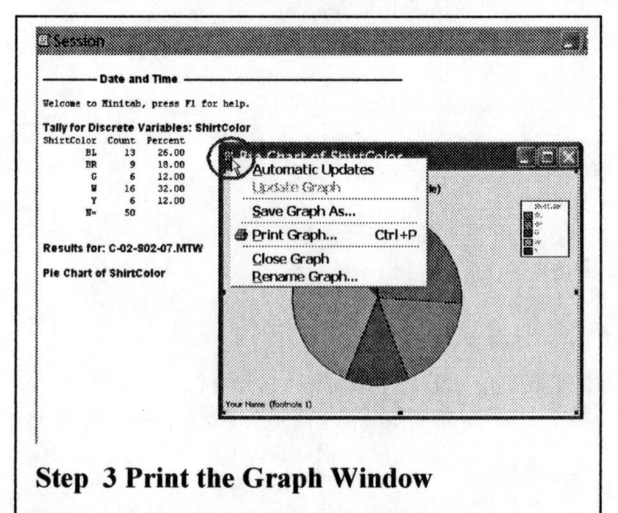

Step 3 Print the Graph Window

Save the Project

Step 4. Select **File>Save Project** or click the Disk icon. Since the project has not been saved, then the Save Project As dialog box opens as shown. If the project has already been saved there will be no dialog box. The file would be saved to its previous location and name (path).

a. The folder should be Projects.

b. The filename should be Intro2. The file extension, MPJ, will be automatic.

c. Click **[Description ...]** and then type **your name, date,** and **a brief description** of the project. This is not the same as the worksheet description but the dialog box is very similar.

d. Click **[Save].** Keep going!

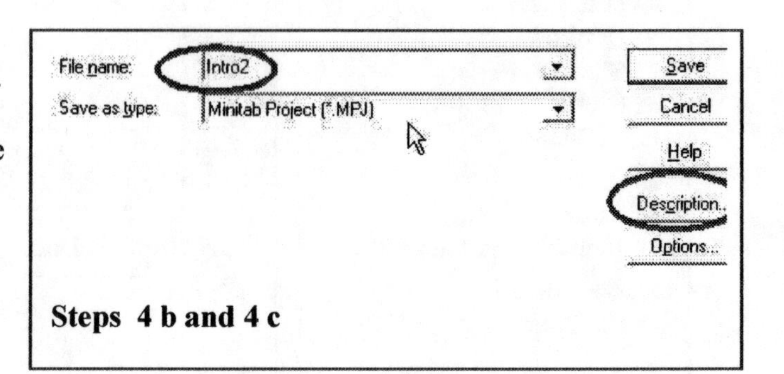

Steps 4 b and 4 c

Construct a Grouped Frequency Distribution

Instructions	***Data/Example/Exercise***
Open the file. Construct a histogram. Calculate statistics/parameters. Print results using the report pad. Start a new project.	Section 2 Exercise 13 Age of Founding Fathers

In these instructions, you will open the file that contains data, in years, of the age of the fifty-five statesmen who signed the Declaration of Independence.

```
41  44  44  35  35  54  52  63  43  46  47  39
60  48  45  40  50  27  46  34  39  40  42  31
53  35  30  34  27  50  50  34  50  55  50  37
69  42  63  49  39  52  46  42  45  38  33  70
33  36  60  32  42  45  62
```

Open the Worksheet File

Step 1. Select **File>Open Worksheet.**
Do not click the File icon!

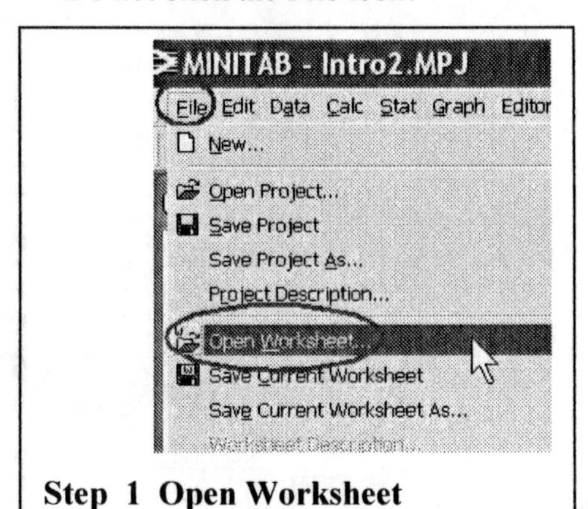

Step 1 Open Worksheet

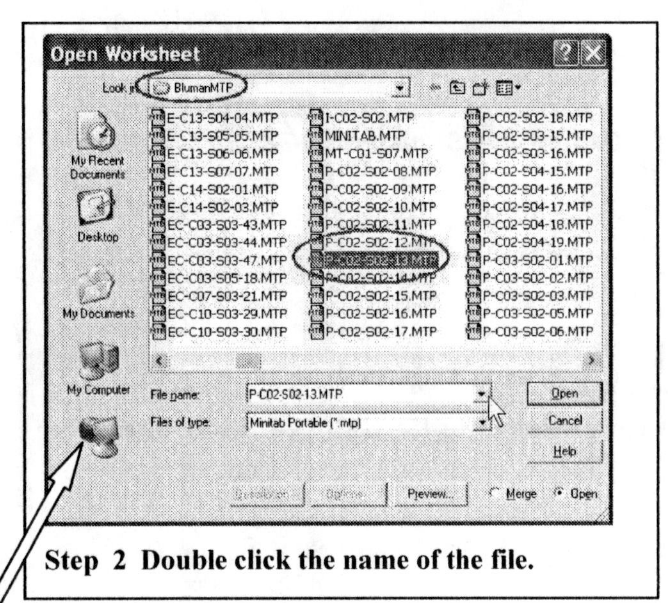

Step 2 Double click the name of the file.

Step 2. Navigate to the directory containing textbook files BlumanMTP.
Step 3. Double click the name of the file.

P-C02-S02-13.MTP

If the file cannot be located, select
File>New>Minitab Worksheet, then type the data in column one of the worksheet.

The data is copied into a second worksheet.
There will now be two worksheets in this project.

The active worksheet has the three asterisks. (.MTP ***)

Only the active worksheet data is used by commands and menus.

Step 3 Two Worksheets, One Project

Calculate Parameters/Statistics

Step 1. Select **Stats>Basic Statistics>Display Descriptive Statistics**.

Step 2. Double click C1 AGE in the variables dialog box.

Step 3. Click **[Statistics]**.

a. Check the boxes of the options you would like calculated. There are 13 items checked in this example. Match them!

b. Click **[OK]** twice.

The Session window will contain the statistics.

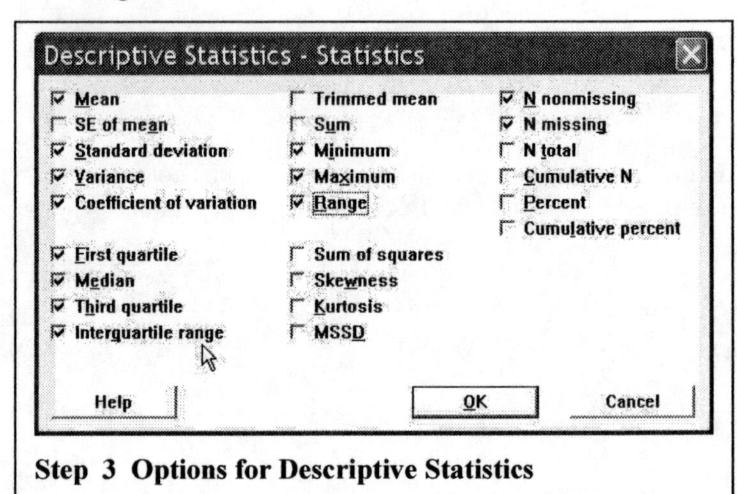

Step 3 Options for Descriptive Statistics

Descriptive Statistics: AGE

```
Variable   N  N*   Mean  StDev  Variance  CoefVar  Minimum     Q1  Median
AGE       55   0  44.51  10.25    105.07    23.03    27.00  36.00   44.00

Variable     Q3  Maximum  Range    IQR
AGE       50.00    70.00  43.00  14.00
```

Construct a Histogram

Step 4. Select **Graph>Histogram...>Simple** then click **[OK]**.

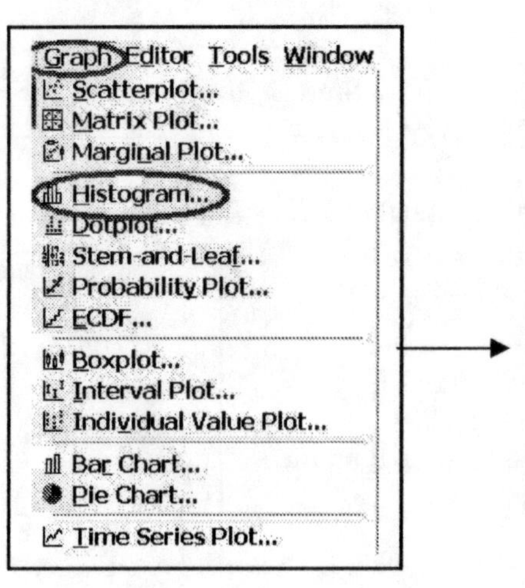

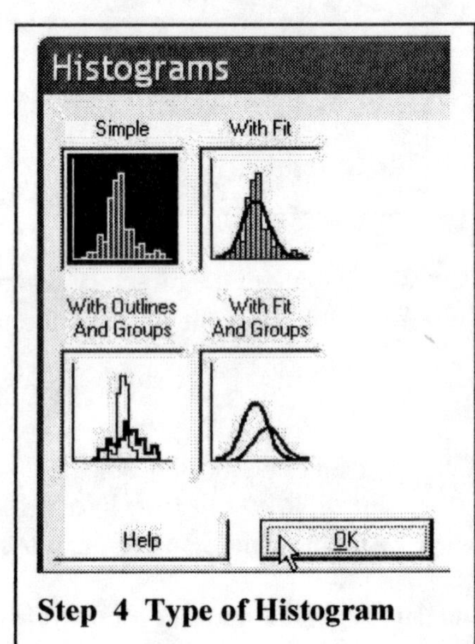

Step 4 Type of Histogram

Step 5. Double click C1 AGE to select it for the variable.
This dialog box is similar for most graphs.

Step 6. Click on **[Scale…]**.

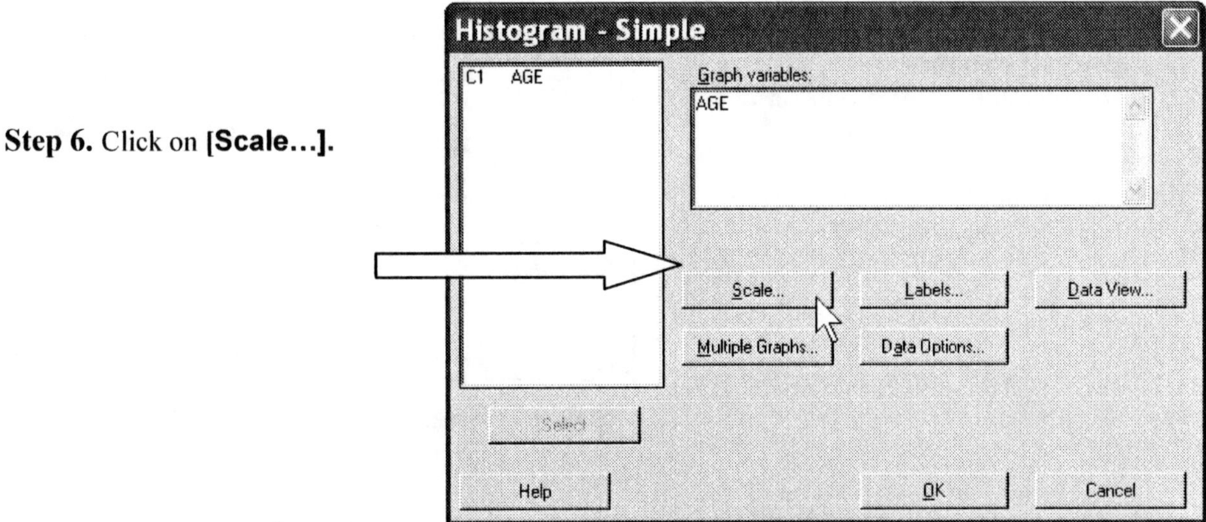

a. Select the Gridlines tab.
   1. Check Y major ticks.
   2. Check X major ticks.

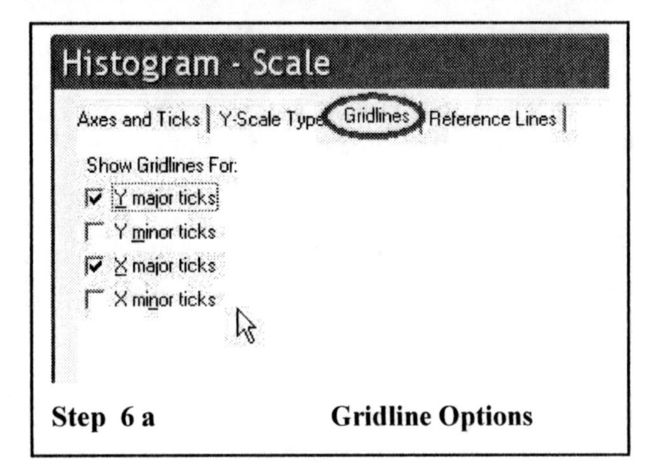

Step 6 a **Gridline Options**

b. Select the Y-Scale Type tab.
c. Check Frequency then click **[OK]**.

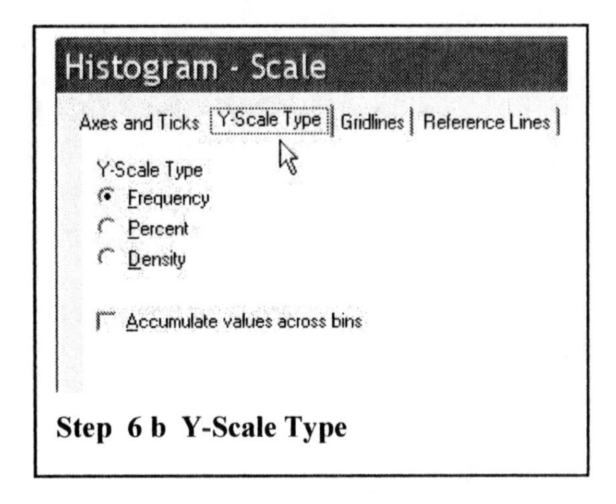

Step 6 b Y-Scale Type

The Simple Histogram dialog box will be open.

Step 7. Click on **[Labels]**.

a. Type an appropriate title
   Age of Statesmen
   who signed the
   Declaration of Independence

b. Type **Your Name** and **Date** in the footnotes then click **[OK]**.

Step 7 Histogram Labels

Step 8. Click on **[Data View]**.

a. Click the tab for Data Display.
b. Check the option for Bars.
c. Click **[OK]**.

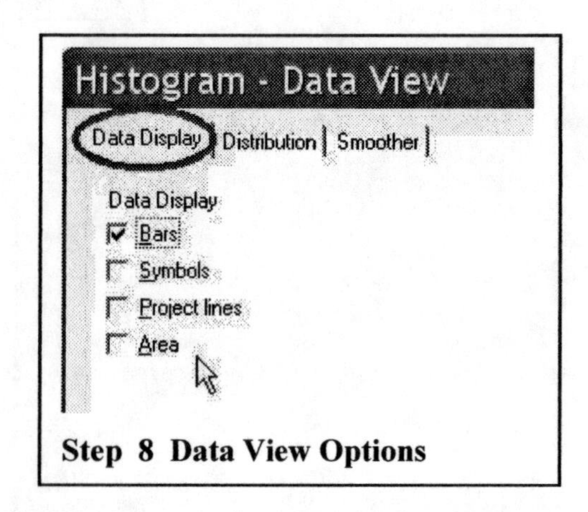

Step 8 Data View Options

The Simple Histogram dialog box will be open.

Step 9. Click **[OK]**.

The histogram will open in a Graph window.

In the next step we will tweak the graph a bit by changing the color of the bars and changing the scale for the horizontal axis.

Edit Graph: Change the Histogram Colors

Step 10. Right click on any bar of the histogram.

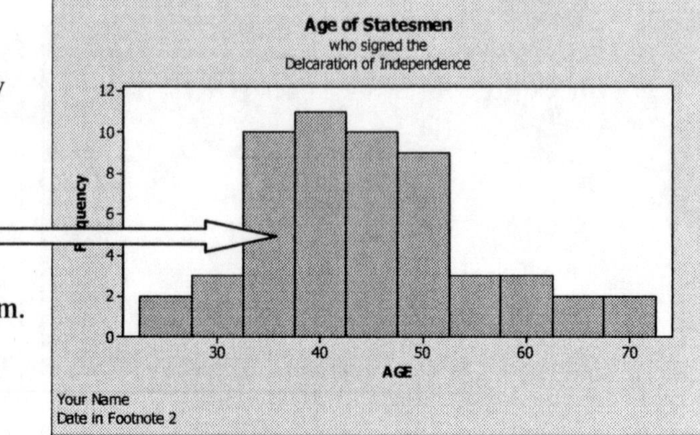

a. The bars will be outlined and the Editor menu will open.

b. Select **Edit Bars**... .
A dialog box opens.

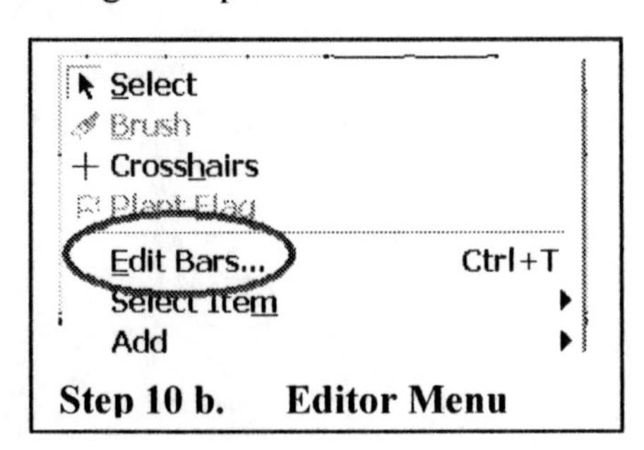

Step 10 b. Editor Menu

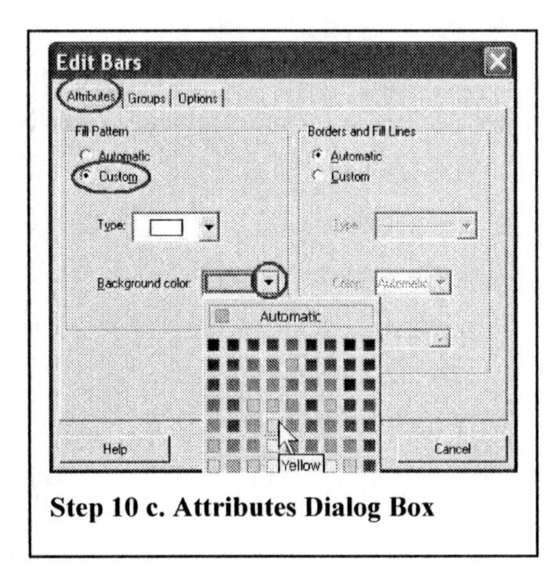

Step 10 c. Attributes Dialog Box

c. Click the tab for **Attributes**.

1. Click the option for **Custom**.
2. Click the drop down arrow for
   Background color:.
3. Click the color of your choice.
   Yellow is selected for the example.

d. Click **[OK]**. Bars will immediately change color.

Edit Graph: Number of Classes/Intervals

Change the number of histogram groups/classes. To make seven classes as the exercise requests, the X-scale has to be edited.

Here is one way to do it. The Graph window must be the active window.

Step 11. Select **Window>Histogram of AGE**.

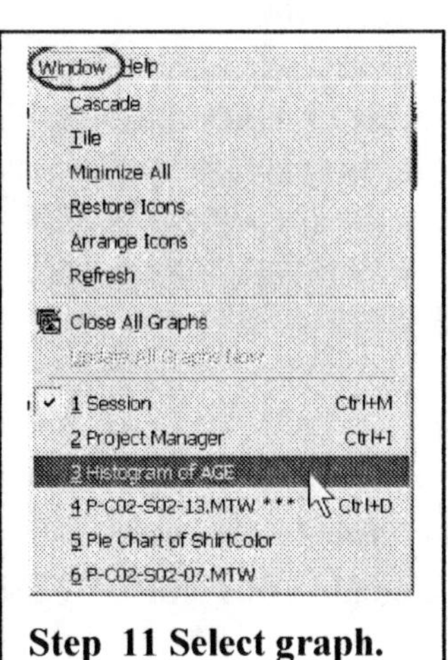

Step 11 Select graph.

Step 12. Select **Editor**.

a. Click on
Select Item>X Scale.

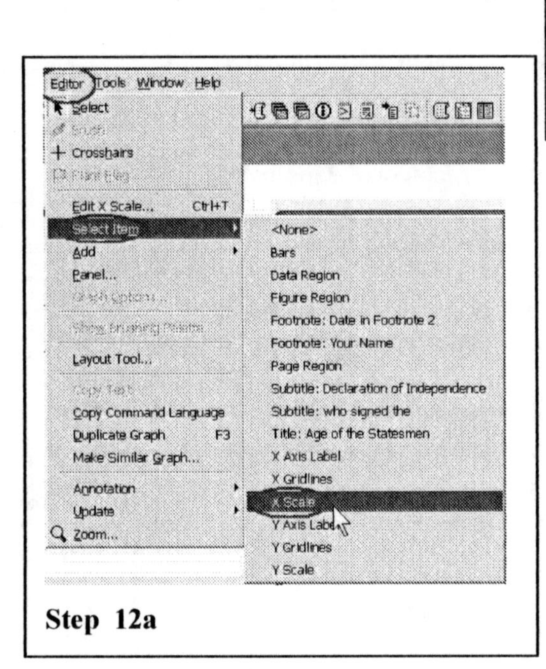

Step 12a

b. Select **Editor>Edit Scale**.

c. Click the tab for Binning, then the option for Midpoint.

d. Click the option for Number of intervals: then type **7** in the text box.

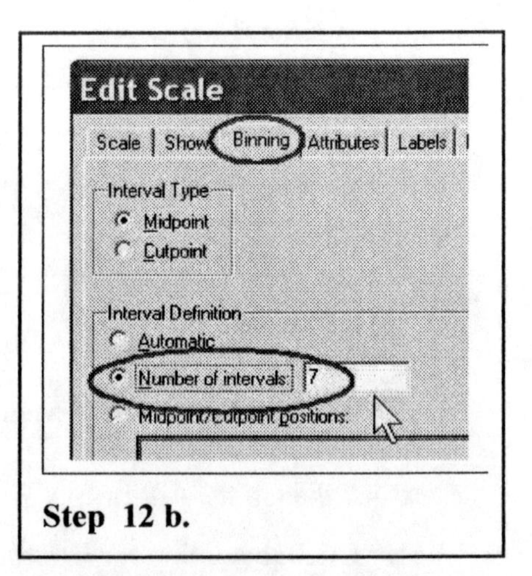

Step 12 b.

Alternately, click the option for Midpoint/cutpoint positions, then type a list of midpoints separated by a space, like this:

24 32 40 48 56 64 72

e. Click **[OK].**

The number of classes will change immediately.

f. Point the mouse at the bar with a midpoint of 48. The frequency and bin (interval) is shown.

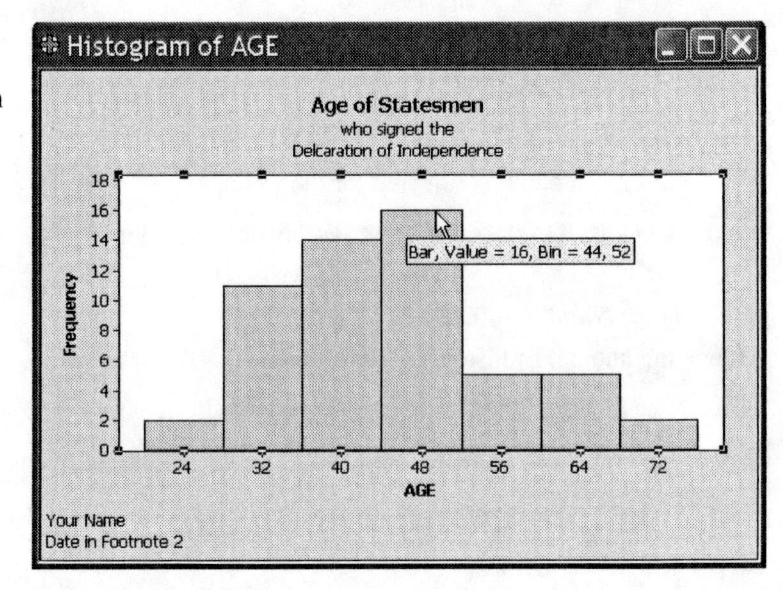

Step 13. Click the Disk icon to save your work. The name of the project should be Intro2.MPJ.

Finish the Project

Complete this project by creating, saving, and printing a report. Here is how.

Create a Report

Step 1. Click the Project Manager icon on the toolbar.

Step 2. Click the Session window folder.

Titles are listed in the right pane. You will select each of these and copy them to the report pad.

a. Hold down <Ctrl> as you click each title. All four should be highlighted.

b. Right click, then select Append to Report.

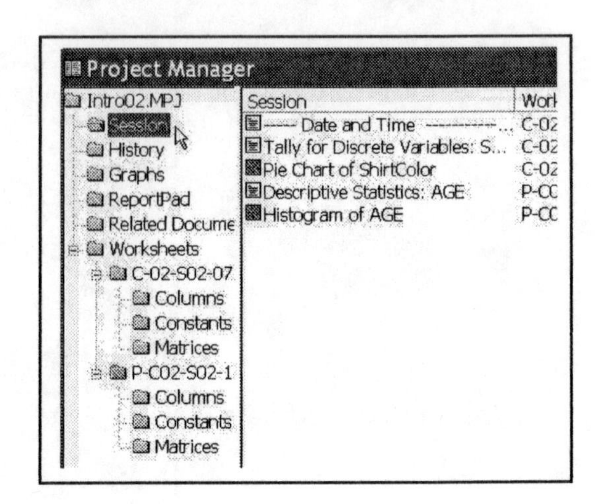

All contents are copied to the ReportPad.

c. Click on the ReportPad folder.
   The document in the right pane will contain all of the items. It is editable.
d. Change the Title of Report to describe the assignment.
e. Type in a comment such as **Your Name** and the **Date**.
f. Right click the ReportPad folder and a brief menu appears.
   Use this menu to save and print the document.

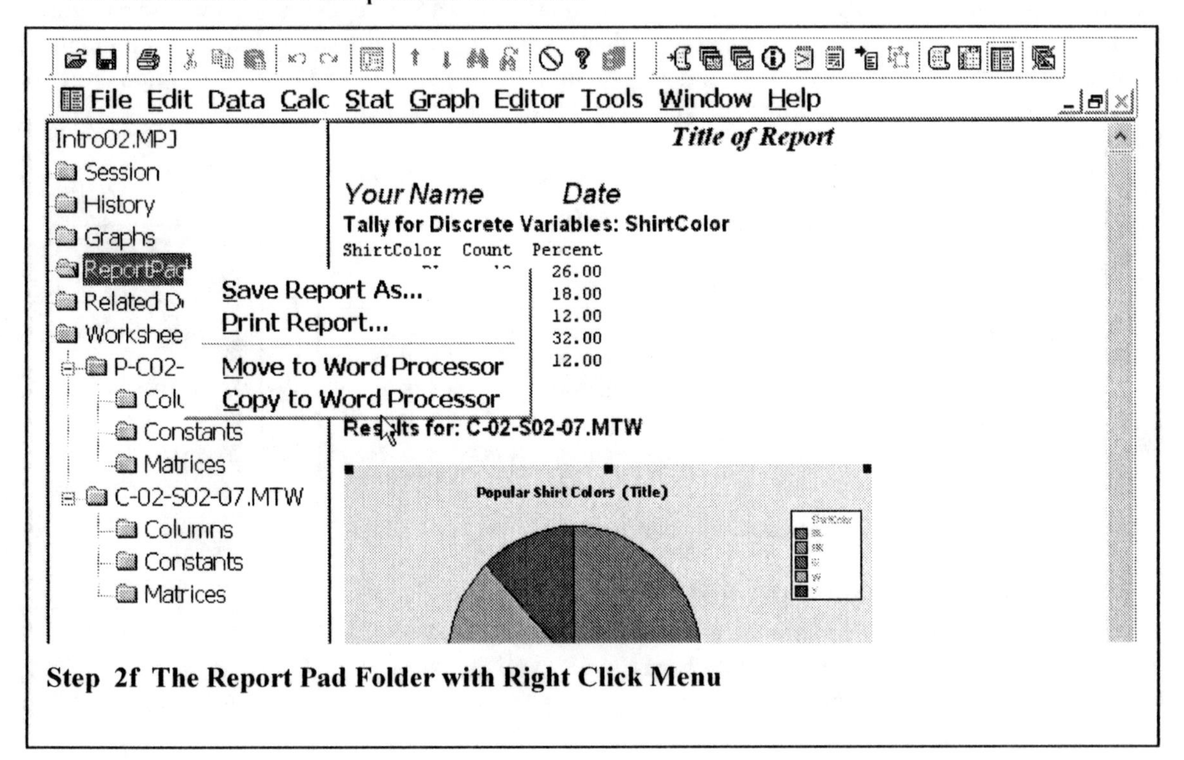

Step 2f The Report Pad Folder with Right Click Menu

Save the Report

Step 3. Click Save Report As... .
   Choose a name such as Intro02. The document will be saved in rich text format, automatically adding the RTF extension.

Print the Report

Step 4. Right click the ReportPad folder again.
   a. Choose Print Report....
   b. Your printer dialog box will open. Click **[OK].**

Copy or Move the Report

Step 5. *Optional:* Skip this step if you aren't sure you have a word processing program installed on your computer. Right click the ReportPad folder then choose Move to Word Processor to copy the document to your favorite word processing program such as Microsoft Word or WordPad. The ReportPad contents are erased. If you choose Copy to Word Processor, the document is copied to your word processing program and the ReportPad contents are unchanged.

Start a New Project

Step 6. To continue with the next section, select **File>New>Minitab Project.**
   All work and data will be cleared. You and MINITAB are ready to start a new project.

2-2 Frequency Distributions (Tables)

Construct an ungrouped distribution — Section 2 Exercise 9: Number of cups of coffee
Construct a grouped distribution. — Section 2 Example 2: Record high temperatures

Construct an Ungrouped Frequency Distribution (Discrete Variable)

Use the same commands to construct an ungrouped table for a discrete quantitative variable that we used in the introduction for a categorical table. This would also work for categorical data whose categories have been coded with integers. For ordinal or higher level data you may want to calculate cumulative counts and percents in addition to the counts and relative frequencies. In exercise 9, the number of cups of coffee consumed with a meal by thirty customers at a restaurant is shown.

Enter the Data

Step 1. Click in the space for the label above the first row in C1.

a. Type the name **Coffee**.

Coffee					
0	2	2	1	1	2
3	5	3	2	2	2
1	0	1	2	4	2
0	1	0	1	4	4
2	2	0	1	1	5

Worksheet	C1
↓	Coffee
1	0
2	2
3	2
4	1

Step 1

b. Read across the list. Type each number down column 1.

Step 2. Select **Stat>Tables>Tally Individual Variables...**
The dialog box may contain the settings from the shirt colors.
Note: Press <F3> to clear a dialog box.

a. Double click on column C1 Coffee.

b. Check the boxes for
Counts
Cumulative Counts
Percent
Cumulative Percents

c. Click **[OK]**.

The results are displayed in the Session window. No one had more than five cups of coffee with a meal. More than three-quarters, 77 percent, had two or less while 7 percent had five. One-third, or ten out of the thirty, had exactly two cups of coffee with a meal.

Tally for Discrete Variables: Coffee

Coffee	Count	CumCnt	Percent	CumPct
0	5	5	16.67	16.67
1	8	13	26.67	43.33
2	10	23	33.33	76.67
3	2	25	6.67	83.33
4	3	28	10.00	93.33
5	2	30	6.67	100.00
N=	30			

Save the Project

Step 3. Click on the Disk icon on the menu bar.
The Save Project As dialog box will open.

a. Change the directory to Projects. Projects

b. Name the file **Project2-2.**

c. The file type should be MINITAB Project with an extension of MPJ.
Continue. Do not exit MINITAB. Click the File icon at anytime to save your work.

Construct a Grouped Frequency Distribution

Open the Worksheet File

The following data represent the record high temperatures for each of the fifty states as of 1994.

TEMPERATURES

112	110	107	116	120	100	118	112	108	113
127	117	114	110	120	120	116	115	121	117
134	118	118	113	105	118	122	117	120	110
105	114	118	119	118	110	114	122	111	112
109	105	106	104	114	112	109	110	111	114

Step 1. Select **File>Open Worksheet**.

a. Navigate to the directory with the files.

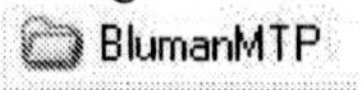

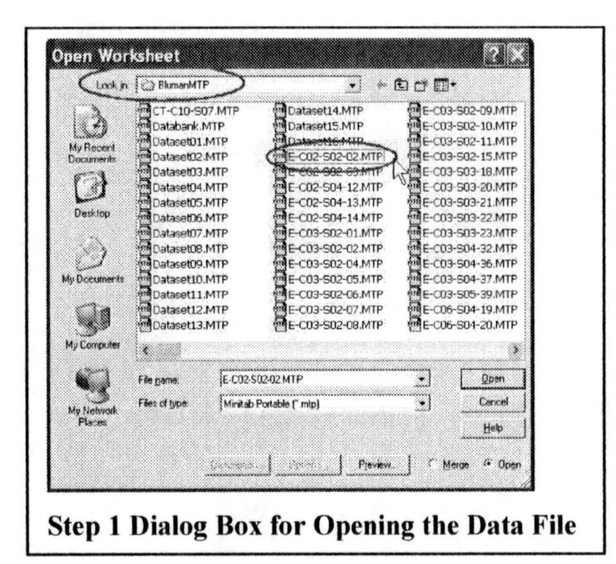

Step 1 Dialog Box for Opening the Data File

b. Double click the name of the file: E-C02-S02-02.MTP.

The data is copied into a new worksheet.
There will now be two worksheets in the project.

There is more than one way to make a grouped frequency table using MINITAB. Here is one of them.

Construct the Character Graph Table

Add the character graph option to the Graph menu.

Step 1. Select **Tools>Customize.**

a. Scroll the list of categories.

b. Click Character Graphs in the command list.

c. Position the mouse over the command.

d. Drag the command up to the Graph menu. The menu will open up.

e. Position the mouse at the top of the menu just above Scatterplot..., then release the mouse.

f. Click **[Close]**. The Graph menu will now have the character graph at the top of the list.

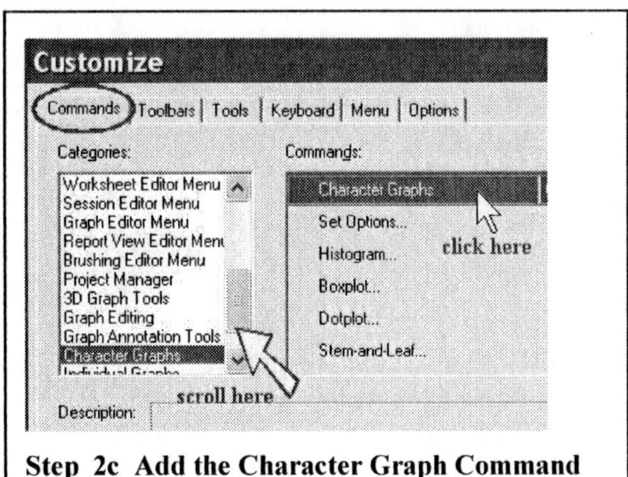

Step 2c Add the Character Graph Command

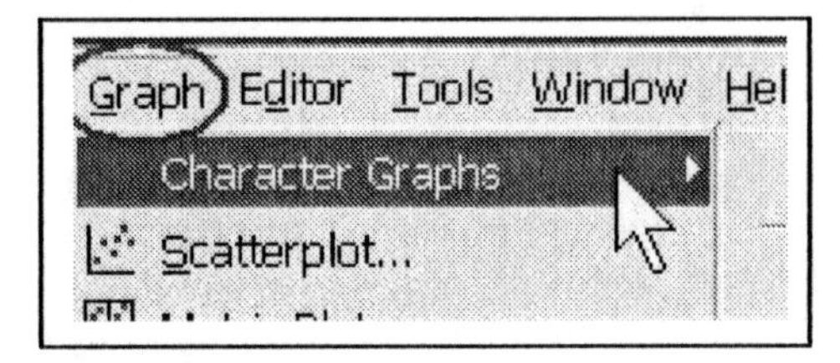

Step 2. Select **Graph>Character Graph>Histogram**.
The cursor should be blinking in the dialog box.

Step 3. Double click **C1 Temperature** in the list box. Do not change any of the dialog or check boxes.

Step 4. Click **[OK]**.

```
Histogram
Histogram of TEMPERAT    N = 50
Midpoint          Count
     100              1  *
     105              6  ******
     110             14  **************
     115             13  *************
     120             14  **************
     125              1  *
     130              0
     135              1  *
```

The table will be displayed in the Session window as shown. It is possible to control the intervals. Here is how.

a. Determine the class width and the first midpoint using the textbook instructions on pages 39 and 40.

   width = 5   first midpoint = 102

b. Click the Edit Last Dialog Box icon on the Standard toolbar.

The settings in the dialog box should not have changed.

Step 5. Change the midpoints and intervals.

a. Type **102** for the First midpoint:.
b. The last midpoint should be **132.**
c. The increment should be **5.**
d. Click **[OK].**

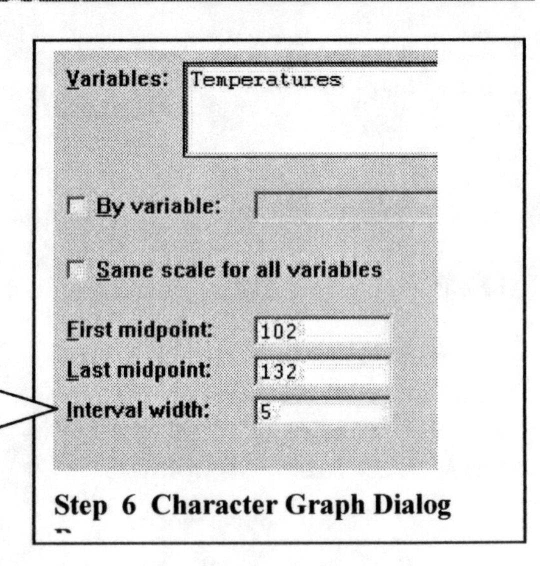

Step 6 Character Graph Dialog

There are now seven classes.
All character graphs are displayed in the session window. You will see the midpoints of each class, the frequencies (counts), and a simple bar graph for each class.

```
Histogram
Histogram of TEMPERAT    N = 50
Midpoint          Count
  102.00              2  **
  107.00              8  ********
  112.00             18  ******************
  117.00             13  *************
  122.00              7  *******
  127.00              1  *
  132.00              1  *
```

Finish the Project

Step 6. Save the project (**Project2-2.MPJ**)

a. Create, save, and print the report.
b. Select **File>New>Minitab Project** or **File>Exit** to quit for now.

2-3 Graphs for Quantitative Data

Graphs commonly used for organizing sets of quantitative data are histograms, frequency polygons, and ogives. All of these graphs are variations of the **Graph>Histogram** command. In the instructions you will make the basic histogram and then change the dialog boxes that will create the relative frequency graphs, frequency polygons, and ogives. The histograms are graphical views of a grouped frequency distribution. They are high resolution graphs displayed in a separate graph window that you can save, print, or copy. The data must be entered in the worksheet from a file or typed in.

New Instructions:	***Data/Example/Exercise***
Construct a Histogram	Section 2 Example 2 Record High Temperatures
Construct a Frequency Polygon	
Construct an Ogive	
Create multiple graphs on the same plot	Section 2 Exercise 18 and Section 3 Exercise 12

Construct a Histogram

Step 1. Type the data into C1 of a MINITAB Worksheet or Open the Worksheet file.

Step 2. Select **Graph>Histogram>Simple** .

Step 3. Double click C1 TEMPERATURES to select it for the Graph variable.

Step 4. Click **[Scale...]**. Select the Gridlines tab.

- **a.** Check Y major ticks.
- **b.** Check X major ticks. (Neither are shown.)
- **c.** Select the Y-Scale Type tab then check Frequency then **[OK]**. *Note* : It is here that you could change the Y axis to relative frequency (percent) instead of counts.

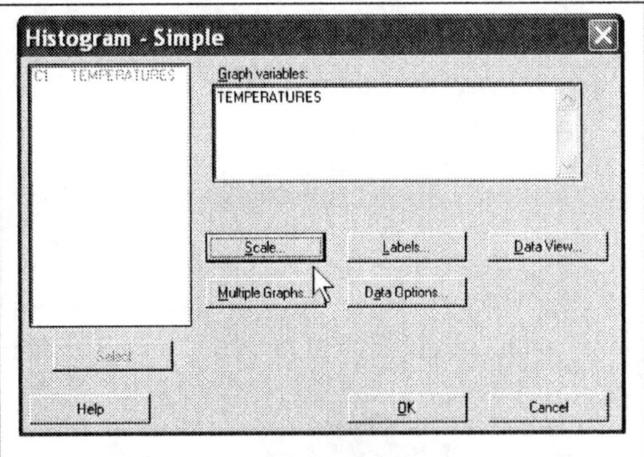

Step 3 Histogram Dialog Box

Step 5. Click **[Labels]**.

- **a.** Type an appropriate title
  Record High Temperatures
- **b.** Type **Your Name** and **Date** in the footnotes then click **[OK]**.

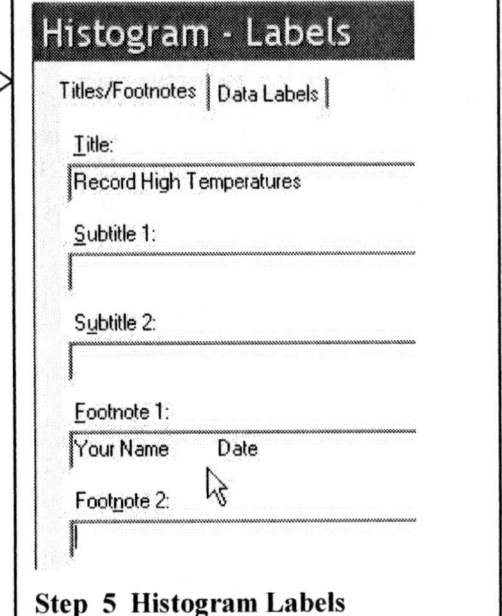

Step 5 Histogram Labels

Step 6. Click **[Data View]**. (not shown)

- **a.** Select the tab for Data Display, then check the box for Bars.
- **b.** Click **[OK]** twice.

Step 7. Click on **[Data Options]**.

- **a.** Click the tab for Group Options.
- **b.** Uncheck both options, then click **[OK]** twice.

The graph is created in a Graph window.

Edit Graph: Change the Histogram Colors

Step 8. Right click on any bar of the histogram.

a. The bars will be outlined and the Editor menu opens.

b. Select Edit Bars… . A dialog box opens.

c. Select the tab for Attributes.

1. Change the Fill pattern option to Custom.
2. Change the color to light aqua.
3. Click **[OK].** Bar colors will change immediately.

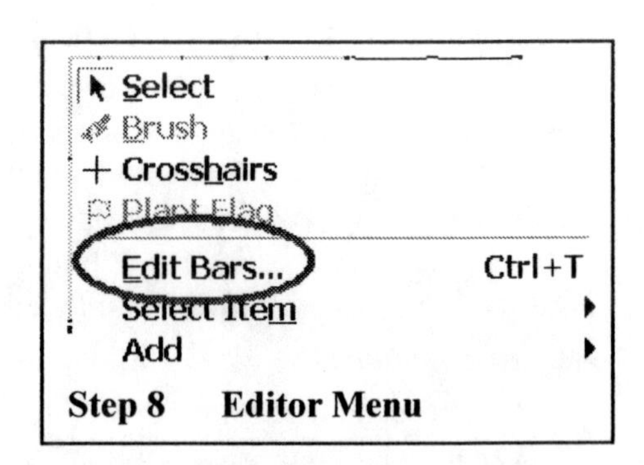

Step 8 Editor Menu

Edit Graph: Number of Classes/Intervals

Change the number of histogram groups/classes. To make seven classes as the exercise requests, the X-scale has to be edited. Here is how. The Graph window must be the active window.

Step 9. Select the Editor menu.

a. Click on **Select Item>X-scale.**

b. Select the **Editor menu>Edit X-scale.**

c. Click the tab for Binning, then the option for Cutpoint.

d. Click the button for Midpoint/Cutpoint positions. Type in the textbox: **99.5:134.5/5.** This is the shortcut for entering the sequence of 99.5 to 134.5 by 5. These are boundaries for each class.

e. Click **[OK].**

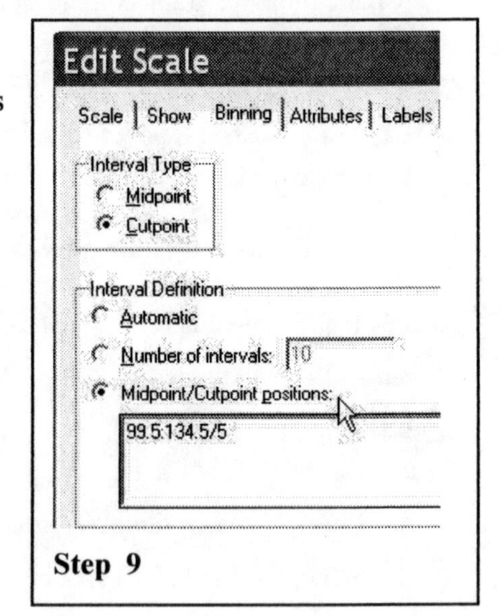

Step 9

The number of classes will change immediately.
Next, change the color of the bars as directed in the introduction.

Step 10. Right click on the green plus to see the Graph window control box.

Step 11. Right click this Control icon.
There are options available for saving, printing, naming, and closing this graph. The Printer icon, for example, will print graphs, not a worksheet or the Session window.

Step 12 Control Box Menu

Step 11 New colors and X-scale

Step 12. Press <Esc> to cancel the menu.

Analyze the graph. The peak is on the interval for temperatures of 109.5 to 114.5 degrees. There are no gaps between the bars.

Step 13. Click the Graph window's minimize button.

Caution: Don't use the Close icon or the graph will be erased.

Construct a Frequency Polygon

Frequency polygons and ogives can be done with another graph type.

A frequency polygon is a line graph.

Step 1. Select **File>New>Minitab Worksheet**.

Step 2. Enter the midpoints and frequencies from the character graph of temperatures into two columns of the worksheet. Use one midpoint higher and lower. They will be needed to anchor the endpoints.

↓	C1	C2
	X	f
1	97	0
2	102	2
3	107	8
4	112	18
5	117	13
6	122	7
7	127	1
8	132	1
9	137	0

Step 3. Select **Graph>Scatterplot>With Connect line**.

Step 4. In the Data Source dialog box, double click C2 f for Y variables and C1 X for X-variables. The buttons are similar to the histogram buttons.

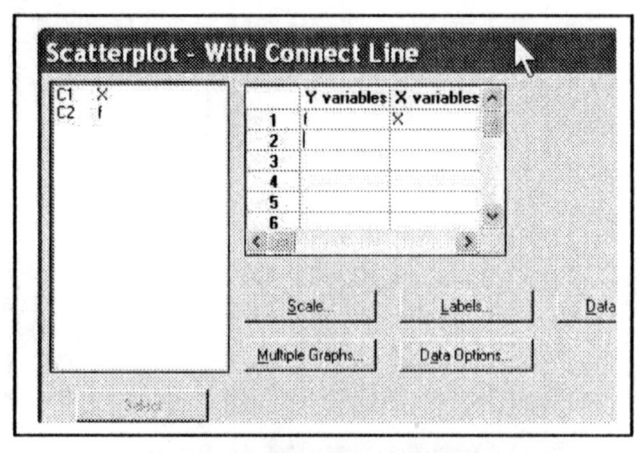

Step 5. Click **[Data View],** then the Data Display tab.

a. Check two options: Symbols and Connect line.

b. Click **[OK].**

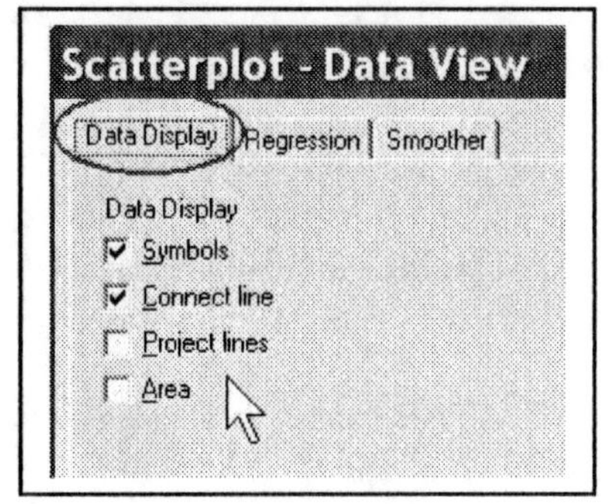

Step 6. Click **[Scale].**

a. Click the tab for Gridlines, then select the major ticks for X and Y.

b. Click the tab for Reference Lines then type a **0** for the Y positions.

c. Click **[OK]** twice.

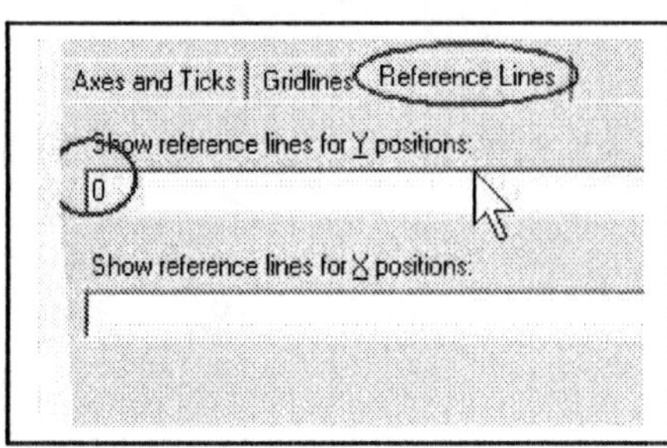

Step 7. Select **Editor>Select Item>X Scale**.

- **a.** In the Scale tab, click the option for Positions of ticks:.
- **b.** Type **97:137/5.**

This abbreviation means
97 to 137 by increments of 5.

- **c.** Click **[OK]** twice.

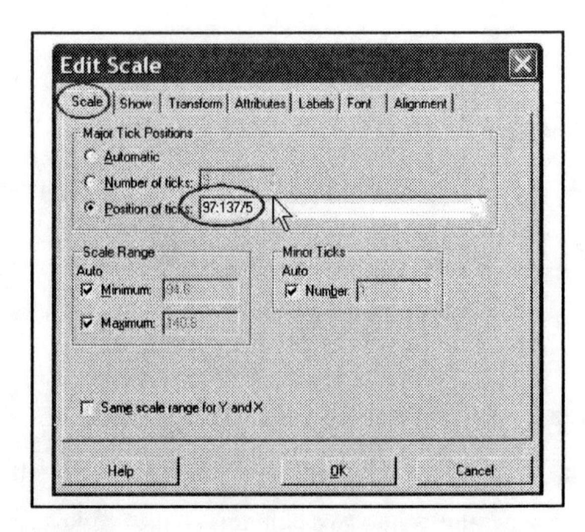

Step 8. Right click the control box, then select **Save Graph As...** .

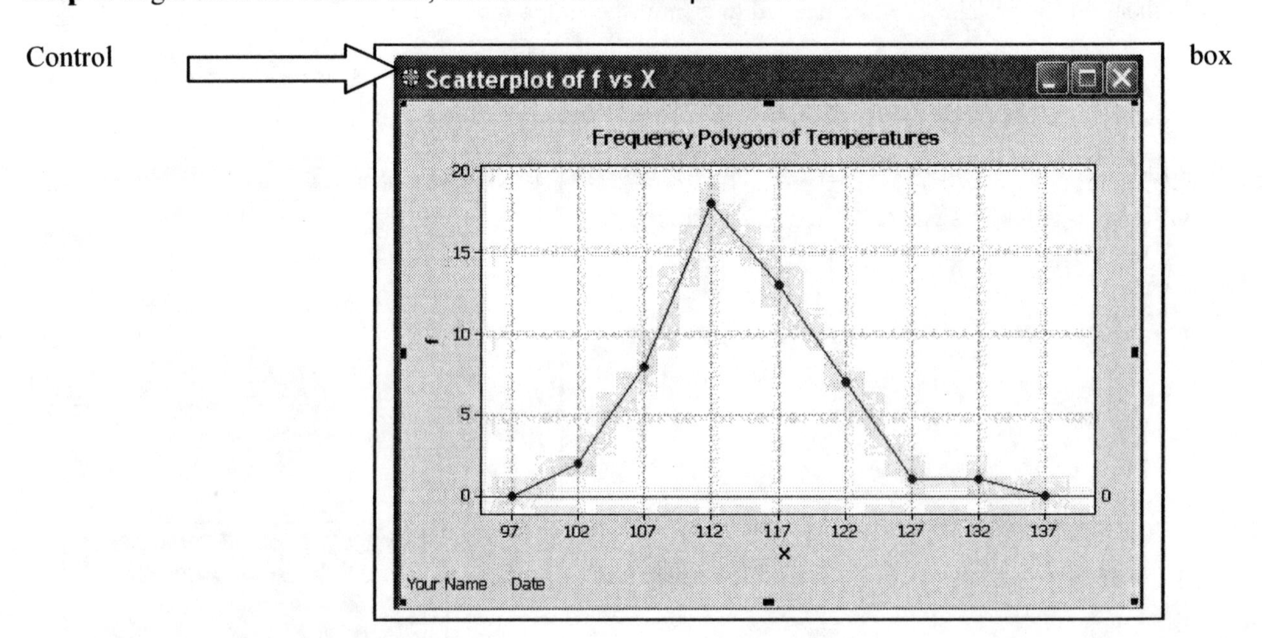

- **a.** Name the graph **FreqPolyTemp.**
- **b.** The file extension MGF will be added to the name. This is the extension for a MINITAB graph file.

The MINITAB graphics file, mgf, takes very little space, only 7Kb. In JPG format, this graph requires about 30Kb of space. The MINITAB format can be inserted as an object in most word processing or presentation programs.

Caution!
A project containing a lot of graphics may not fit on a floppy disk.

Step 9. Use **File>Save Project As** and save it as **Project2-3.MPJ.** Keep going.

Construct an Ogive

Repeat the steps for a frequency polygon with two minor changes:

Step 10. Change the distribution values in the worksheet.

a. Change the X values to the class limits from **100** to **135** by **5.**

b. Change the frequencies to cumulative frequencies.

C1	C2
X	f
100	0
105	2
110	10
115	28
120	41
125	48
130	49
135	50

Step 11. Select **Graph>Scatterplot>With Connect line.** Don't change the settings.

Step 12. Click **[Labels].** Change the title.
Ogive of Record High Temperatures

Step 13. Click **[OK]** twice. It's a beauty!

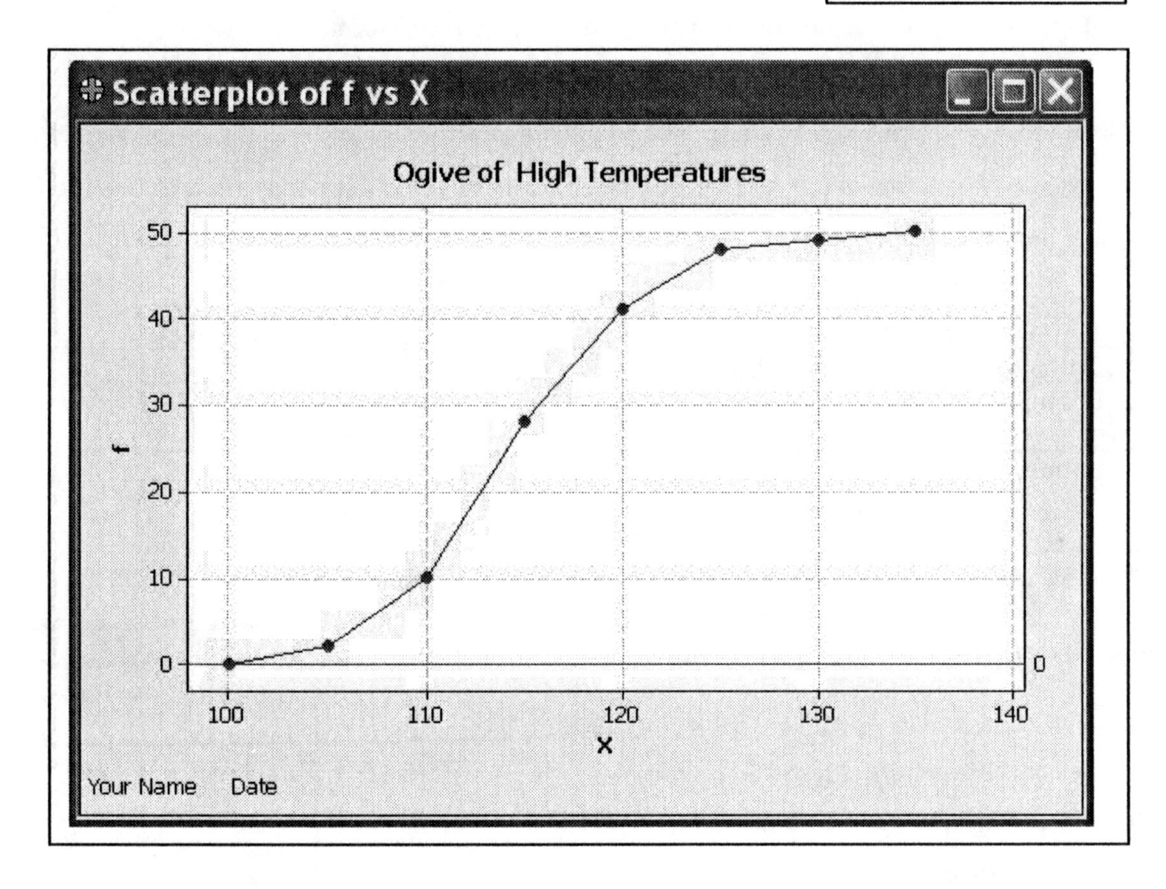

Step 14. Save the project as **Project2-3.MPJ.**

Step 15. Create and print a report including the Session window items and the graphs.

Step 16. Exit MINITAB.

Construct a Histogram from a Frequency Distribution

New in MINITAB Release 14, many graphs can be made from data in a frequency table. Exercise 3 in section 3 contains a frequency distribution for the scores of the Ladies Professional Golf Association tournament at the Giant Eagle Tournament during the 2002 season. The scores are summarized in a table. To make the graph, use the midpoints for each class and the frequencies.

C1	C2
Xm	f
203	2
206	7
209	16
212	26
215	18
218	4

→

↓	C1	C2
	LPGA Scores	f
1	203	2
2	206	7
3	209	16
4	212	26
5	215	18
6	218	4

Step 1. Enter the data into the first two columns of a MINITAB worksheet.

Step 2. Select **Graph>Histogram>Simple,** then click **[OK].**

Step 3. In the Histogram Simple dialog box, double click LPGA Scores as the graph variable. Use each button to make the settings summarized in the table. You may need to click **[OK]** as you complete each button.

Histogram Buttons	Tab Name	Settings and Check Boxes
1. **[Data Options]**	Frequency	**C2 f**
2. **[Scale]**	Y-Scale type	frequency
	Gridlines	Check all four options.
3. **[Labels]**	Title	LPGA Scores
	Subtitle	Giant Eagle 2002
	Footnote1	**Your Name** and **Date**
4. **[Data View]**	Data Display	Bars, the default

Step 4. Click **[OK]** so the Graph window opens with the histogram.

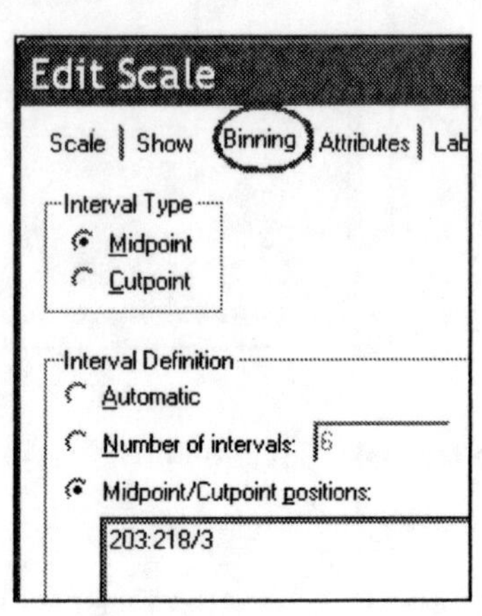

Step 5. The X-Scale must be adjusted using the Graph Editing menu.

a. Select **Editor>Select Item>X-Scale.**

b. Select **Editor>Edit X-Scale,** then click the tab for Binning and click Midpoint for Interval Type.

c. Click the Midpoint/Cutpoint positions:, then inside the box type

203:218/3

d. Click **[OK].**

Step 6. Right click the control box then select Print graph.

Step 7. Select **File>New>Minitab Project,** then save the project as **LPGA.MPJ.**

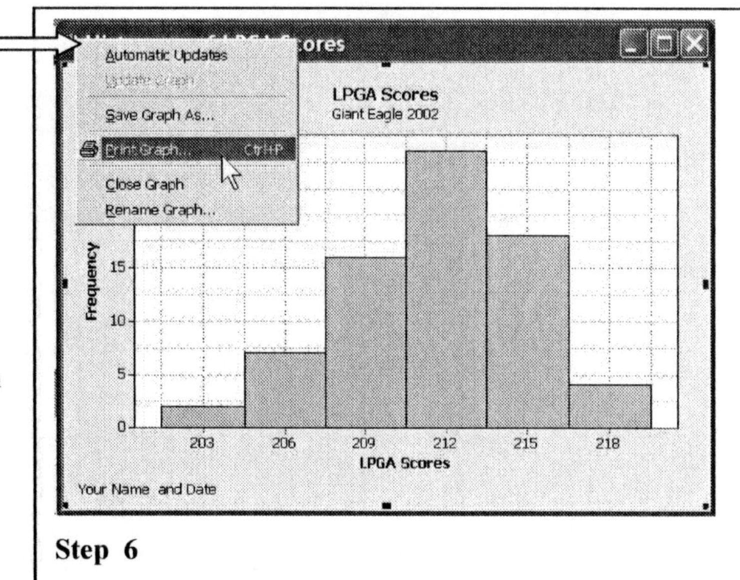

Step 6

2-4 Other Graphs

Bar charts, pie charts, and Pareto charts are used to summarize sets of qualitative data, while a time series plot shows how some quantitative variable changes over time.

New Instructions:	*Data/Example/Exercise*
Construct a bar chart and Pareto Chart	Super Bowl Snack Foods Section 4 Example 10
Construct a time series plot.	Drive-In Theaters Section 4 Example 9

Construct a Bar Chart from a Frequency Distribution

Step 1. Enter the data into the first two columns of MINITAB.

a. Name C1 **Snack.** They are entered here in alphabetical order.

b. Name C2 **f**.

c. Select **File>Save Current Worksheet As....**

d. Save the worksheet as **E-C02-S04-10.MTW.**

↓	C1-T	C2
	Snack	f
1	Peanuts	2.5
2	Popcorn	3.8
3	Potato Chips	11.2
4	Pretzels	4.3
5	Tortilla Chips	8.2
6		

Step 2. Select **Graph>Bar Chart.**

a. Click on the drop-down list in Bars represent:, then select Values from a table.

b. Click on the Simple chart then click **[OK].**

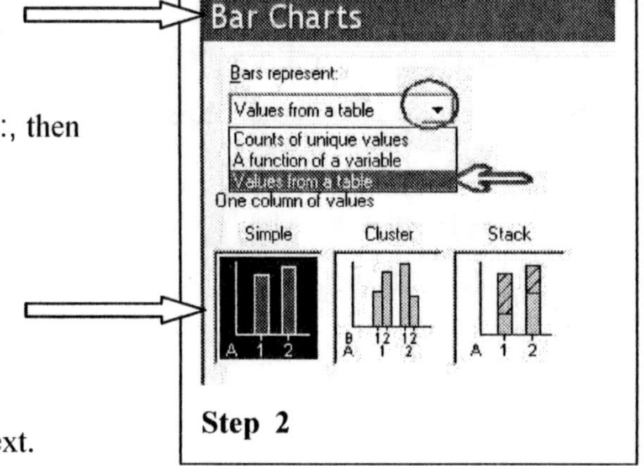

Step 2

The settings needed for the Bar Chart dialog box are next.

c. Double click the frequency column, C2 f for Graph variables:.

d. Press <Tab> to move the cursor to the Categorical variable text box.

e. Double click Snack for the Categorical variable.

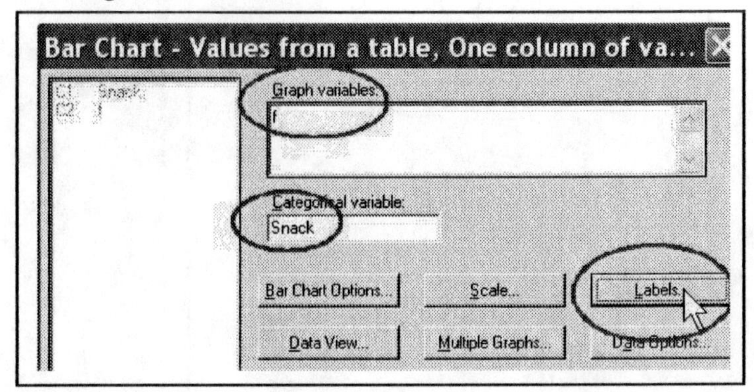

Step 3. Click on **[Labels]**. The Titles/Footnote tab should be open.

Step 4. Type the title in: **1998 Super Bowl Snacks**. Click **[OK]**.

Step 5. Click the button for **[Bar Chart Options],** then click the option to show Y as a percent.

a. Click **[OK]** twice.

The chart will open in a Graph window. The bars are in the order the data was entered in the rows of the worksheet.

b. Right click any bar then

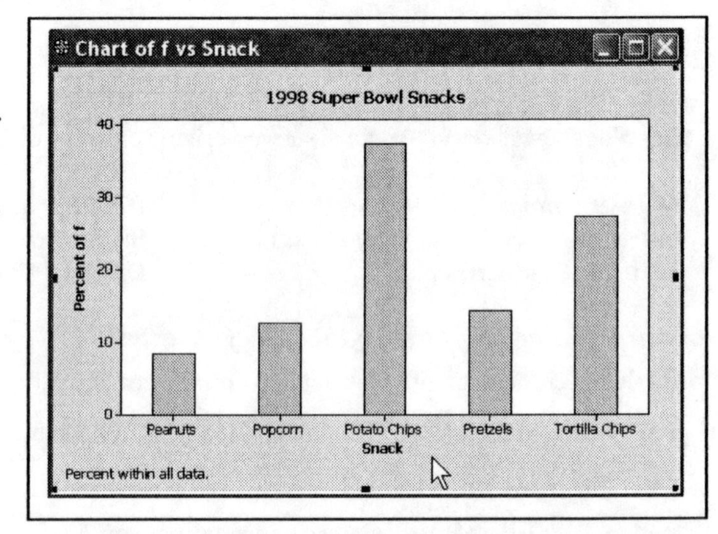

Construct a Pareto Chart

Pareto charts are a quality control tool. They are similar to a bar chart with no gaps between the bars and the bars are arranged by frequency.

Step 1. Select **Stat>Quality Tools>Pareto**.

a. Click the option to Chart defects table.

b. Click in the box for the Labels in: and type **Snack.**

c. In the Frequencies in: box, type **f.**

Step 2. Click on **[Options]**.

a. Type in the title, **1998 Super Bowl Snacks**.

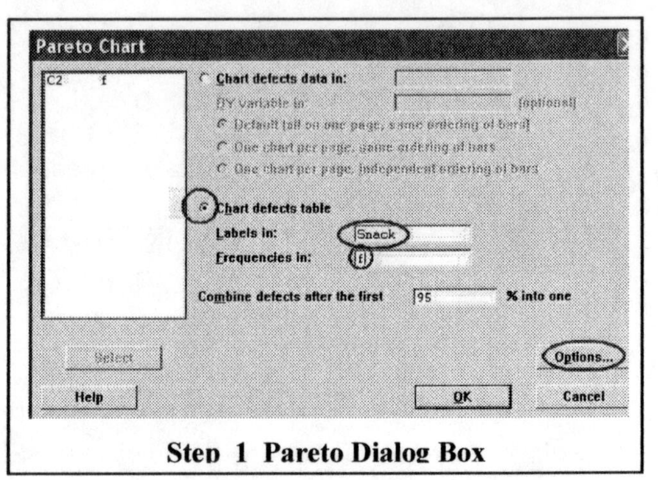

Step 1 Pareto Dialog Box

b. Click **[OK]** twice.

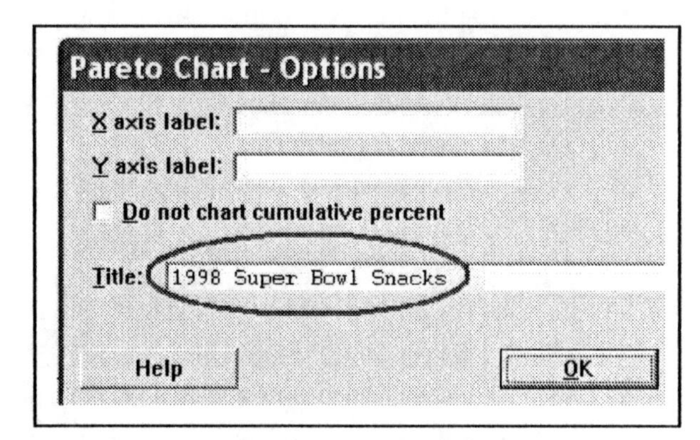

The chart is complete.

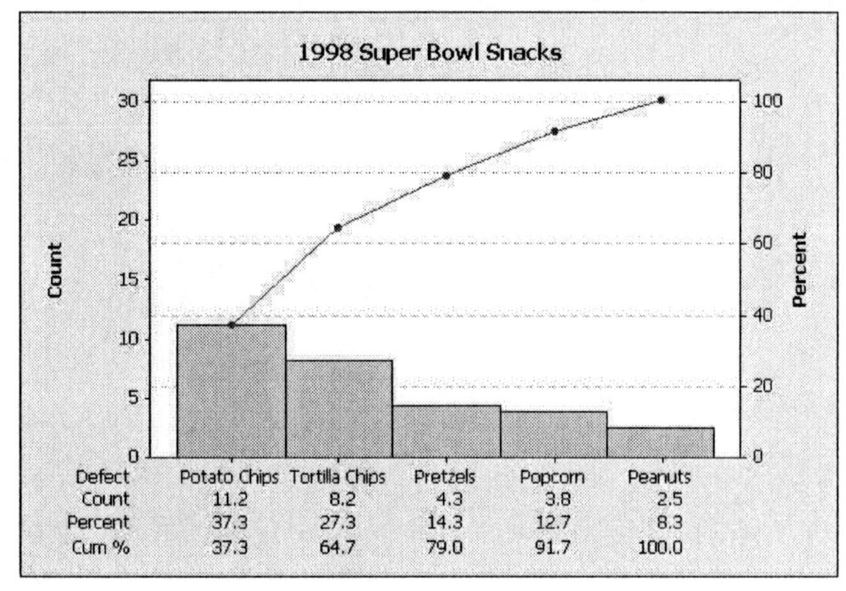

Construct a Time Series Plot

The data used is from Section 4 Example 9, the number of drive-in theaters operating from 1988 to 2000. The data was collected in order and should be kept in sequence. That's what makes it a time series.

Step 1. Add a blank worksheet to the project by selecting **File>New>Minitab Worksheet**. Click **[OK]**. Here is a shortcut for entering a sequence of values, the years from 1988 to 2000.

Step 2. Select **Calc>Make Patterned Data>Simple Set of Numbers**.

a. Type Year in the text box for Store patterned data in:.

b. From first value: should be **1988**.

c. To last value: should be **2000.**

d. In steps of: should be **2** (for every other year).

e. The last two boxes should be **1**, the default value.

f. Click **[OK]**. The sequence from 1988 to 2000 will be entered in C1 whose label will be Year.

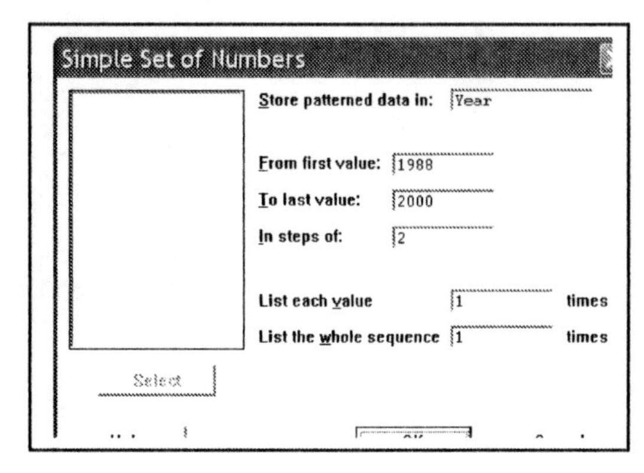

Step 3. Type the number of drive-ins into C2 then name the column **Drive-Ins.** This data is the time series, recorded every other year for twelve years.

↓	C1 Year	C2 Drive-Ins
1	1988	1497
2	1990	910
3	1992	870
4	1994	859
5	1996	826
6	1998	750
7	2000	637

Step 4. To make the graph, select **Graph>Time Series Plot,** then Simple, and press **[OK].**

Step 5. For Series, select Drive-ins, then click **[Time/Scale].**

a. Click the Stamp option and select Year for the Stamp column.

b. Click the Gridlines tab and select all three boxes, Y major, Y minor, and X major.

c. Click **[OK]** twice.

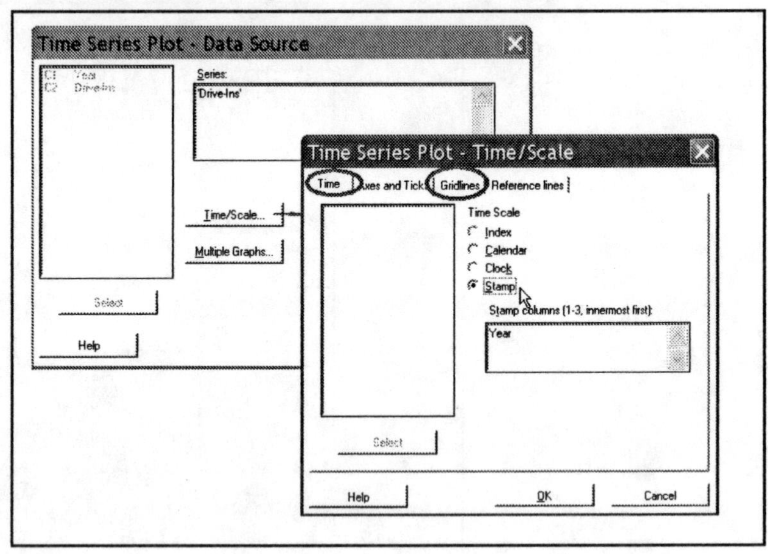

A new window will open containing the graph.

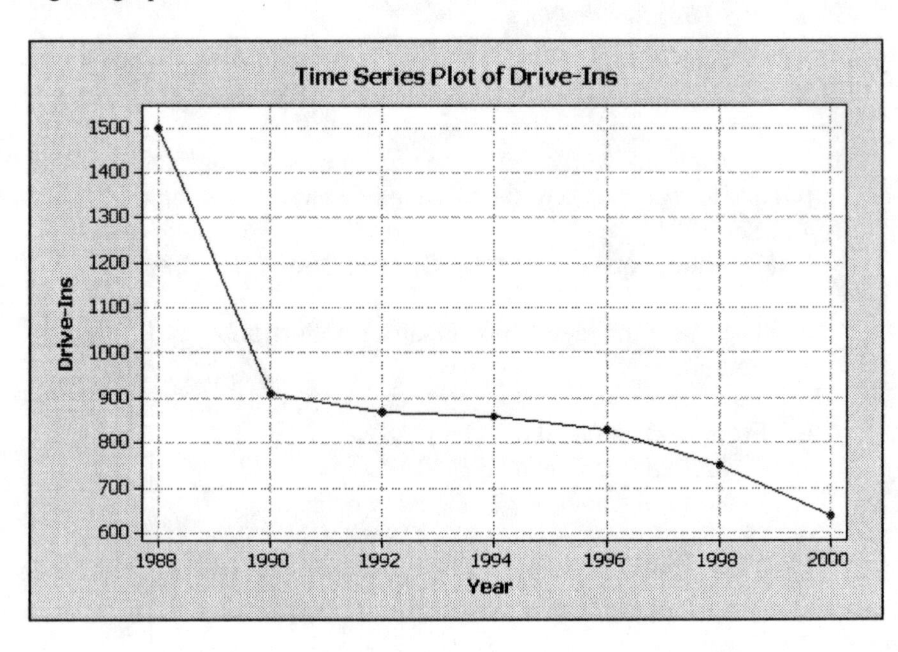

Make a Time Series Graph Using Scatterplot

MINITAB assumes the data was collected at equal time intervals. If the time sequence is irregular you must make the graph using **Graph>Scatterplot.** This MINITAB graph can be used to plot any two quantitative variables. Date variables can also be used as a numeric variable. In this example, the years will be entered as a date type. This type of data can behave much differently than numeric data. The data is from Exercise 15 in the Review and Practice for chapter 2, Minimum Wage from 1960 through 2000.

Step 1. Select **File>New>Minitab Worksheet.**

If prompted, click **[Yes]** to add the worksheet to the project. Next, enter the sequence of years as a date variable and the minimum wage in column 2. Do not use the dollar sign!

Step 2. Select **Calc>Make Patterned Data>Simple Set of Date/Time Values... .**

a. Type **Yr** in the text box for storing the data.

b. Use a start date of **1/1/1960** and an end date of **12/31/2000.**

c. Select Year from the drop-down list for Increment: and then type **5** in by:.

d. Click **[OK].**

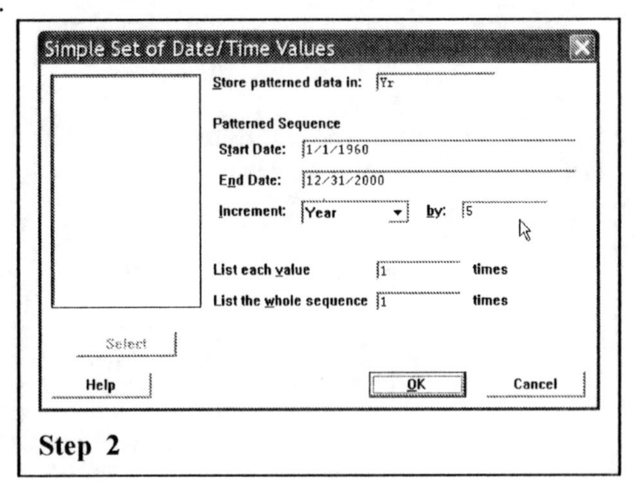

Step 2

Step 3. Enter the wage data in column 2.

a. Name the column **MinWage.**

b. Type **1/1/2004** for the year and **5.15** in the tenth row for the 2004 minimum wage.

Next, create the scatterplot for the time series.

Step 4. Select **Graph>Scatterplot>Simple.**

Step 5. Select C2 MinWage for the Y-variable and C1 Yr for the X-variable.

↓	C1-D	C2
	Yr	MinWage
1	1/1/1960	1.00
2	1/1/1965	1.25
3	1/1/1970	1.60
4	1/1/1975	2.10
5	1/1/1980	3.10
6	1/1/1985	3.35
7	1/1/1990	3.80
8	1/1/1995	4.25
9	1/1/2000	5.15
10	1/1/2004	5.15

Sten 3

Step 6. Click **[Scale]** then the tab for gridlines.

a. Select the major ticks' options for the X and Y scale.

b. Click **[OK].**

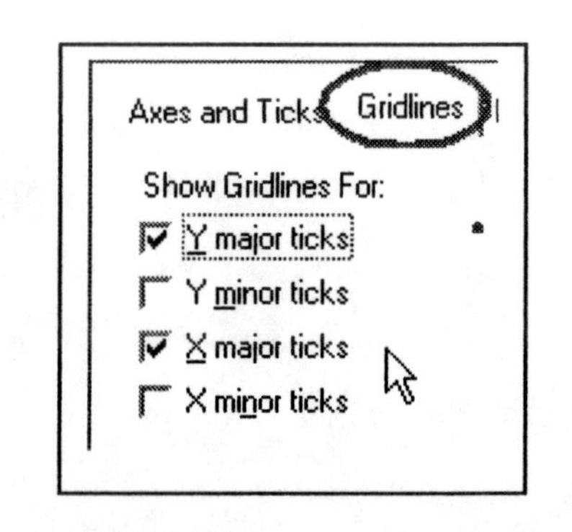

c. Click **[Labels].**

1. Type the title **Minimum Wage.**
2. Type **your name** in the footnote text box.
3. Click **[OK].**

d. Click **[Data View].**

1. Select the Data Display tab.
2. Check the options for **Symbols** and **Connect line.**
3. Click **[OK]** twice.
   The Graph window with the plot will open.

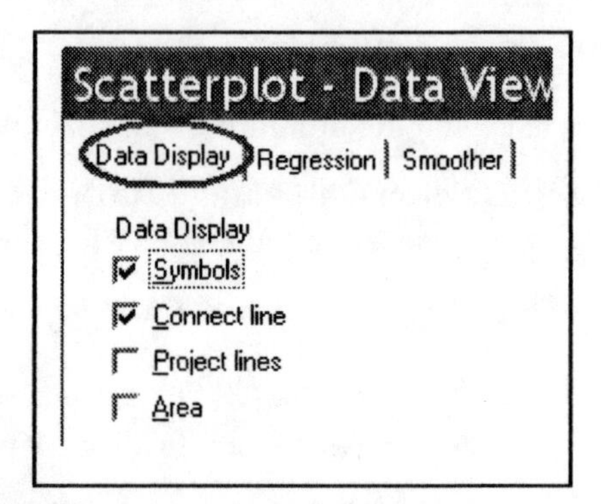

Step 7. Select **Editor>Select item.**

a. Select X-scale in the menu.

b. Select **Editor>Edit X Scale.**

c. Click the Alignment tab. Change the text angle to **60.**

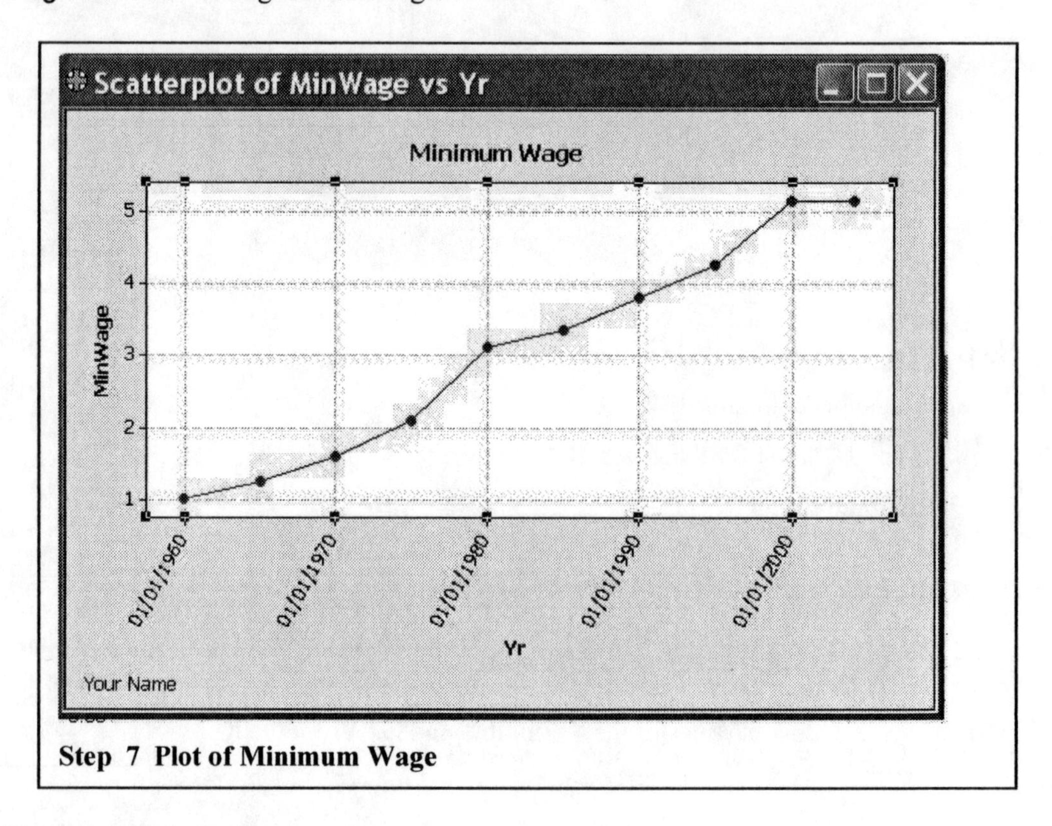

Step 7 Plot of Minimum Wage

d. Click **[OK].** The graph is displayed as shown.

Construct a Stem-and-Leaf Plot

Step 1. Enter the data for Example 13 in Chapter 2 Section 4 into a column of a MINITAB worksheet. Type it from the list below or open the file from the data file.

a. Select **File>Open worksheet,** then navigate to the BlumanMTP file folder.

b. Double click E-C02-S04-13.MTP.

c. Label the column **CarThefts.**

```
CarThefts
   52   58   75   79   57   65   62   77   56   59   51   53   51   66   55
   68   63   78   50   53   67   65   69   66   69   57   73   72   75   55
```

Step 2. Select **STAT>EDA>Stem-and-Leaf.** This is the same as **Graph>Stem-and-Leaf.**

a. Double click on C1 CarThefts in the column list.

b. The check mark for Trim outliers should be removed.

c. Click in the Increment text box, and enter the class width of **5.**

d. Click **[OK].**

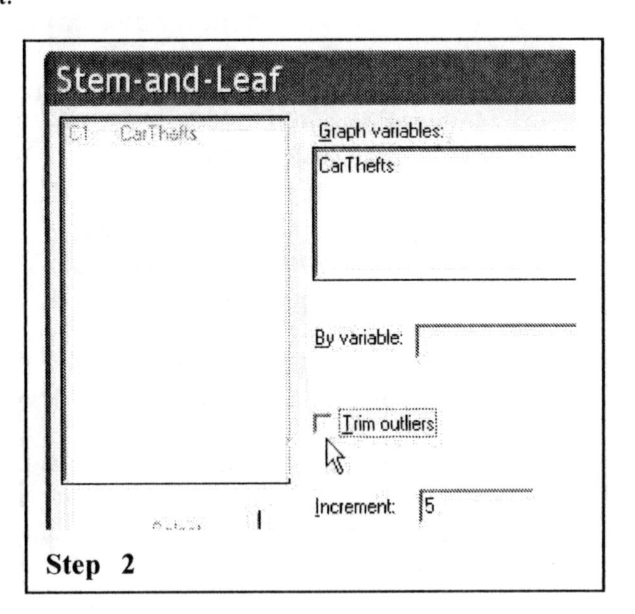

Step 2

Stem-and-Leaf Display: CarThefts

```
Stem-and-leaf of CarThefts  N = 30
Leaf Unit = 1.0
 6   5  011233
 13  5  5567789
 15  6  23
 15  6  55667899
 7   7  23
 5   7  55789
```

This Character graph will be displayed in the Session window.

Step 3. Save the project. Select **File>Save Project As** and name the project **Project2-4.**

Create and print a report

Step 4. Click the Program Manager icon then click the Session window folder.

a. One at a time, select the titles of each item in the Session window, then right click and choose Append to Report.

b. Right click the ReportPad folder, then scroll through the report and add any comments that you choose.

c. Right click the ReportPad folder, then choose Print the report.

Step 5. Click the Close icon in the upper right-hand corner of the window to exit MINITAB.

2-5 Data Analysis

Project 1: From the Databank choose one of the following variables: age, weight, cholesterol level, systolic pressure, IQ, or sodium level. The following abbreviated instructions leave it up to you to fill in some detail--when to click **[OK]**, for example!

Step 1. Choose a variable, weight.
Select at least thirty values. For these values, construct a grouped frequency distribution. Draw a histogram, frequency polygon, and ogive for the distribution. Use eight classes. Briefly describe the shape of the distribution.

Select a random sample of thirty-five. Use any number that is thirty or more.

Step 2. Use **File>Open Worksheet** to open the Databank file.
Select **Calc>Random Data>Sample from Columns.**
Sample thirty-five rows from C6 Weight.
Store the sample in a new column named **Weight35.**

Make a grouped table using default classes.

Step 3. Use **Graph>Character Graph>Histogram.**
The variable should be Weight35.
Leave the rest of the boxes empty.

Make the graphs for quantitative data.

Step 4. Make a histogram. Use **Graph>Histogram** and the instructions of Section 2-3 to make a histogram or frequency polygon. The variable should be Weight35.

Note: Your random sample will be different so your graphs may be different.

Describe the weights of the individuals in your sample.

Project 2: From the Databank, select at least thirty subjects, and construct a categorical distribution for their marital status. Draw a pie chart and briefly describe the findings.

Step 5. Select a random sample of thirty-five.
Select thirty (or more) rows from the column for Marital Status.
Store the sample in a new column named **Marital30.**

Step 6. Select **Calc>Random Data>Sample from Columns**.

a. Use the column for Marital Status as the "from" variable.
b. Name the new column **Marital35.**

Make the frequency distribution with counts and percents.

Step 7. Use **Stat>Tables>Tally Individual Variables** for the new column, Marital35.
Check the boxes for counts and percents.

Optional : Select **Data>Code>Text to Text** to change from the single letter codes to words.
The dialog box is shown. It may seem strange that the From and Into columns have the same name. It is correct. The data is taken from the original, converted and put back into the same column replacing the originals.

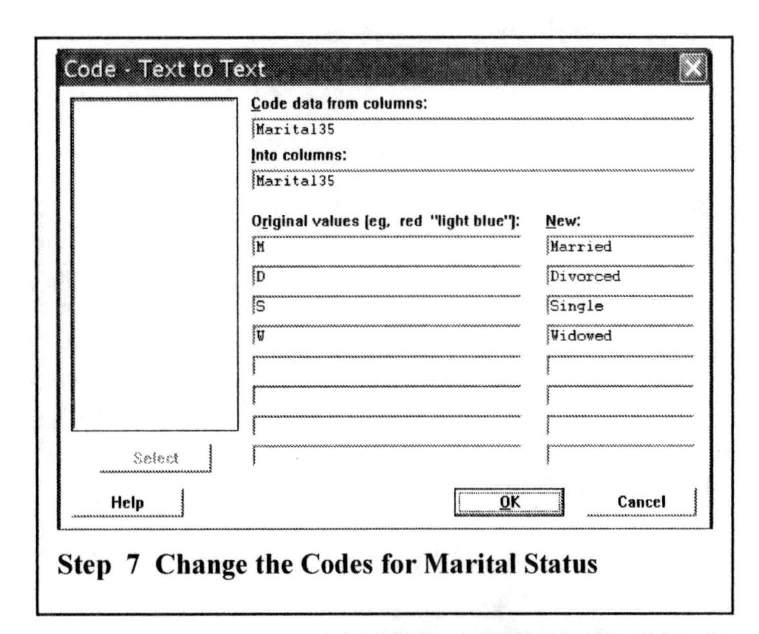

Step 7 Change the Codes for Marital Status

Step 8. Select **Graph>Pie Chart** for the data in Marital35.

- **a.** Click **[Labels…],** then type an appropriate title with your name in the footnote in the **Titles** tab.
- **b.** Click the tab for Slice Labels.
  1. Check all of the option boxes.
  2. Click **[OK]** twice.

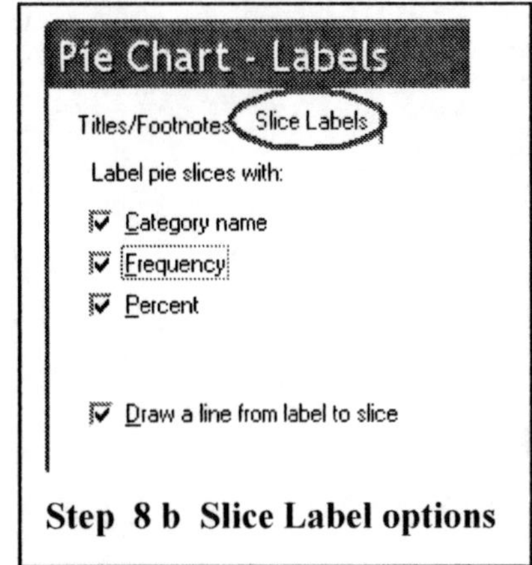

Step 8 b Slice Label options

Interpret the results:

More than half of the sample is married (54%). Only three of the thirty-five are widowed. Twenty percent are single and nearly 9 percent are divorced. Do not expect your random sample to be the same!

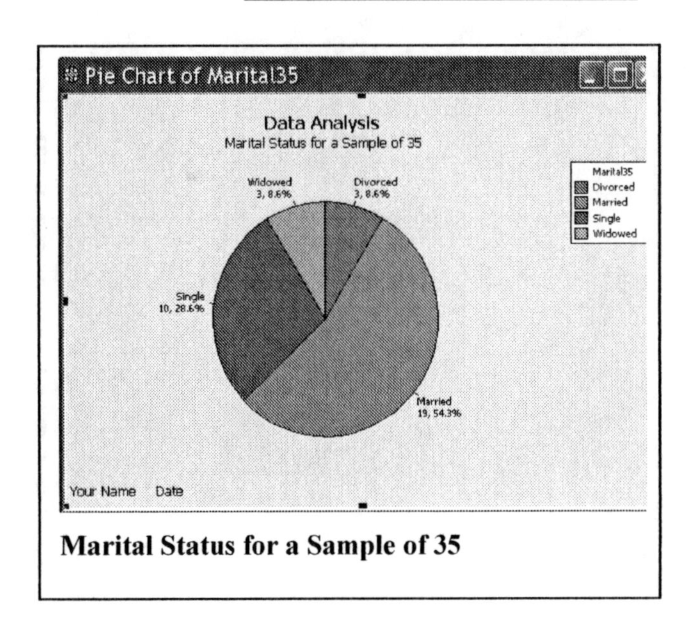

Marital Status for a Sample of 35

Finish the Project

Create and print a report.
Save the project as **DataAnalysis02.MPJ.**
Exit the program.

Chapter 2 Textbook Problems Suitable for MINITAB

Page 43	2.2	7 - 18	
Page 57	2.3	1 - 18	
Page 77	2.4	1-13, 15 – 19	
Page 73	Extend the Concepts	20 - 23	
Page 86		7 – 11, 9 -12	
Page 89	Data Analysis	1 - 3	Databank
		4	Dataset IV
		5	Dataset XII
		6	Dataset IX
		7	Dataset V
Page 93	Data Projects	1 - 3	Original data

Chapter 2: EndNotes

Chapter 3 Data Description

3-1 Introduction: The Least You Need to Know About Data Description

Use tables and graphs to describe the spread and concentration of a set of data. Use measures of central tendency to describe the typical value. Use measures of variation to describe how much difference there is in a set of data.

Calculate Measures of Central Tendency and Variation for Raw Data

Step 1. Enter the data from Example 3-23 into C1 of MINITAB.

11.2 11.9 12.0 12.8 13.4 14.3

a. Name the column **AutoSales.**

b. Save the worksheet as **E-C03-S03-23.MTP**

Step 2. Select **Stat>Basic Statistics>Display Descriptive Statistics.**

The cursor will be blinking in the Variables text box.

a. Double click C1 AutoSales.

b. Click **[Statistics]** to view the statistics that can be calculated with this command.

c. Check the option boxes for Mean, Standard deviation, Variance, Coefficient of variation, Median, Minimum, Maximum, and N nonmissing.

d. Remove the checks from other options.

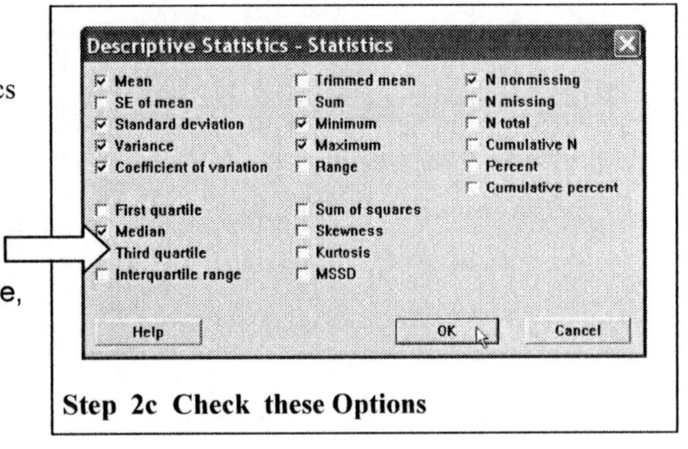

Step 2c Check these Options

Step 3. Click **[OK]** twice. The results will be displayed in the Session window as shown.

Descriptive Statistics: AutoSales

```
Variable    N*    Mean  StDev  Variance  CoefVar  Minimum  Median  Maximum
AutoSales    0  12.600  1.130     1.276     8.97   11.200  12.400   14.300
```

Step 4. Finish the project.

a. Type **your name** and **date** at the top of the Session window.

b. Print the Session window.

c. Save the project as **Intro03.MPJ.**

d. Start a new project using **File>New>Minitab Project.**

3-2 Measuring Central Tendency and Variation

The instructions for descriptive statistics in Step 2 above, calculate and display several statistics all at once. In this section, each statistic will be calculated and stored, one at a time.

Calculate and Store the Mean

Step 1. To open the worksheet for AutoSales that is part of Intro03.MPJ select **File>Open Worksheet**.

- **a.** Navigate to the directory containing your projects and double click Intro03.MPJ.
- **b.** Click **[Preview]**. A window will open that displays the contents of the file. (not shown)
- **c.** Click **[OK]** twice.

  The data will be extracted from the project and copied into your new project. If you can't find this file, type the data into the worksheet from the previous page.

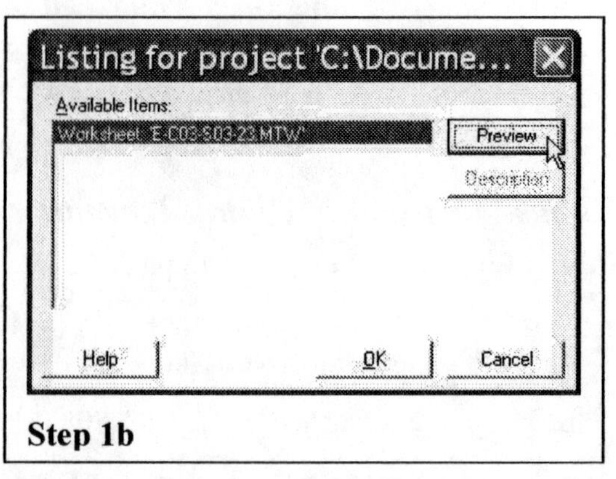

Step 1b

Step 2. Select **Calc>Column Statistics... .**

- **a.** Click the button for Mean.

  Only one of the buttons can be selected at a time. Notice the choices. The median, standard deviation, range, and more can be calculated and stored using this command.
- **b.** Click in the box for Input variable: then double click on C1 AutoSales in the list box.

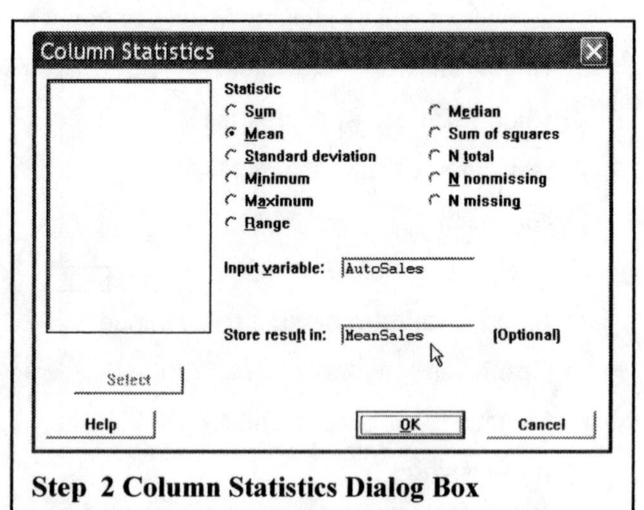

Step 2 Column Statistics Dialog Box

- **c.** Click in Store result in: then type **MeanSales.** Column statistics are stored in a constant, K1.
- **d.** Click **[OK].**

Calculate the Standard Deviation

Step 3. Select **Edit >Edit Last Dialog** or click the Edit Last Dialog icon on the Standard toolbar.

The Column Statistics dialog box will open.

- **a.** Check the button for Standard deviation.
- **b.** The Input variable should still be AutoSales.
- **c.** Press <Tab> to the Store result in: textbox then change the name to **s.**
- **d.** Click **[OK].** The standard deviation will be stored in K2 and named **s.**

  The value will also be displayed in the Session window.

Count and Store the Sample Size

Step 4. Click the Edit Last Dialog icon.

a. Check the box for N nonmissing and name the constant **n.** This is the sample size.

b. Click **[OK].**

Step 5. Select **File>Save Project As... ,** then save your work as **Project3-2.MPJ**

In the Session window you will see the result of each calculation as shown. The mean amount of sales is 12.6 million dollars with a standard deviation of 1.1 million. When **Calc>Column Statistics** is used, results will be stored in constants if you choose to store them. MINITAB can store up to 1,000 constants, each containing a number or text string. They are listed in Project Manager as K1, K2, K3, ..., K1000 along with their name.

Minitab assigns the values of the missing value code *, e, and pi to the last three stored constants (K998 = *, K999 = 2.718, and K1000 = 3.141). To save space, these values are not shown in the Project Manager.

A MINITAB variable is a column not a constant. A result stored in a variable will be placed in a column such as C1 or C2. View details about constants and columns in the Project Window.

Step 6. Select **Window>Project Manager** or click the Project Manager icon.

Project Manager

Untitled, Session, History, Graphs, ReportPad, Related Docume, Worksheets, E-C03-S03-2, Columns, Constants, Matrices

Name	Id	Value
MeanSales	K1	12.6
s	K2	1.1296
n	K3	6

Step 7 View Constants

Step 7. Click the File icon by Constants in the worksheet. The panel on the right side displays information about all the constants associated with this worksheet.

Calculate the Midrange.

The formula for the midrange is MidRange = (max(C1) + min(C1))/2.

Step 8. Select **Calc>Calculator.**

a. Type **K4** in Store result in variable:.

b.

The default storage location is a MINITAB column, a variable. If a label is typed the results are stored in the first row of the first empty column of the worksheet. K4 is a constant. Typing **K4** forces MINITAB to save the result in the constant instead of a column.

c. In the dialog box type the right side of the equation. **(min(C1) + max(C1))/2**

d. Click **[OK].**

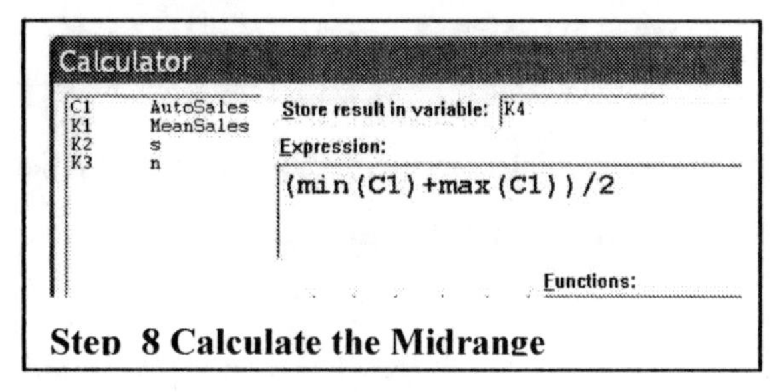

Step 8 Calculate the Midrange

Step 9. Select **Window>Project Manager,** then click the Constants Folder icon.

a. Right click on Unnamed, then select Rename.

b. Type in **MidRange.**

c. Click the Disk icon to save the project.

Name	Id	Value
MeanSales	K1	12.6
s	K2	1.1296
n	K3	6
*** Unnam		

Set Description...
Rename

Calculate Population Variance (σ^2)

Use the data for the ages of the patriots who signed the Declaration of Independence.

Step 1. Select **File>Open Worksheet... .**

a. Navigate to the folder with the textbook data files. BlumanMTP

b. Double click P-C02-S02-13.MTP. The ages should be in C1 of the worksheet.

Step 2. Calculate the deviations from the mean.

a. Select **Calc>Calculator.**

b. Type **Deviations** in the text box for Store the result in:.

c. Press <Tab>, then type the right side of the formula in the text box for Expression.

C1-Mean(C1)

d. Click **[OK].** A column of differences is added to the worksheet.

Step 3. Calculate the squared deviations.

a. Click the Edit Last Dialog icon.

b. Type **Deviations^2** for Storing the result. Press <Tab>. The formula should be highlighted.

c. Double click C2 Deviations to replace the first formula then type **Deviations**2**.

The double asterisk is the MINITAB symbol for exponents. Type two asterisks.

d. Click **[OK].** This dialog box is shown with the first few lines of the worksheet after completing this step.

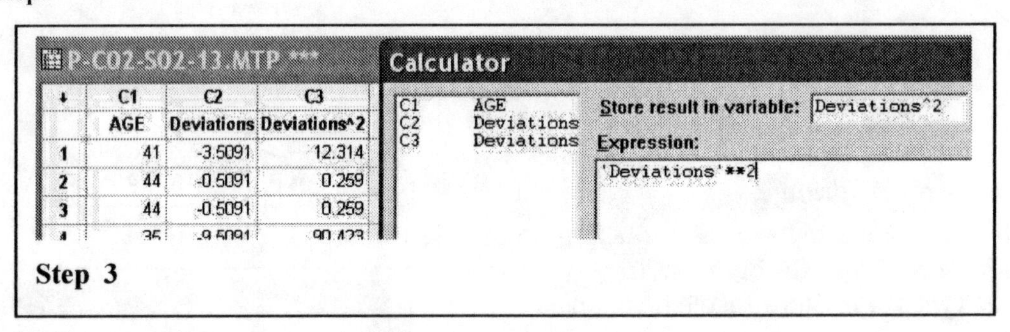

Step 3

Step 4. Calculate variance. The squared deviations will be added then divided by N.

a. Select **Calc>Column Statistics.**

b. Check the option for SUM then choose C3 Deviations^2 for the Input variable and **SST** for the name. Click **[OK].** This result will be stored in a constant.

c. Select **Calc>Column Statistics** again, then the option for **N nonmissing**. Change SST to **N** in the box for Store result in:. Click **[OK].**

d. Select **Calc>Calculator,** then type **K3** for storage. In the Expression box type the formula **SST/N.** The variance is 103.159.

Calculate Population Standard Deviation (σ)

The standard deviation for the population, (σ, sigma) is the square root of the variance.

Step 5. Select **Calc>Calculator,** then type **K4** for storage.

a. The expression should be SQRT(K3).

b. Click **[OK].** The standard deviation is 10.1567.

Step 6. To view the results, click the Project Manager icon, and click the constants tab for this worksheet. Some of the results are in constants and some are in the columns.

a. Right click K3, choose rename, then type **Variance.**

b. Right click K4, choose rename, and type **Sigma.**

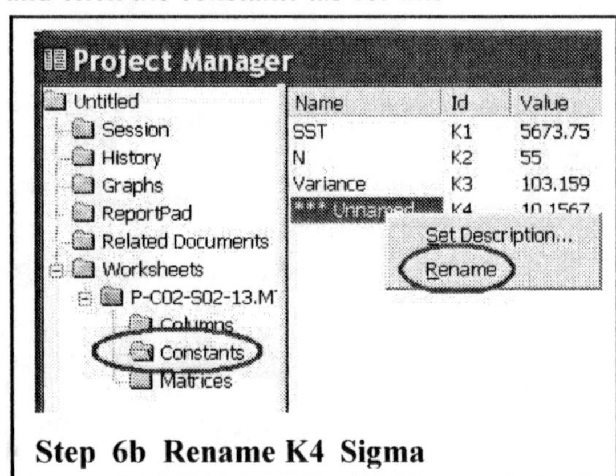

Step 6b Rename K4 Sigma

Step 7. ***Finish the Project***

a. Create and print a report.

b. Save the project as **Project3-2.MPJ.**

c. To continue, select **File>New>Minitab Project,** or to quit, click the Close icon.

3-3 Measures of Central Tendency and Variation for Frequency Distributions

If the table is an ungrouped frequency distribution, the mean and standard deviation should be exactly the same as the results calculated from the raw data. If the data is not available but a grouped frequency distribution is given, it is still possible to estimate some of the statistics such as the mean, variance, and standard deviation.

The distribution of tournament golf scores in Exercise 13 in Section 3 is a frequency distribution for an LPGA event. You may recall making a histogram from this data back in chapter 2.

Multiple menu selections must be used to calculate the statistics from a table. Here is how.

Step 1. Here is a shortcut for entering the midpoints into C1. They are a sequence of integers.

a. Select **Calc>Make Patterned Data>Simple Set of Numbers.**

b. Type **X** to name the column.

c. Type in **203** for the First value; **218** for the Last value; and for Steps, type in a **3.**

d. Click **[OK].**

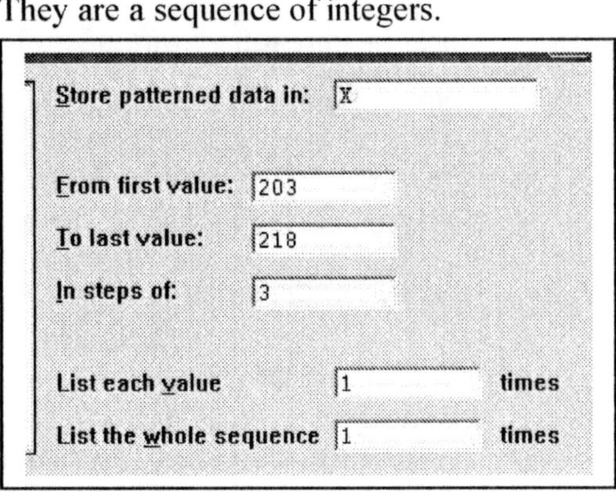

Step 2. Type the frequencies in C2. Name the column **f.**

The worksheet is shown.

↓	C1	C2	C3
	X	f	
1	203	2	
2	206	7	
3	209	16	
4	212	26	
5	215	18	
6	218	4	
7			

Calculate the Products ***($f{\cdot}X$ and $f{\cdot}X^2$)***

Step 3. Select **Calc>Calculator.**

a. Type in **fX** for the variable and **f*X** in the Expression dialog box, then click **[OK].**

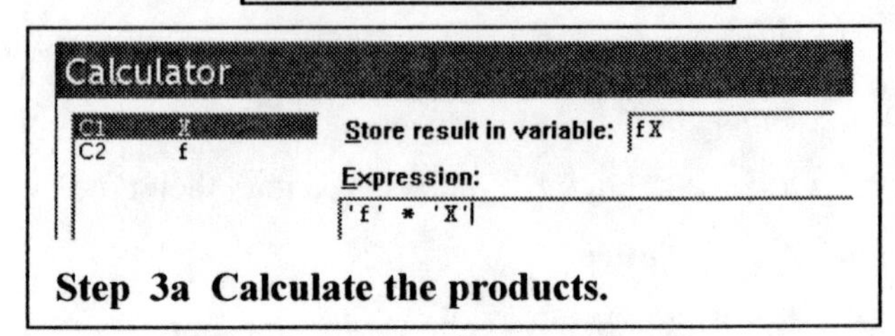

Step 3a Calculate the products.

Note: If you double click the columns to select them instead of typing them, MINITAB inserts the single quotes around the column names. The quotes are not required unless the column name has spaces.

b. Select **Edit>Edit Last Dialog** and type in **fX2** for the variable and **f*X**2** for the expression.

↓	C1	C2	C3	C4
	X	f	fX	fX2
1	203	2	406	82418
2	206	7	1442	297052
3	209	16	3344	698896
4	212	26	5512	1168544
5	215	18	3870	832050
6	218	4	872	190096

Step 3c Worksheet with calculated columns.

c. Click **[OK].** There are now four columns in the worksheet.Calculate the Column Sums

Step 4. Select **Calc>Column Statistics.** This command stores results in constants, not columns. Click **[OK]** after each step.

a. Click the option for Sum then select C2 f for the Input column and type **n** for Store result in:.

b. Select **Edit>Edit Last Dialog,** then select C3 fX for the column, and type **sumX** for storage.

c. Edit the last dialog box again. This time select C4 fX2 for the column, then type **sumX2** for storage.

Statistic
- (•) Sum
- ( ) Mean
- ( ) Standard deviation
- ( ) Minimum
- ( ) Maximum
- ( ) Range
- ( ) Median
- ( ) Sum of squares
- ( ) N total
- ( ) N nonmissing
- ( ) N missing

Input variable: fX2

Store result in: sumX2 [Optional]

Step 4 c Calculate the Sum of X^2

d. The sums are 73, 15446, and 3,269,060.

e. To verify the results, click the Project Manager icon, then the **Constants** folder of the worksheet. The last sum is so large, it is displayed in scientific notation. The decimal point should be +006, six places to the right!

Name	Id	Value
n	K1	73
sumX	K2	15446
sumX2	K3	3.26906E+006

Step 4e Project Manager Constants

Calculate the Mean, Variance, and Standard Deviation

Step 5. Select **Calc>Calculator.**

a. Type **K4** for the variable, then press <Tab> then type **sumX/n** in the **Expression** box.

b. Click **[OK].** The mean is calculated and stored. Continue.

c. Click the Edit Last Dialog icon and type **K5** for the variable.

d. In the expression box, type in **(sumX2-sumX**2/n) / (n-1)**

e. Edit the last dialog box and type **K6** for the variable, then press <Tab>. The previous expression is highlighted.

f. Scroll the function list until **Square root** is visible. Double click it. Parentheses will surround the highlighted text and the SQRT function will be inserted at the beginning of the line.

g. Click **[OK].**

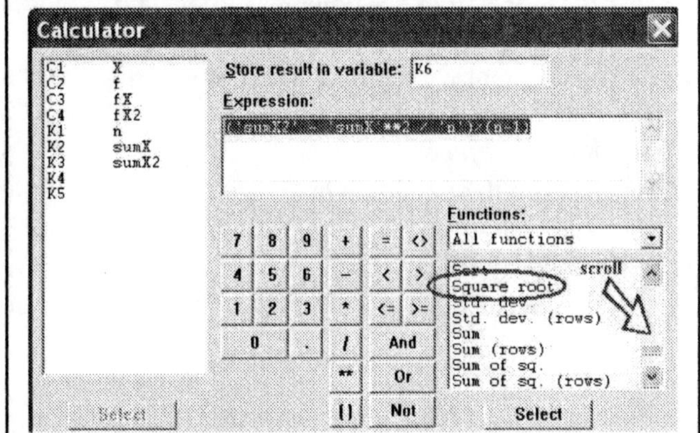

h. Click the Project Manager icon; then the **Constants** folder; and right click on **K4**, an unnamed constant. Choose rename, then type **Mean.** Repeat for **Variance** and **s**, the standard deviation.

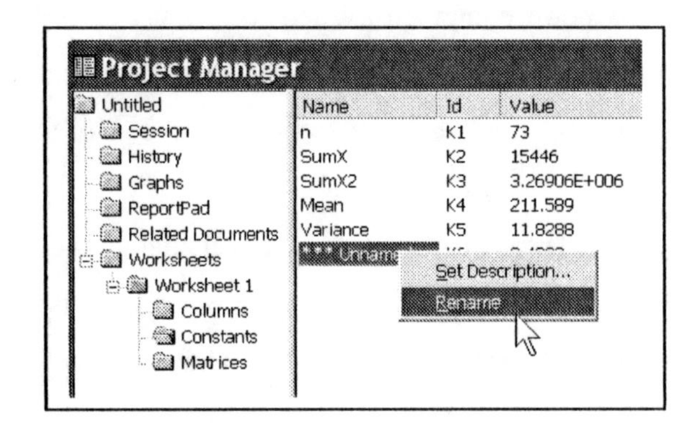

Display Results

Step 6. Select **Data>Display Data.**

a. Highlight all columns and constants in the list.

b. Click **[Select]** then **[OK].**

The Session window will display all of our work! This histogram was created in Chapter 2-2.

Data Display

```
n           73.0000
sumX        15446.0
sumX2       3269056
Mean        211.589
Variance    11.8288
s           3.43930

Row    X    f    fX       fX2
  1  203    2   406     82418
  2  206    7  1442    297052
  3  209   16  3344    698896
  4  212   26  5512   1168544
  5  215   18  3870    832050
  6  218    4   872    190096
```

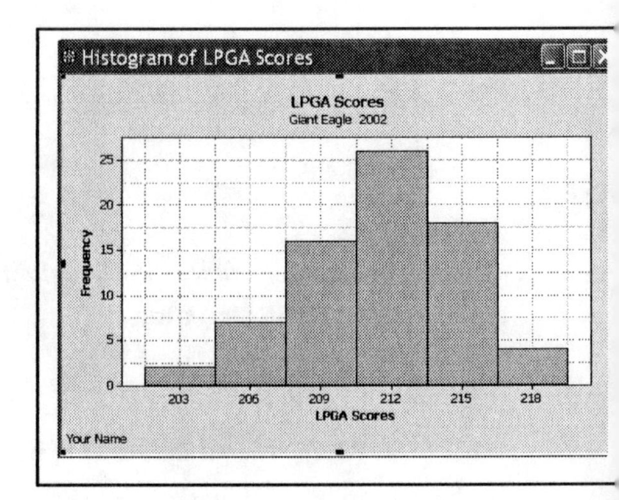

Delete Columns and Constants from the Worksheet.

Columns of data may be deleted from the worksheet by clicking the column ID and pressing <Delete>. Constants can only be deleted by using the **Data** menu.

Step 1. Select **Data>Erase Variables.**

Step 2. Double click the columns or constants you erase.
For example, double click on Variance K5 11.8288, then click **[OK].**
The constant will be erased, and it will disappear from the Project Manager window.

Finish the project.

Step 3. Print the Session window.

Step 4. Print the History window. Click the History folder in the Project Manager window. Right click the History folder, then choose Print History.

Step 5. Select **File>New>Minitab Project.** Save this project as **Project3-3.mpj.**

Using a Macro (Optional; skip to 3-4 Measures of Position)

It is possible to automate the process above by creating a macro, a text file that stores the Session window commands. When needed, use **File>Other files>Run an Exec…** to perform the task. The macro for calculating the mean, variance, and standard deviation for a frequency table has been saved in a file named **FreqDistr.mtb.** The contents of this file are shown in Appendix A.

The sequence of steps for calculating the mean and standard deviation for a frequency distribution leaves a lot of room for errors. When there is a process like this that you might like to repeat, save the History window commands as a macro file. When the file is executed, the commands in the file are carried out. Macros are a tool for automating a process. They are also very sensitive to errors.

To use this macro, midpoints and frequencies must first be entered into the first two columns of a worksheet. Then select **File>Other files>Run an Exec…** and locate the file. Here goes!

Chapter 2 Section 3 Example 24: Find the mean, variance, and standard deviation for this frequency distribution. The table summarizes the number of miles run by twenty randomly selected joggers in one week.

Step 1. Enter the midpoints in C1 and the frequencies in C2 into a new worksheet.

C1	C2
midpoint	**f**
8	1
13	2
18	3
23	5
28	4
33	3
38	2

Step 2. Select **File>Other files>Run an Exec.**

a. Click **[Select File].**

b. Find, then double click the macro file FreqDistr.mtb then click **[OK].**

The Session window results:

Data Display

```
n             20.0000
SumX          490.000
SumX2         13310.0
Mean          24.5000
Variance      68.6842
s             8.28759
```

The mean number of miles run by the joggers each week is 24.5, with a standard deviation of 8.3 miles.

Finish the project.

Step 3. Create and print a report.

Step 4. Save the project as **Project3-2.MPJ.**

Step 5. To quit, exit MINITAB, or to continue, start a new project.

Calculate a Weighted Mean (Optional)

The instructions for calculating a weighted mean is nearly the same as calculating from a frequency table. The only difference is that in step 6c, divide by n instead of n-1. The macro WtMean.MTB does this for us.

↓	C1	C2
	Grades	Credits
1	4	3
2	2	3
3	3	4
4	1	2

Step 1. Enter the table for Chapter 3 Section 3 Example 17 in a worksheet. Don't worry about the column labels. The macro changes them!

Step 2. Run the macro WtMean.MTB. The mean is 2.7 with a standard deviation of 1.03.

```
Mean        2.66667
Variance    1.05556
s           1.02740
```

Continue.

3-4 Measures of Position

Determine the Quartiles and the Interquartile Range

Section 4 Example 38: A stockbroker recorded the number of clients she saw each day over an 11 day period.

```
Clients
33   38   43   30   29   40   51   27   42   23   31
```

A previous command, **Stat>Basic Statistics>Display Descriptive Statistics** can be used to calculate the minimum, maximum, quartiles, and interquartile range. The statistics are displayed in the Session window.

Step 1. Enter the data in C1 of a MINITAB worksheet.

a. Name the column **Clients.**

b. *Optional*: Sort the data. It is not necessary to sort the data before calculating the statistics.

1. Select **Data >Sort.**
2. Double click C1 Clients for the Sort column.
3. Press <Tab> to jump the cursor to the By column: text box.
4. Double click Clients again.
5. Check the option by the last text box so the sorted list will be placed in a new column of this same worksheet.
6. Click **[OK].**

   A new column labeled X is created with the sorted list.
7. Select **Data>Display Data** then double click C2 X. The sorted list will be displayed in the Session window. `23   27   29   30   31   33   38   40   42   43   51`

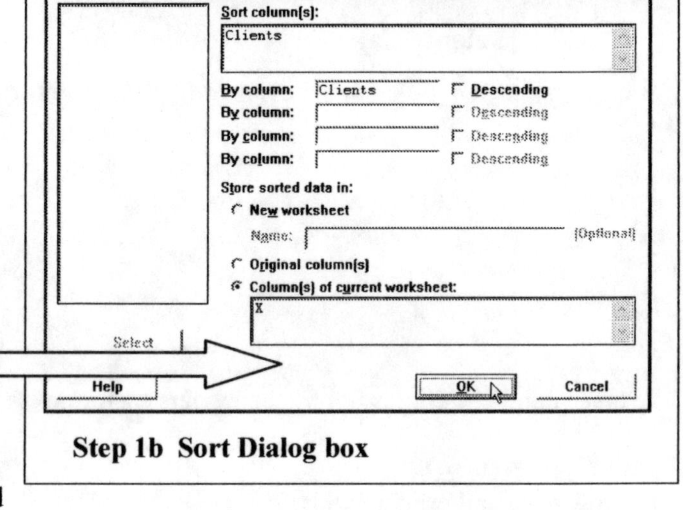

Step 1b Sort Dialog box

Step 2. Select **Stat>Basic Statistics>Display Descriptive Statistics.**

a. Double click C1 Clients then **[Statistics].**

b. In the window there is a list of statistics with check boxes.

Check the quartiles, median and Interquartile range, minimum, maximum, and N nonmissing and N missing as shown. Remove the checks from other options.

c. Click **[OK].**

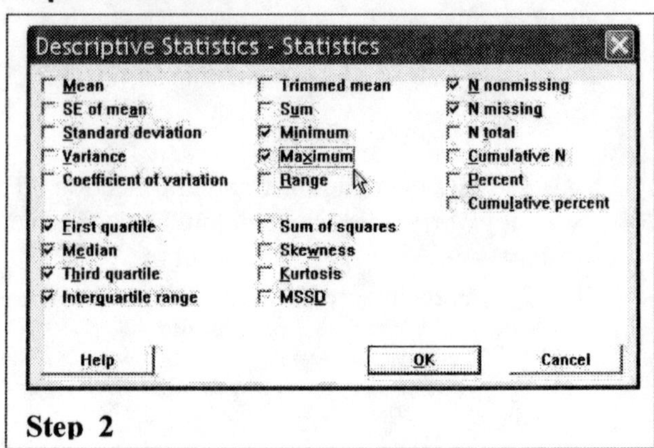

Step 2

Step 3. The statistics are displayed in the Session window. The five-number summary is highlighted.

Descriptive Statistics: Clients

```
Variable    N  N*  Minimum      Q1  Median      Q3  Maximum    IQR
Clients    11   0    23.00   29.00   33.00   42.00    51.00  13.00
```

Calculate Standard Scores

It is easy to calculate the standard scores, or Z-values, for all of the data in a column.

$$Z = \frac{(X - \overline{X})}{S}$$

Step 1. Select **Calc>Standardize.**

a. Double click C1 Clients to select it for the variable.

b. Click in the box for Store results in:, then type **Z.**

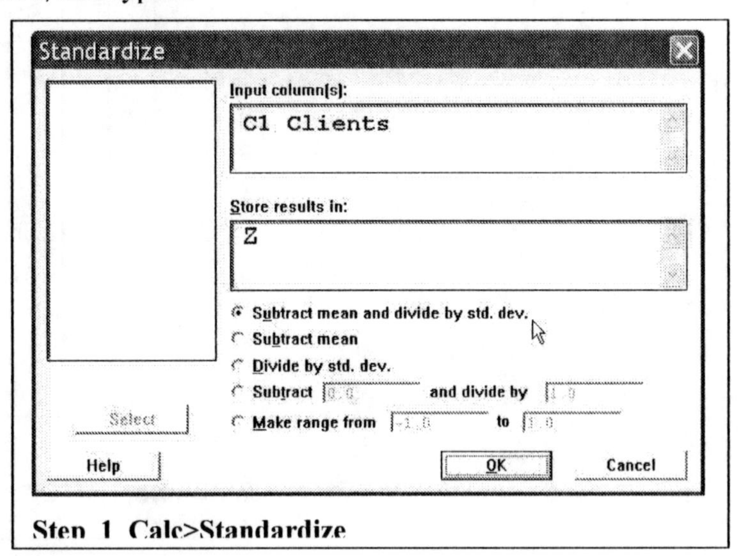

Step 1 Calc>Standardize

c. The default option is Subtract mean and divide by the std dev. That is exactly the formula required. Don't change it.

d. Click **[OK].**

In the first available column, the Z score for the values in each row of clients will be calculated and stored in the same row of a new column as shown. The sorted column X is ignored.

In the first row there were 33 clients. The Z score for 33 is -.26235. Thirty-three is .26235 standard deviations below the mean or about ¼ of a standard deviation below average.

↓	C1	C2	C3
	Clients	X	Z
1	33	23	-0.26235
2	38	27	0.33887
3	43	29	0.94008
4	30	30	-0.62308
5	29	31	-0.74332
6	40	33	0.57935
7	51	38	1.90203
8	27	40	-0.98381
9	42	42	0.81984
10	23	43	-1.46478
11	31	51	-0.50284

Step 2d Calculate Z scores

Finish the project

Step 2. Print the worksheet.

Step 3. Save the project as **Project3-4.MPJ.**

Continue to the next section.

3-5 Exploratory Data Analysis

Construct a Stem-and-Leaf Plot.

Step 1. Select **STAT>EDA>Stem-and-Leaf.**

- **a.** Double click C1 Clients.
- **b.** Remove the check for Trim outliers option.
- **c.** Click in the Increment text box and enter the class width of **5.**
- **d.** Click **[OK].** This character graph will be displayed in the Session window.

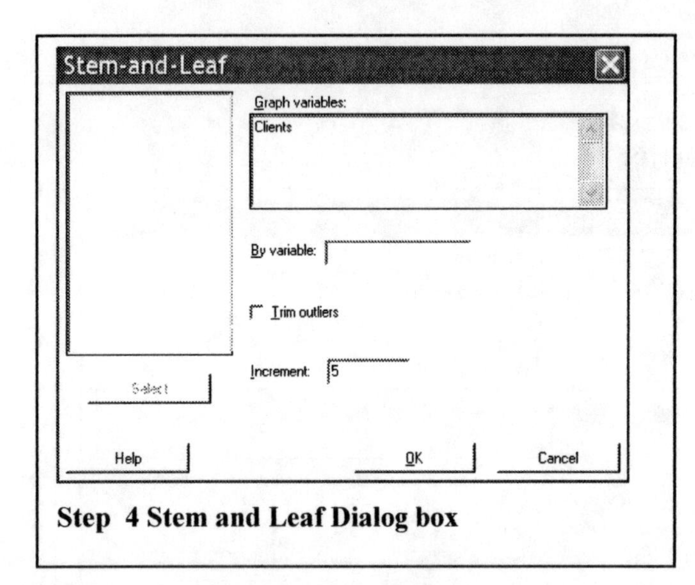

Step 4 Stem and Leaf Dialog box

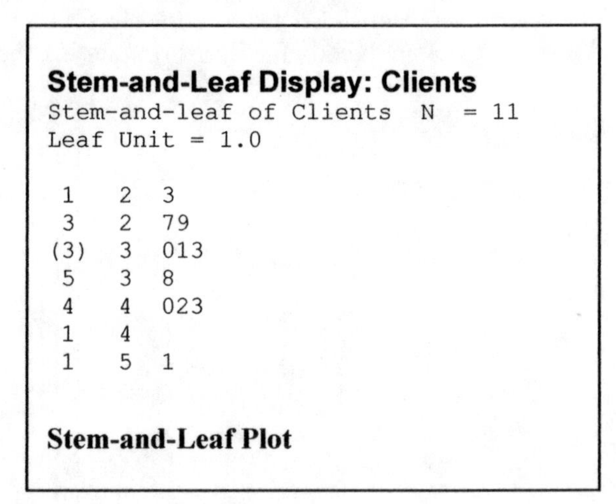

Stem-and-Leaf Display: Clients

```
Stem-and-leaf of Clients  N  = 11
Leaf Unit = 1.0

 1   2  3
 3   2  79
(3)  3  013
 5   3  8
 4   4  023
 1   4
 1   5  1
```

Stem-and-Leaf Plot

Construct a Boxplot.

Step 2. Select **Stat>EDA>Box Plot,** then **Simple,** and **[OK].**

- **a.** Double click C1 Clients to select it for the Graph variable.
- **b.** Click **[Labels]**, then type an appropriate title such as **Clients Per Day.**
- **c.** Type **your name** and **date** in the footnote.

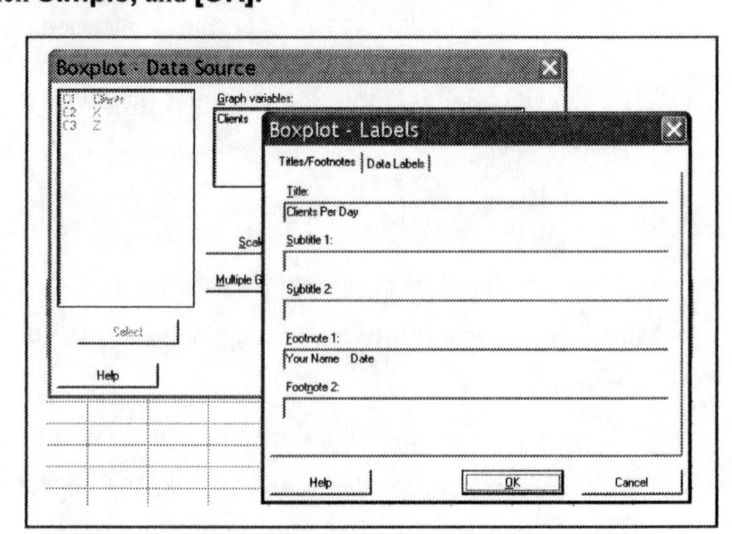

Step 3. Click **[OK]** twice. The graph will be displayed in a Graph window.

Optional: Help Menu
Step 4. Click the Edit Last Dialog icon.
Step 5. Click **[Help]** or press <F1>.

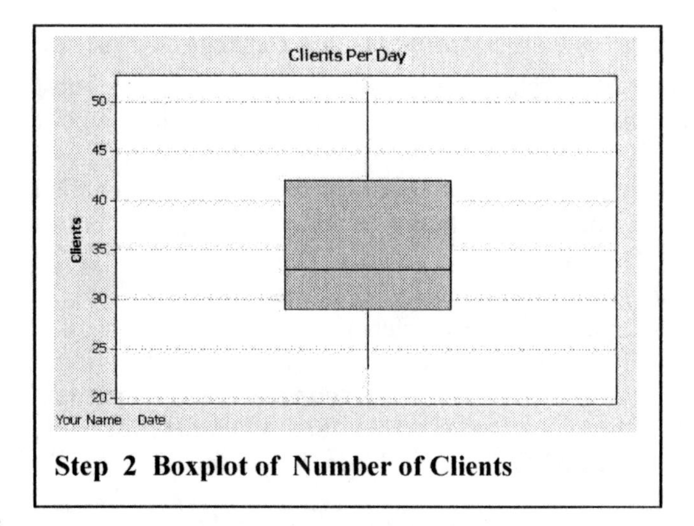

Step 2 Boxplot of Number of Clients

You may need to click main topics at the top of the window.
Step 6. Scroll down the Help window until you see the display shown.

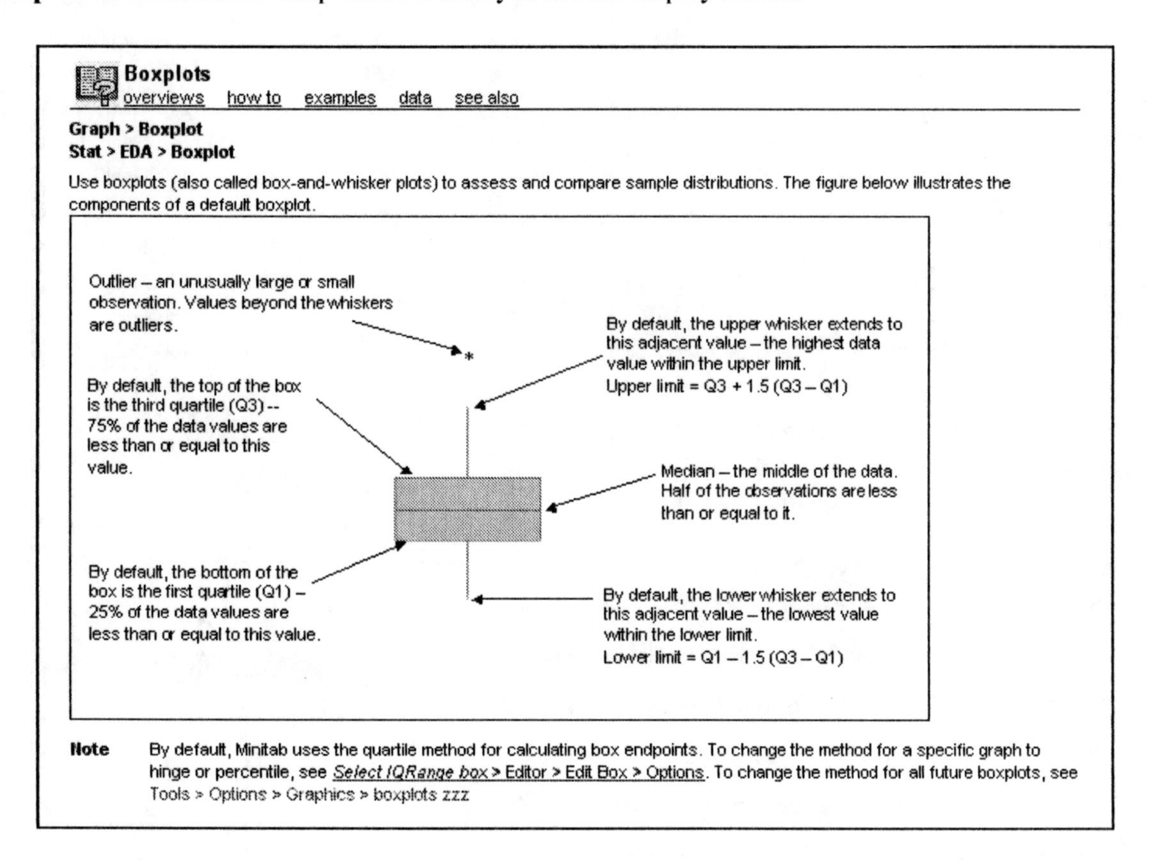

Boxplots
overviews how to examples data see also

Graph > Boxplot
Stat > EDA > Boxplot

Use boxplots (also called box-and-whisker plots) to assess and compare sample distributions. The figure below illustrates the components of a default boxplot.

Note By default, Minitab uses the quartile method for calculating box endpoints. To change the method for a specific graph to hinge or percentile, see *Select IQRange box* > Editor > Edit Box > Options. To change the method for all future boxplots, see Tools > Options > Graphics > boxplots zzz

The five number summary is used to make the high resolution box plot.

Any data value outside the interval limits would be considered outliers or extreme values.

A symmetrical distribution will have a boxplot with the box in the center of the range. The median is in the middle of the box and whiskers are about the same length. The boxplot for this sample indicates a distribution that is symmetrical.

Compare Two Distributions Using Descriptive Statistics and Boxplots

A sample of real cheese and another sample of a cheese substitute are tested for their sodium levels in milligrams. The data is used in Section 5 Example 39 in of the textbook.

Step 1. Select **File>Open Worksheet,** and locate the file E-C03-S05-39.MTP.
To make a boxplot for the same measurement for two groups, the data must be stacked. Stack the data.

Step 2. Select **Data>Stack>Columns.**

- **a.** Double click each column.
- **b.** Click the option for Column of current worksheet: and type the name **Sodium** in the text box.
- **c.** Press <Tab>, then type **Group** in the subscripts text box.
- **d.** The option Use variable names in subscript column should be checked.
- **e.** Click **[OK].** The worksheet is shown.

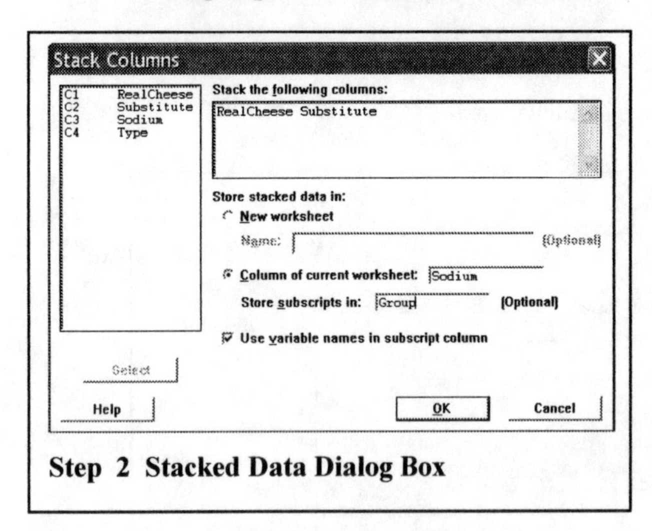

Step 2 Stacked Data Dialog Box

↓	C1 RealCheese	C2 Substitute	C3 Sodium	C4-T Group
1	310	270	310	RealCheese
2	220	130	220	RealCheese
3	420	180	420	RealCheese
4	240	260	240	RealCheese
5	45	250	45	RealCheese
6	180	340	180	RealCheese
7	40	290	40	RealCheese
8	90	310	90	RealCheese
9			270	Substitute
10			130	Substitute
11			180	Substitute
12			260	Substitute
13			250	Substitute
14			340	Substitute
15			290	Substitute
16			310	Substitute

When stacked, the data will look like C3 and C4-T. C3 contains all the sodium data and C4 has codes that identify the type of cheese (qualitative data).

Here is an easy way to get everything we need!

Step 3. Select **Stat>Basic Statistics>Display Descriptive Statistics.**

- **a.** Press <F3> to reset the defaults for the dialog boxes.
- **b.** Double click C3 Sodium for the variable. Press <Tab>.
- **c.** Double click C4 Group for By variables:.
- **d.** Click **[Graphs…]**, then check the option for Boxplot of data and then **[OK].**

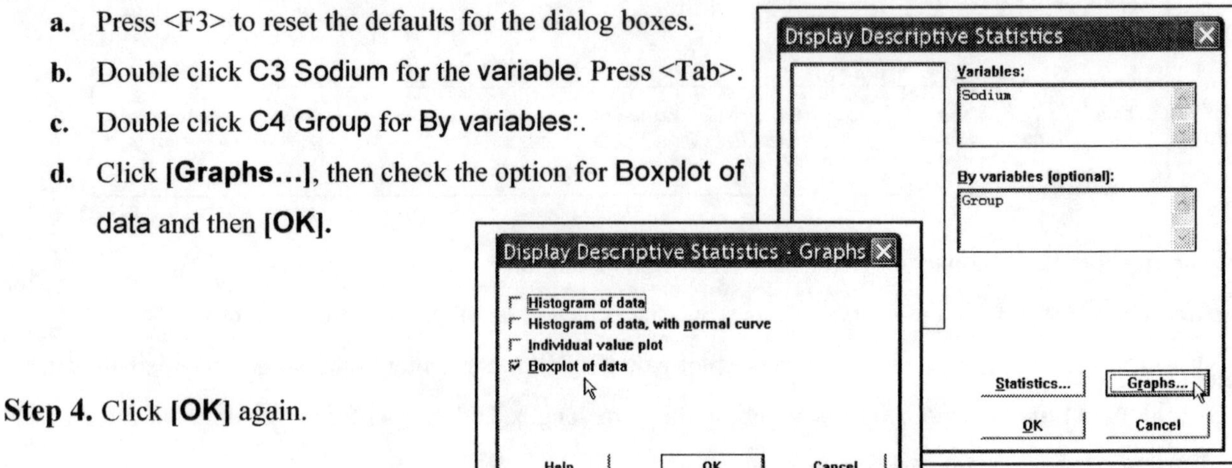

Step 4. Click **[OK]** again.

The boxplot shows there is more variation in the sodium levels for the real cheese. The median level is lower for the real cheese than the median for the sodium in the cheese substitute. The longer whisker for the real cheese indicates higher sodium levels for the real cheese type. The distribution is skewed right. The sodium levels for the substitutes are symmetrical and not as spread out.

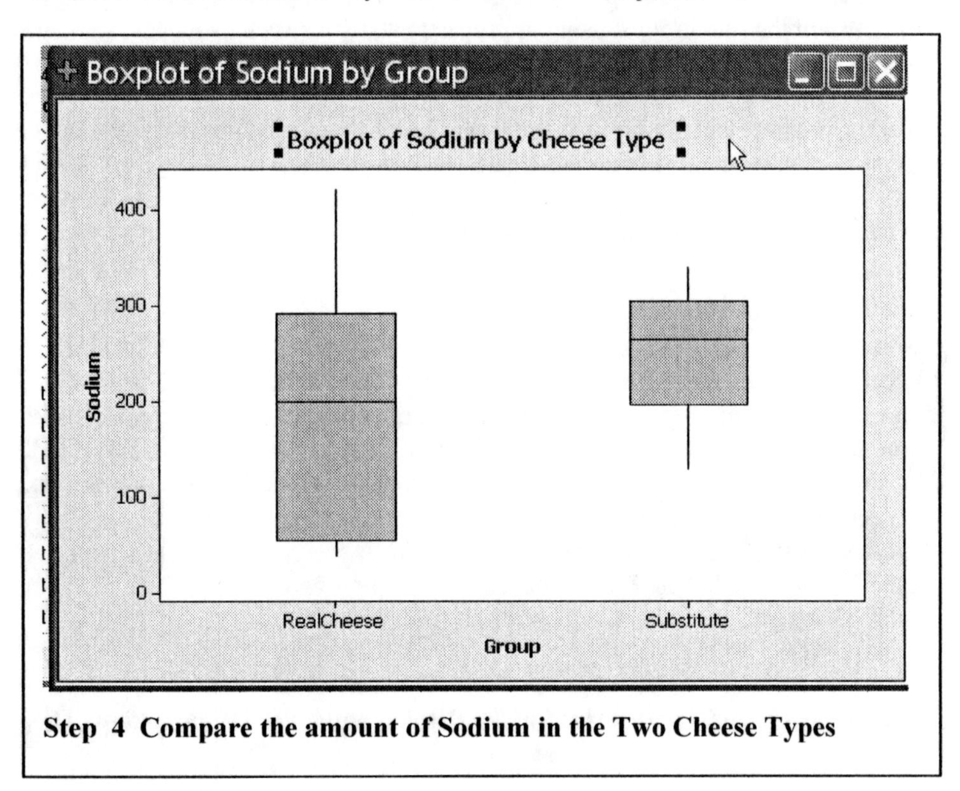

Step 4 Compare the amount of Sodium in the Two Cheese Types

The Session window will contain the descriptive statistics for sodium for each cheese type as shown. The mean amount of sodium for real cheese is 193.1 milligrams compared to 253.8 for the substitute. Even though the mean is smaller for the real group, the standard deviation is almost double indicating more variation in the sodium levels for the real cheese.

Descriptive Statistics: Sodium

```
Variable  Group        N  N*   Mean  SE Mean  StDev  Minimum     Q1  Median
Sodium    RealCheese   8   0  193.1     47.1  133.2     40.0   56.3   200.0
          Substitute   8   0  253.8     24.3   68.6    130.0  197.5   265.0

Variable  Group           Q3  Maximum
Sodium    RealCheese   292.5    420.0
          Substitute   305.0    340.0
```

Finish the project.

Step 5. Create and print a report.

Step 6. Save the project as **Project3-5.MPJ.**

Step 7. Quit MINITAB or start a new project to continue.

Data Project

Project 1: Select a variable and collect about ten values for two groups. Define the variable and the populations. Describe how the samples were selected. Write a paragraph describing the similarities and differences between the two groups, using appropriate statistics such as means, standard deviations, and so on.

In this project, we will select a sample of ten from the Databank file using Height and Gender. Surely, men weigh more than women! Let's investigate.

Calculate the Descriptive Statistics and Make a Boxplot

Step 1. Open the Databank file.

Step 2. Select a random sample of twelve from two columns: Weight and Gender.

a. Select **Calc>Random Data>Sample from Columns...**

b. Type **12** for the number of rows, then select both columns, Weight and Gender.

c. Store the samples in two new columns, Weight12 and Gender12. Click **[OK].**

Remember, this is a random sample. Your sample will be different from the one shown.

Step 3. Select **Stat>Basic Statistics>Display Descriptive Statistics.**

a. Select Weight12 for the variable.

b. Select Gender12 for the BY variable, the grouping variable.

c. Click **[Graphs]**, then check the option for boxplots.

d. Click **[OK]** twice.

The descriptive statistics will be displayed in the Session window. The boxplot will be in a Graph window.

Descriptive Statistics: Weight12

```
Variable  Gender12  N  N*    Mean  SE Mean  StDev  Minimum      Q1  Median
Weight12  F         4   0  137.8      12.8   25.6    109.0   113.0   138.0
          M         8   0  178.13     9.74  27.56   143.00  162.25  172.00

Variable  Gender12      Q3  Maximum
Weight12  F          162.3    166.0
          M         191.75   234.00
```

The men in the sample weighed 40 pounds more than the women on average. The weight of the women ranged from 109 to 166 while the men's weight ranged from 143 to 234 pounds.

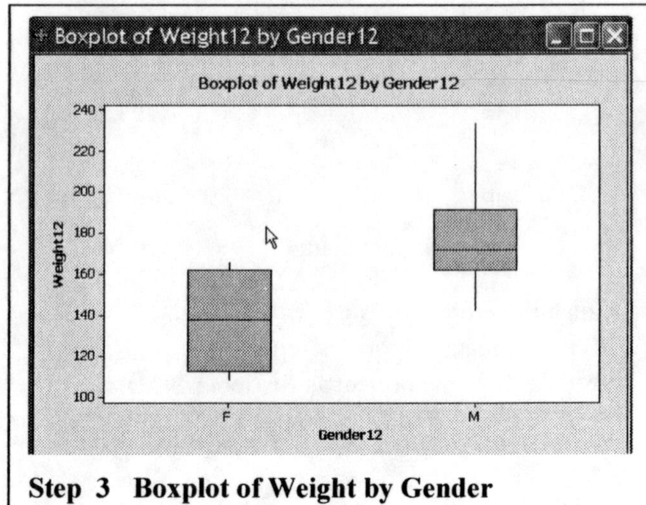

Step 3 Boxplot of Weight by Gender

Finish the project.

Step 4. Create and print a report.

Step 5. Save the project as **DataProject03.MPJ.**

Step 6. Exit MINITAB.

Chapter 3 Textbook Problems Suitable for MINITAB

Page	109	3.2	1 - 31, 26 - 31, 38 a-e
Page	126	3.3	6 – 27, 48
Page	141	3.4	none
Page	153	3.5	1 – 6, 11 - 17
Page	159	Review	1 - 9, 14, 21
Page	161	Data Analysis	1 - 5
Page	165	Data Projects	1 , 2

Chapter 3: EndNotes

Chapter 4 Probability

4-1 Introduction: The Least You Need to Know About Probability

Probability is an essential element of inferential statistics. Understanding basic probability concepts and how probabilities are determined is essential for understanding sampling theory. Classical probabilities use sample spaces to determine the numerical probability that an event will occur. Empirical probabilities are relative frequencies, proportions (fractions or percents). MINITAB calculates empirical probabilities.

Section 2 Example 14: Hospital records indicated that maternity patients stayed in the hospital for the number of days shown in the distribution. Find these probabilities:

a) A patient stayed exactly five days.
b) A patient stayed less than six days.
c) A patient stayed at most four days.
d) A patient stayed at least five days.

C1	C2
X	f
3	15
4	32
5	56
6	19
7	5

Frequency distribution

4-2 Calculate Relative Frequencies (from a distribution)

The random variable, X, represents the number of days patients stayed in the hospital from Example 14 in Section 2.

1. In C1 of a worksheet, type in the values of X. Name the column **X.**

2. In C2 enter the frequencies. Name the column **f.**

Calculate the relative frequencies and store them in a new column named Px.

3. Select **Calc>Calculator.**

a. Type **Px** in the box for Store result in variable:.

b. Click in the Expression: box then double click C2 f.

c. Type or click the division operator.

d. Scroll down the function list to Sum, then click **[Select].**

e. Double click C2 f in the list to select it.

f. Click **[OK].**

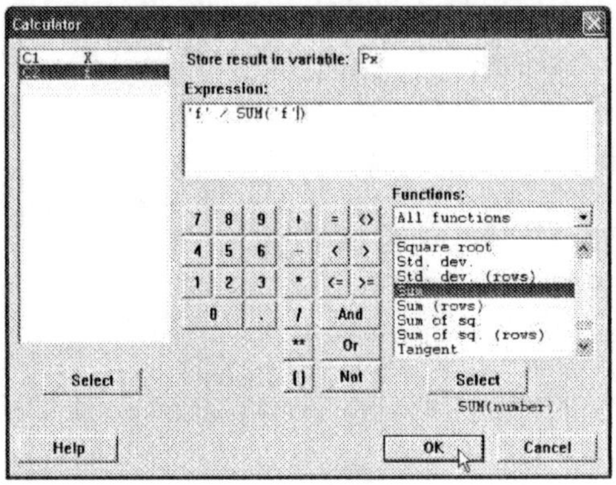

The proportions are displayed. MINITAB does the calculation with fifteen decimal places of precision but only displays six places.

Step 3 Use Calc>Calculator

Format the column of probabilities so they are rounded to three decimal places.

4. Click in any cell of C3 Px.

a. Right click, select Format columns in the pop-up menu, then Numeric.

b. Click on **Fixed decimal with** and type **3** in the box for number of **decimal places.**

This only rounds the values for display. Internally all fifteen decimal places are stored so you can change the number of decimal places at any time.

c. Click **[OK].**

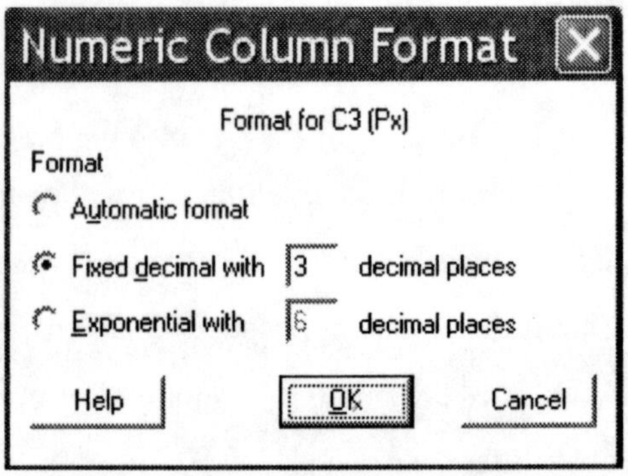

Numeric Column Format

Format for C3 (Px)

Format

Automatic format

Fixed decimal with 3 decimal places

Exponential with 6 decimal places

Help OK Cancel

Step 4 Format Column

4-3 Addition Rule for Probability

The completed worksheet is shown. Relative frequencies are empirical probabilities. For example, the probability that a patient will stay exactly three days is .118.

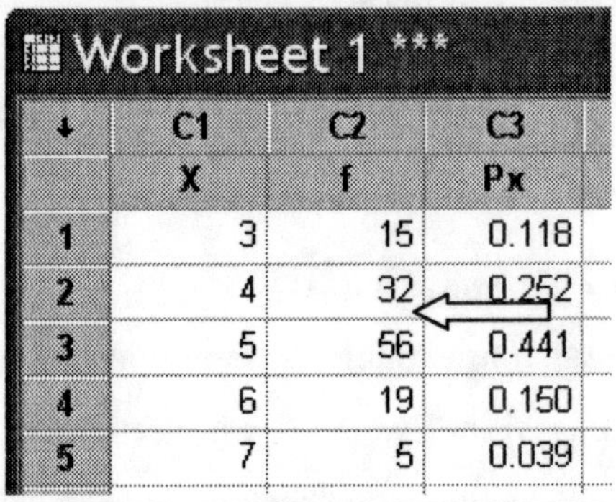

Worksheet 1 ***

↓	C1 X	C2 f	C3 Px
1	3	15	0.118
2	4	32	0.252
3	5	56	0.441
4	6	19	0.150
5	7	5	0.039

Solutions:

a) P(exactly five) = .441
This is the value in the Px column for X=5.

b) P(less than six) = .811
Add the probabilities for 3, 4, and 5, the values of X that are less than 6.
The addition rule for probability is applied.
.118 + .252 + .441

c) P(at most four) = .37
Sum of the values for 3 and 4, the values of X that are or less than or equal to 4.
.118 + .252

d) P(at least five days) = .63
Sum the probabilities for 5, 6, and 7.
.441 + .150 + .039

MINITAB table commands are used to calculate the observed counts and empirical probabilities from data in a worksheet.

Calculate Relative Frequencies from Data

Step 1. Open the Databank.MTP file.

a. Use **File>Open worksheet.**

b. Make sure the file type is Minitab Portable [*.mtp] then double click the Databank.MTP file.

Step 2. Select **Stat>Tables>Tally Individual Variables.**

You can select more than one variable in this text box.

A separate table will be made for each variable selected.

a. Double click Gender and Smoking Status to select them for the Variables.

b. Check the boxes for Counts, Percents, and Cumulative percents.

c. Click **[OK].** Look for the tables in the Session window.

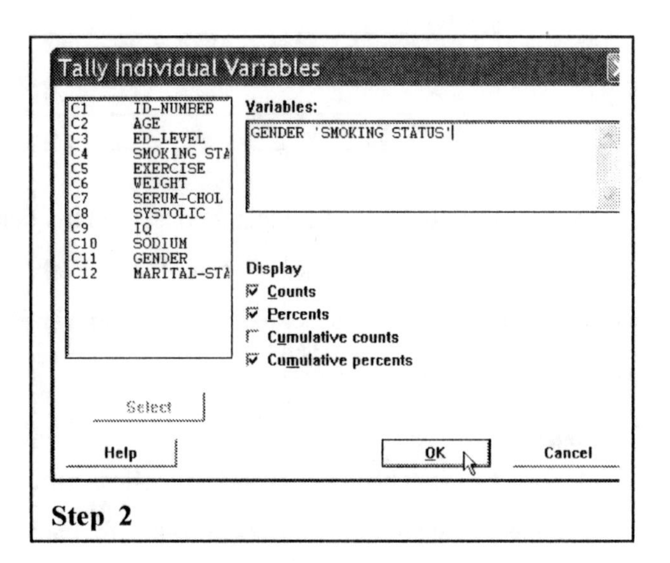

Step 2

Percentages are empirical probabilities.

Tally for Discrete Variables: GENDER, SMOKING STATUS

GENDER	Count	Percent	CumPct
F	50	50.00	50.00
M	50	50.00	100.00
N=	100		

SMOKING STATUS	Count	Percent	CumPct
0	47	47.00	47.00
1	37	37.00	84.00
2	16	16.00	100.00
N=	100		

Fifty percent of this sample is females. Forty-seven percent don't smoke. The probability of randomly selecting an individual from this group who don't smoke is .47. The probability that an individual smokes is .53, the sum of .37 and .16.

e) Smoking levels are mutually exclusive. Each of the 100 individuals is counted in one category of smoking. The genders, male or female, are also mutually exclusive. Each individual is male or female, not both. Smoking levels and gender are not mutually exclusive. It is possible for one individual to smoke a pack or more of cigarettes a day *and* be male. A descriptive technique, cross-tabulation, produces a contingency table that summarizes two variables. The setup for the table is explained first.

Sample Spaces: Contingency Tables (Two Qualitative Variables)

A table will be made with one row for each smoking level (3) and one column for each gender (2). The squares where two characteristics cross are called cells of the table. There are six cells in this table. Labels and totals are not counted as a row or a column. The cell for females who don't smoke is highlighted.

Gender Smoking	F	M
0		
1		
2		

The highlighted cell will contain an observed frequency, the number of individuals in this sample who are female and don't smoke. To complete this table manually, you would have to count all the rows of data that have a zero in the smoking column and an F in the gender column. Try it! There should be twenty-five. MINITAB can easily do this for us.

Construct a Contingency Table

Step 1. Select **Stat>Tables>Crosstabulation and Chi-Square.** This option sets up the table as described, then tallies the observed counts for each cell and the row and column totals.

Step 2. Double click SMOKING STATUS for rows and GENDER for columns.
Step 3. Click the display box for Count.
Step 4. Click **[OK].**

There is a row for each value of the smoking variable and a column for each gender. At the end of the table the contents of each cell is described. The Session window title is Tabulated Statistics.

Tabulated Statistics: SMOKING STATUS, GENDER

```
Rows: SMOKING        Columns: GENDER
            F          M        All

0          25         22         47
1          18         19         37
2           7          9         16
All        50         50        100

  Cell Contents --
                     Count
```

Seven women smoke one pack or more and nine men smoke a pack or more.
There are fifty men and fifty women.
The sample size is 100.
Thirty-seven smoke less than one pack a day.

There are twenty-five women who don't smoke.

Find the following marginal probabilities:

P(male) $= \frac{50}{100} = .5$

P(doesn't smoke) $= \frac{47}{100} = .47$

P(smoke one pack or more) $= \frac{16}{100} = .16$

MINITAB calculates marginal and joint probabilities if you check the box for total percents in the Cross-tabulation dialog box.

Step 5. Choose **Stat>Tables>Crosstabulation** or click Edit Last Dialog Box icon.

Step 6. Check the button for Total percents, then **[OK].**

The legend at the bottom of the table displays the contents of each cell. These probabilities are pretty uninteresting because the sample size is 100 and so the counts and percents are the same!

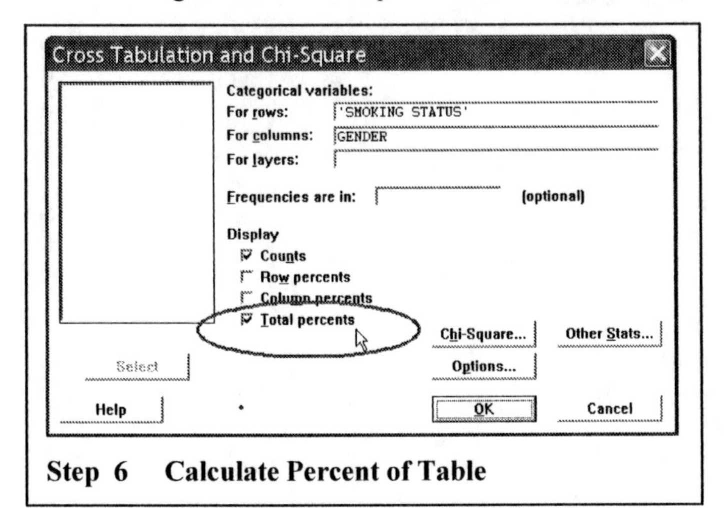

Step 6 Calculate Percent of Table

Rows: SMOKING Columns: GENDER

	F	M	All
0	25	22	47
	25.00	22.00	47.00
1	18	19	37
	18.00	19.00	37.00
2	7	9	16
	7.00	9.00	16.00
All	50	50	100
	50.00	50.00	100.00

Cell Contents --
Count
% of Tbl

Find the following joint probabilities. They are the cell count divided by the sample size.

P(male *and* doesn't smoke) $= \frac{22}{100} = .22$ or 22%

P(female *and* smokes one pack or more) = $\frac{7}{100} = .07$ or 7%

P(male *or* doesn't smoke) =
P(male) + P(doesn't smoke) - P(male *and* doesn't smoke) = $\frac{50}{100} + \frac{47}{100} - \frac{22}{100} = \frac{75}{100}$ or .75.

The reason the probability of both is subtracted is that these twenty-two are counted twice, once for the males and again for the number who don't smoke. Because of double counting, this number must be subtracted.

Conditional Probability: Row and Column Percents

Create a table for smoking and marital status. Calculate column percents and row percents.

Step 7. Select **Stat>Tables>Crosstabulation**.

- **a.** Double click C12 Marital Status for rows.
- **b.** Double click C4 Smoking Status for columns.
- **c.** Click the buttons for Counts, Row percents and Column percents, then click **[OK]**.

The table will be displayed in the Session window.

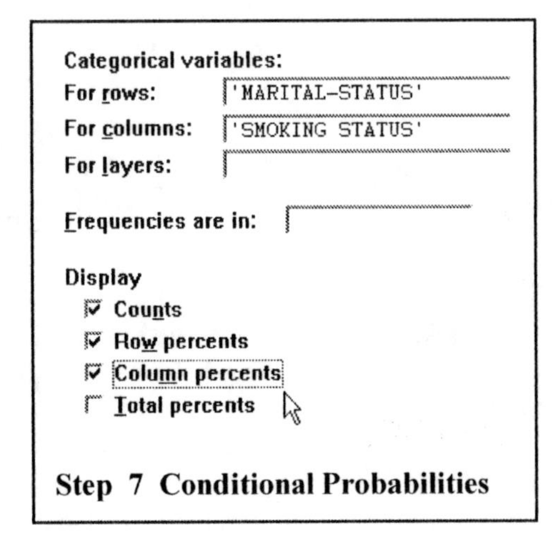

Step 7 Conditional Probabilities

Codes for Marital Status:
D = divorced
M = married
S = single
W = widow or widower

Smoking Codes:
0 doesn't smoke
1 smokes less than one pack
2 smokes one pack or more

Tabulated Statistics:

Rows: MARITAL- Columns: SMOKING

	0	1	2	All
D	3	12	1	16
	18.75	75.00	6.25	100.00
	6.38	32.43	6.25	16.00
M	31	16	6	53
	58.49	30.19	11.32	100.00
	65.96	43.24	37.50	53.00
S	8	7	7	22
	36.36	31.82	31.82	100.00
	17.02	18.92	43.75	22.00
W	5	2	2	9
	55.56	22.22	22.22	100.00
	10.64	5.41	12.50	9.00
All	47	37	16	100
	47.00	37.00	16.00	100.00
	100.00	100.00	100.00	100.00

Cell Contents --
Count
% of Row
% of Col

The column and row percents are the conditional probabilities.
Find P(doesn't smoke | D) is a conditional probability. P(doesn't smoke | D) is the probability an individual doesn't smoke given he/she is divorced. Two more ways of asking for a conditional probability:

1--What is the probability that an individual doesn't smoke if you know he/she is divorced?

2--What percent of those who are divorced don't smoke?

The solution:
P(doesn't smoke | D) = $\frac{3}{16} = .1875$. There are only sixteen divorcees and three of them don't smoke.

Another example:
P(doesn't smoke | M) = $\frac{31}{53} = .5849$. Fifty-nine percent of the married individuals don't smoke.

The highlighted values in the tables are a column of row percents, the percent of each marital status who don't smoke. These percents are not the same as the 47% of all. Smoking is not independent of marital status. They are related or dependent characteristics. A larger percent of those who are married don't smoke.

Calculate row percents. Compare them down each column.
Calculate column percents. Compare them across each row.

There are two ways to find joint probabilities in a table. P(M *and* 0) = probability that an individual is married and doesn't smoke.

1-- Get the frequency or percents directly from the table. P(M *and* 0) = $\frac{31}{100}$ = .31 (total %)

2-- Use the multiplication rule: P(M and 0) = P(M) * P(0 | M)

$$\frac{53}{100} * \frac{31}{53} = \frac{31}{100} \text{ or } .31.$$

It does not work to multiply P(M) * P(0) because the two events--married and doesn't smoke--are not independent! This idea is used again in chapter 11 for the chi-square test of independence.

Complementary events:

$P(\overline{D})$ = 1 - P(D) = 1 - .16 = .84. Use complementary rule. OR

$P(\overline{D})$ = P(S) + P(M) + P(S) = $\frac{84}{100}$ Use cumulative probabilities.

4-5 Crosstabulation and a Control Variable

In the previous dialog box, two variables were entered: Marital Status and Smoking Status. A third variable would be used as a control variable. The best way to explain is to show an example using gender as the control variable.

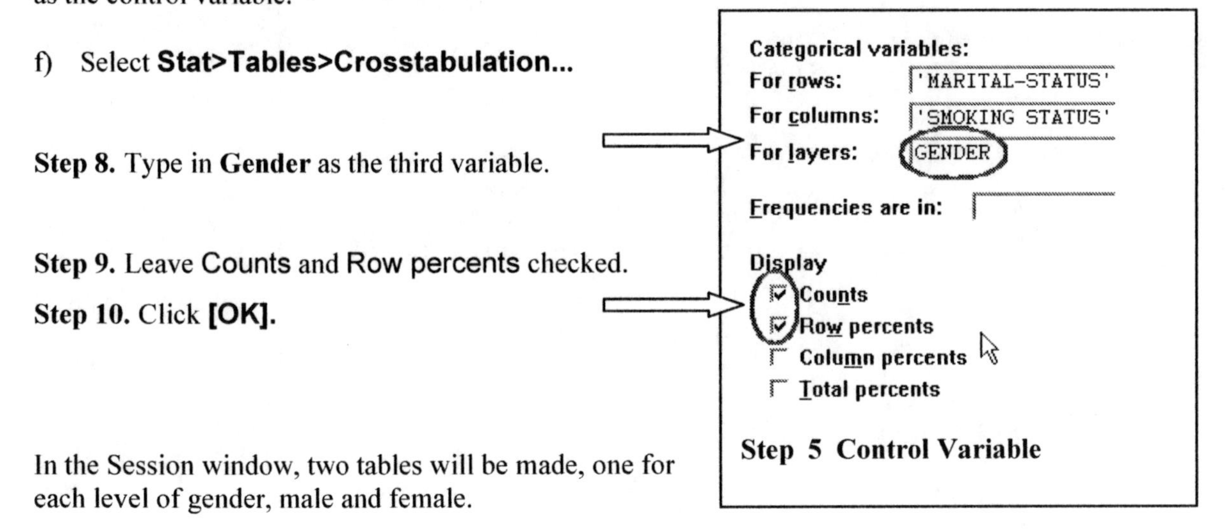

f) Select **Stat>Tables>Crosstabulation...**

Step 8. Type in **Gender** as the third variable.

Step 9. Leave Counts and Row percents checked.

Step 10. Click **[OK].**

In the Session window, two tables will be made, one for each level of gender, male and female.

Step 5 Control Variable

The highlighted row percent is the percent of married women who don't smoke (70.83%) compared to 48.28 percent of married men who don't smoke. It appears smoking is dependent on gender as well as marital status.

See the table on the next page.

Results for GENDER = F

```
Rows: MARITAL-STATUS   Columns: SMOKING STATUS

           0       1       2     All

D          0       8       0       8
        0.00  100.00    0.00  100.00

M         17       6       1      24
       70.83   25.00    4.17  100.00

S          4       3       6      13
       30.77   23.08   46.15  100.00

W          4       1       0       5
       80.00   20.00    0.00  100.00

All       25      18       7      50
       50.00   36.00   14.00  100.00

Cell Contents:      Count
                    % of Row
```

Results for GENDER = M

```
Rows: MARITAL-STATUS   Columns: SMOKING STATUS

           0       1       2     All

D          3       4       1       8
       37.50   50.00   12.50  100.00

M         14      10       5      29
       48.28   34.48   17.24  100.00

S          4       4       1       9
       44.44   44.44   11.11  100.00

W          1       1       2       4
       25.00   25.00   50.00  100.00

All       22      19       9      50
       44.00   38.00   18.00  100.00

Cell Contents:      Count
                    % of Row
```

Let's explore a bit more. This will create a table and calculate descriptive statistics for a quantitative variable for each cell of the table. In this example we use weight.

g) Select **Stat>Tables>Descriptive Statistics.**

a. In For rows:, double click Marital-Status.

b. In For columns:, double click Smoking Status.

c. Clear the contents in For layers:.

d. Click on **[Associated Variables].**

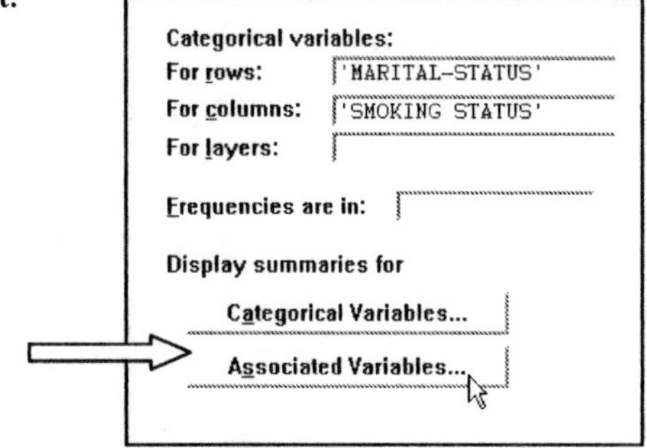

e. Double click Weight.

f. Check the buttons for Means, Standard deviations, and N nonmissing.

g. Click **[OK]** twice.

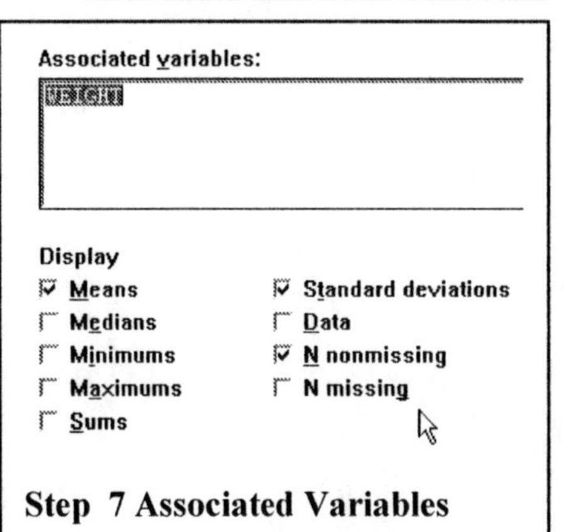

Step 7 Associated Variables

The new table includes the frequencies for each cell and the mean and standard deviation of the weights for that group. The divorced individuals who don't smoke weigh, on average, 159.7 pounds with a standard deviation of 8.33 pounds.

Tabulated statistics: MARITAL-STATUS, SMOKING STATUS

Rows: MARITAL-STATUS   Columns: SMOKING STATUS

	0	1	2	All
D	159.7	151.6	179.0	154.8
	8.33	34.98	*	30.96
	3	12	1	16
M	147.7	158.6	176.3	154.2
	24.22	36.22	25.81	29.47
	31	16	6	53
S	132.1	133.6	124.9	130.3
	24.49	20.20	35.13	26.15
	8	7	7	22
W	159.6	161.5	183.0	165.2
	9.71	7.78	32.53	17.01
	5	2	2	9
All	147.1	151.7	154.8	150.0
	23.48	32.87	39.18	29.84
	47	37	16	100

Cell Contents: WEIGHT : Mean
WEIGHT : Standard deviation
WEIGHT : Nonmissing

The legend at the bottom of the table displays the contents of each cell.

The mean weight of all the nonsmokers is 147.1 pounds compared to 150.7 pounds for all 100 in the sample. There is so much you can do with two-way tables.

Step 11. Finish the project.

a. Type **your name** and **date** at the top of the Session window, then print the Session window.

b. Save this project as **Project04.mpj.**

c. Close MINITAB.

Chapter 4 Textbook Problems Suitable for Minitab

Page 182	4.2	31
Page 187	4.3	19 – 21, 24

Chapter 4: EndNotes

Chapter 5 Discrete Probability Distributions

5-1 Introduction: The Least You Need to Know About Probability Distributions

The menus and dialog boxes encountered when calculating binomial probabilities is typical for most probability distributions. The example is used to explain how to use the dialog box.

Calculate a Probability from a Known Distribution

From Section 4 Example 19: It is known that 5 percent of the population is afraid of being alone at night. If a random sample of twenty Americans is selected, what is the probability that exactly five of them are afraid?

No data needs to be entered in the worksheet.

Step 1. Select **Calc>Probability Distributions>Binomial.**

The dialog box is in three sections.

Step 2. Click the option for Probability. Given a value of x, calculate P(x). x = 5

One of these options must be selected.

Cumulative probability would find the probability that $X \leq 5$.

Inverse probability finds X if you know the cumulative area.

Step 3. Click in the text box for Number of trials:. This section is unique to binomial distributions. In this example, n = 20 and p = .05.

Step 4. Type in **20** then <Tab> to Probability of success: and type **.05**.

Step 5. Click the option for Input constant:.

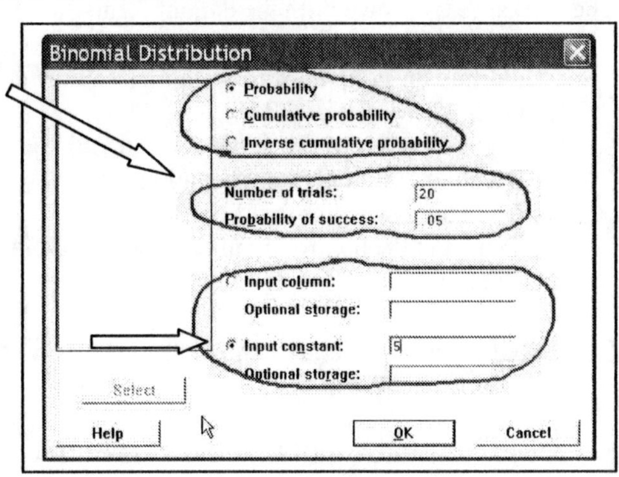

These two choices determine whether you want to calculate a lot of values at the same time or if you want to calculate one value. In this example, calculate P(5), the probability that exactly five out of twenty are afraid.

a. Type in **5** in the text box for Input constant:.

b. Leave the text box for Optional storage: empty. The result will be displayed in the Session window. If you type in a constant such as **K1,** the result is stored and not displayed.

c. Click **[OK].** Use the delete key to erase extra lines in the output.

Probability Density Function

```
Binomial with n = 20 and p = 0.05

x  P( X = x )
5   0.0022446
```

The probability that five out of twenty would be a success is small, .002 .

d. Select **File>New>Minitab Worksheet.**

5-2 Calculate Empirical Probabilities from a Frequency Distribution

Given a frequency distribution, use MINITAB to calculate the relative frequencies. Relative frequencies are empirical probabilities.

Section 2 Example 5-3: X represents the number of chain saws a supplier rents each day.
Step 1. In C1 of the worksheet type in the values to be used for X from 0 to 2.
Step 2. Name the column **X**.
Step 3. In C2 enter the frequencies. Name the column **f**.

Calculate the relative frequencies and store them in a new column named Px.
Step 4. Select **Calc>Calculator.**

- **a.** Type **Px** in the box for Store result in variable:.
- **b.** Click in the Expression: box then type **f/SUM(f)** .
- **c.** Click **[OK]**.

Worksheet 1 ***

↓	C1	C2	C3
	X	f	Px
1	0	45	0.500000
2	1	30	0.333333
3	2	15	0.166667

Step 4 Empirical probabilities

Each frequency is divided by the total frequency; then the result is stored in the same row of the new column, Px.

The worksheet is shown with the default formatting for C3 at six decimal places.

Step 5. Format the column of probabilities so they are displayed to three decimal places.
- **a.** Click in any cell of C3 Px.
- **b.** Right click anywhere in the column with the probabilities.
- **c.** Select Format column, then Numeric.
- **d.** Click the option for fixed decimal.
- **e.** Type **3** in the box for number of decimal places.
- **f.** Click **[OK]**.

Step 6. Click the Disk icon on the toolbar, then save this project as **Project5-2.mpj.**
Continue.

5-3 Mean, Variance, and Standard Deviation of a Discrete Probability Distribution

Use the formulas $\mu = \sum X * P(X)$ and $\sigma^2 = \sum X^2 * P(X) - \mu^2 \quad \sigma = \sqrt{\sigma^2}$

Continue to use the distribution for Example 5-3.

Calculate the Mean, Variance, and Standard Deviation from Probabilities

Step 7. Select **Calc>Calculator** to multiply the columns.

- **a.** Type **X*Px** for the storage variable label.
- **b.** Enter the Expression to multiply the two columns: **X*Px,** then click **[OK]**.

c. Edit the last dialog box then type **X2*Px** for the variable and **'X'**2*'Px'** for the Expression. Both dialog boxes are shown.

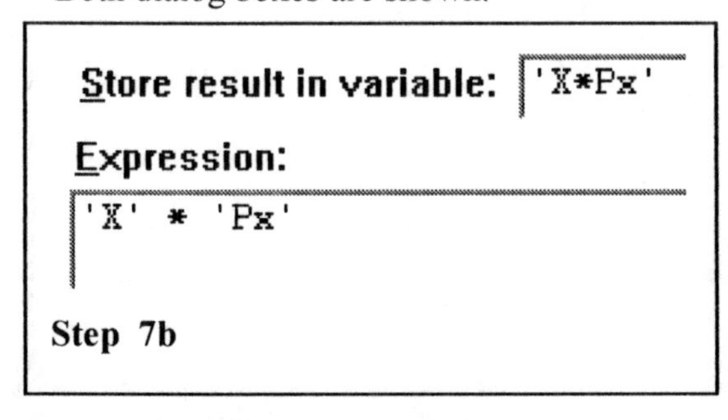

Step 7b

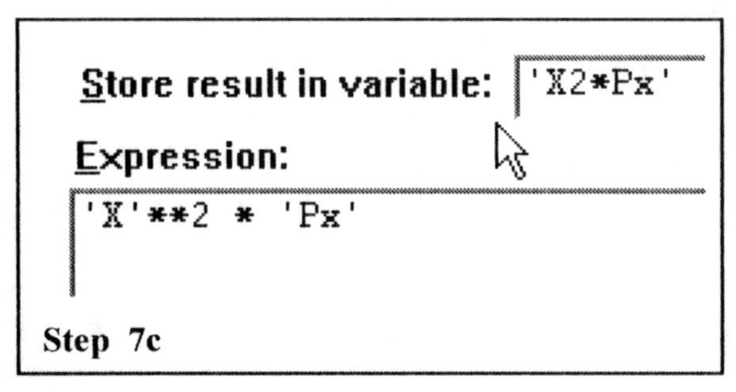

Step 7c

Two new columns are created in the worksheet.

C1	C2	C3	C4	C5
X	f	Px	X*Px	X2*Px
0	45	0.500000	0.000000	0.000000
1	30	0.333333	0.333333	0.333333
2	15	0.166667	0.333333	0.666667

The mean is the sum of X * Px.

Step 8. Select **Calc>Column Statistics.**

a. Click the option for **Sum.**

b. Click inside the box for Input variable:, then select X*Px.

c. Press <Tab>.

d. Type **MeanX** to name the constant.

e. Click **[OK]**.

This sum is the mean of the distribution.

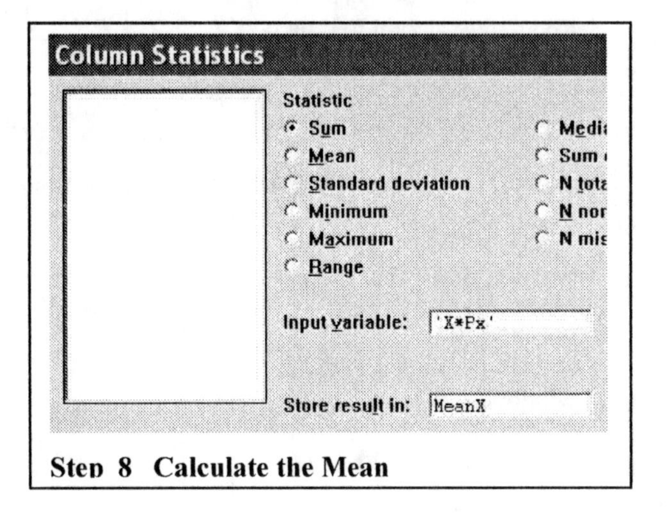

Step 8 Calculate the Mean

Step 9. To calculate the variance, select **Calc>Calculator.**

a. Type **K2** in the variable box.

b. Enter the formula to calculate the variance as shown.

SUM('x2*Px')-'MeanX'2**

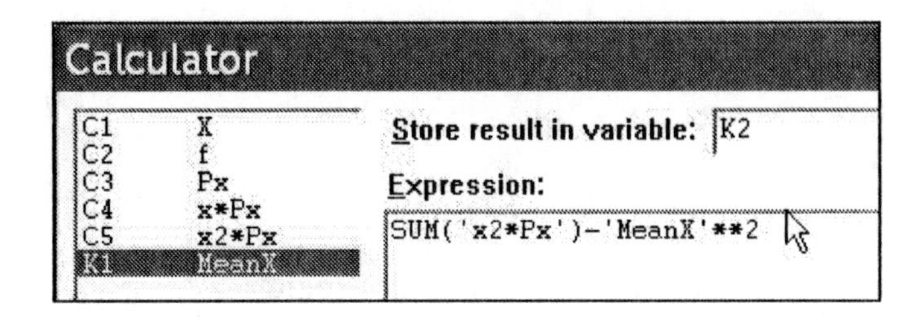

c. Click **[OK]**.

To calculate the standard deviation, find the square root of the variance.

Step 10. Select **Calc>Calculator.**

- **a.** Type **K3** for the variable.
- **b.** In the Expression box type **SQRT(K2)** then click **[OK].**

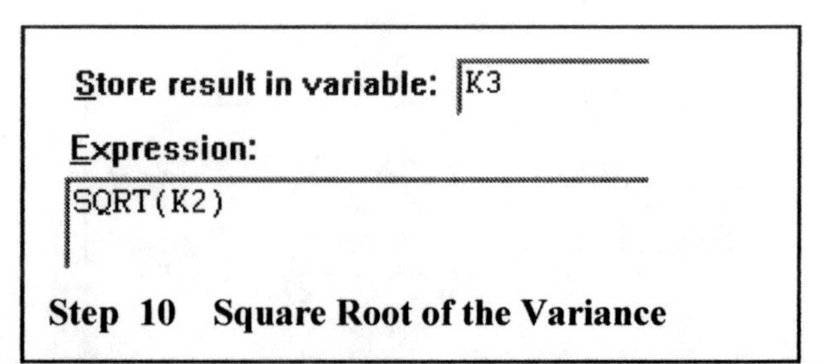

Step 10 Square Root of the Variance

Step 11. Click the icon for the Project Manager window on the Standard toolbar.

The Project Manager window will open.

- **a.** Click the Constants folder icon. The list of constants will appear in the Tree View of the right pane.
- **b.** Right click the name for K3 then choose Rename. Type in **Sigma.** Repeat instruction to rename K2 Variance.

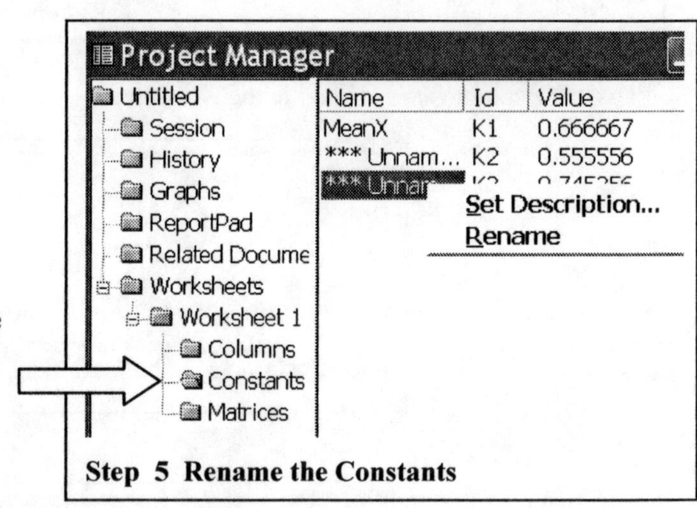

Step 5 Rename the Constants

Step 12. To display the results in the Session window select **Data>Display data.**

- **a.** Drag the mouse over the columns and constants.
- **b.** Click **[Select].**
- **c.** Click **[OK].**

The results will be printed in the Session window as shown.
The mean of the distribution is .667 with a standard deviation of .745, just as we expected!

Data Display

```
MeanX       0.666667
Variance    0.555556
Sigma       0.745356

Row  X   f     Px       X*Px      X2*Px
  1  0  45  0.500  0.000000   0.000000
  2  1  30  0.333  0.333333   0.333333
  3  2  15  0.167  0.333333   0.666667
```

Step 13. Save the project, then continue.

- **a.** Type **your name** and **date** at the top of the Session window.
- **b.** Save the project by clicking the Disk icon.
  The project is already named Project5-2.MPJ.

Example 5-11. The probability that 0, 1, 2, 3, or 4 people will be placed on hold when they call a radio talk show is shown in the distribution. Find the variance and standard deviation for the data. The radio station has four phone lines. When all lines are full, a busy signal is heard.

X	0	1	2	3	4
P(X)	.18	.34	.23	.21	.04

Step 1. Use **File>New>Minitab Worksheet** to create a fresh worksheet.

- **a.** Enter the data into two columns of a MINITAB worksheet.
- **b.** Name the columns **X** and **Px.**

Construct a Graph for the Discrete Distribution

Step 2. Select **Graph>ScatterPlot,** then **Simple**.

- **a.** Double click C3 Px to select the probabilities for the Y-variable.
- **b.** The X variable is C1. Use the slider to see that you have selected these two variables.
- **c.** Click **[Data View],** then the tab for Data Display.
- **d.** Click Project lines.
- **e.** Remove the check for Symbols.
- **f.** Click **[OK]**.

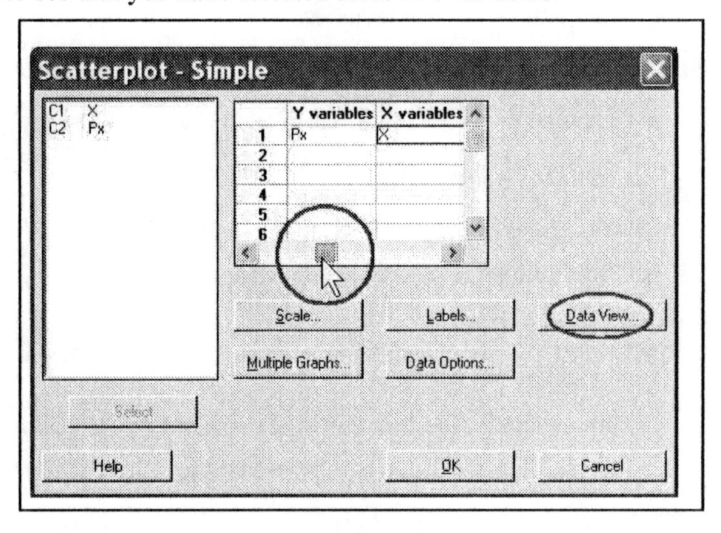

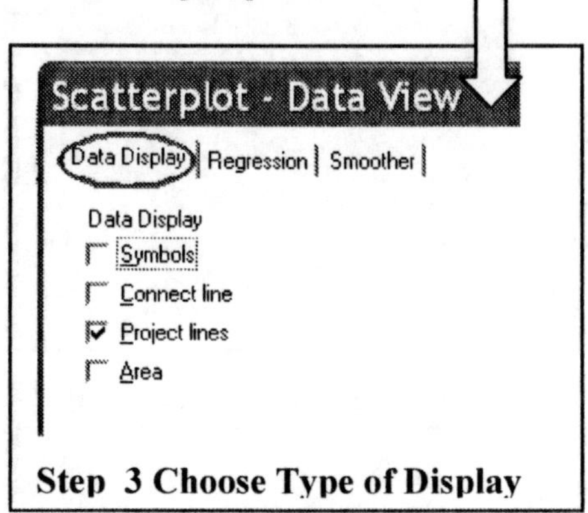

Step 3 Choose Type of Display

The Scatterplot - dialog box will be visible.

Step 3. Click **[Labels]**.

- **a.** Type in the title, **Callers on Hold.**
- **b.** Click **[OK]**.

Step 4. Click **[Scale],** then the tab for Gridlines.

- **a.** Select all four.
- **b.** Click **[OK]** twice.

The graph will open in a Graph window.

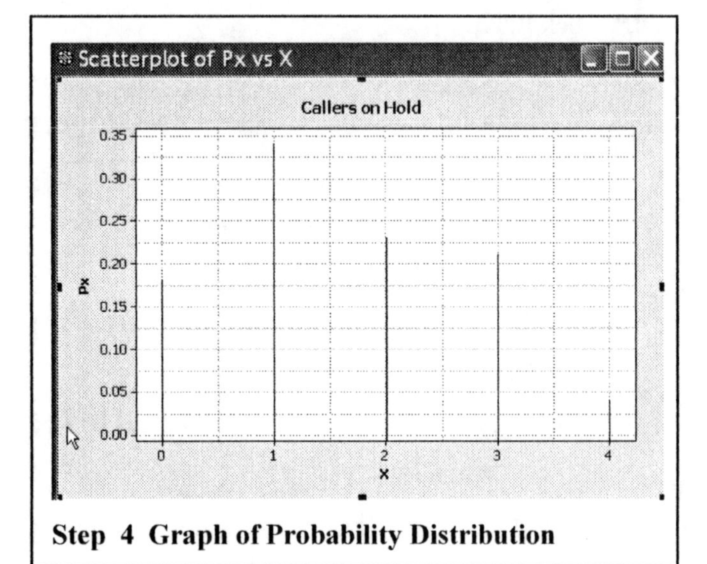

Step 4 Graph of Probability Distribution

Step 5. Calculate the mean.

a. Select **Calc>Calculator.** Press <F3> to clear the dialog box.

b. Enter **K1** for the variable.

c. Type **SUM(X* Px)** in the Expression box.

d. Click **[OK].**

Step 6. Calculate the variance, but a little different this time.

a. Click the Edit last dialog icon.

Change the variable to **K2**.

b. Type **SUM(X**2 *Px) - K1**2** for the Expression:, then click **[OK].**

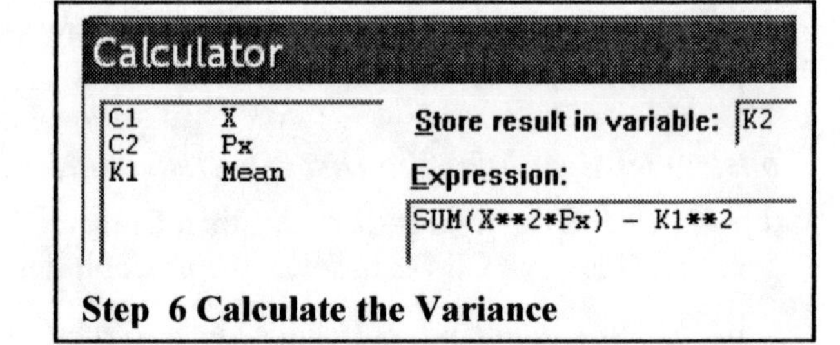

Step 6 Calculate the Variance

Calculate the standard deviation.

Step 7. Click the Edit last dialog icon.

a. Change the constant to **K3.** Press <Tab>. The entire expression for variance will be highlighted.

b. Scroll down the function list until you see Square root.

c. Double click the square root function. Immediately, the function will surround the previous expression: **SQRT(SUM ('X'**2 * 'Px') - K1**2))**

d. Click **[OK].**

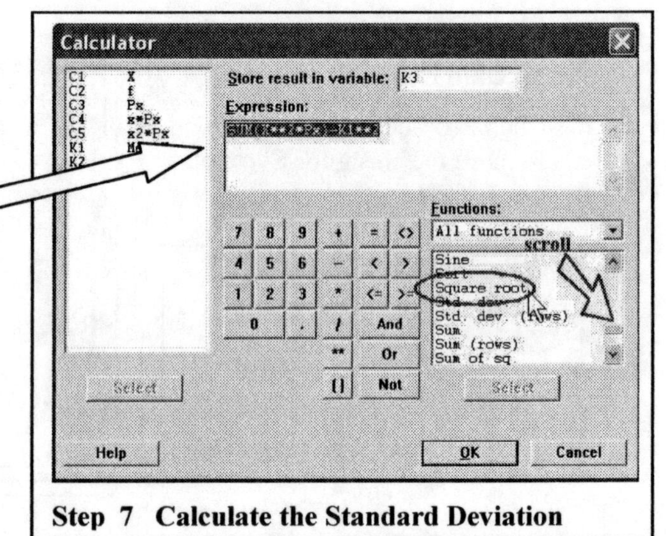

Step 7 Calculate the Standard Deviation

Step 8. To rename the constants, click the Project Manager icon then the Constants folder for Worksheet 2.

a. Right click each constant and rename them, **Mean, Variance** and **StDev.**

b. Select **Data>Display Data.**

c. Drag the mouse over the columns and constants of this worksheet.

d. Click **[Select],** then **[OK].**

The values stored in the columns and constants will be displayed in the session window.

Data Display

```
Mean       1.59000
Variance   1.26190
StDev      1.12334
```

Create and Print the Report.

Step 9. In the Project Manager window, Worksheet 2 folder should be green.

a. Click the Session folder.

b. Right click on the Scatterplot of Px vs 'X', then click Append to Report.

c. Right click Data Display for Worksheet 2, then Append to Report.

d. Click ReportPad in the left pane.

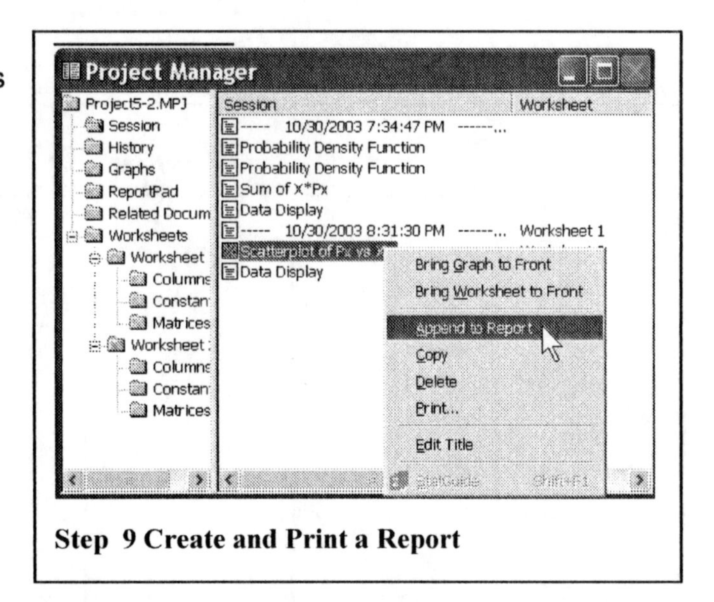

Step 9 Create and Print a Report

e. Maximize the window. Type **your name,** the **date** and a **brief description** of the project at the top of the report. At the end, type a brief paragraph that summarizes the result.

This report can be printed, saved as Rich Text for exporting to other programs, or moved to your word processor for inclusion in the body of a document.

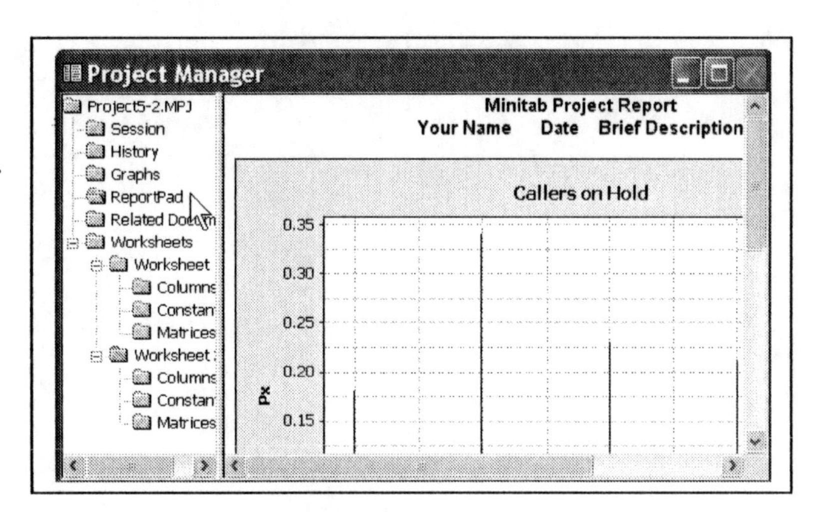

f. Right click the ReportPad folder again and select Print the Report.

See the ReportPad output on the next page.

g. Right click the ReportPad folder once again then select Save Report as. You will be prompted for a name. Type **Project05-2 ,** for the file name then click **[SAVE]**. The extension for Rich Text Files (.RTF) will be appended to the file.

Step 10. Select **File>New>Minitab Project.** Save this project as **Project5-2.MPJ.**

Minitab Project Report
Your Name and Date

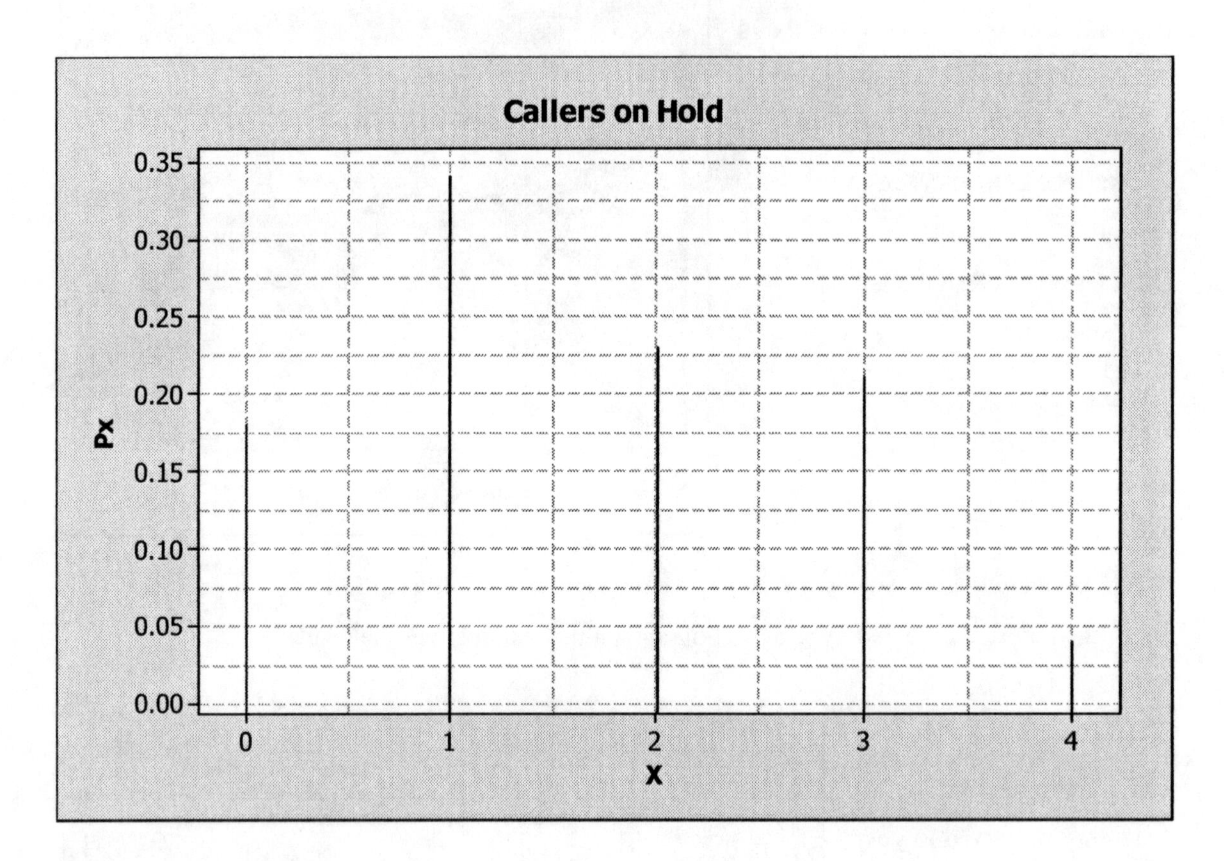

Data Display

```
Mean         1.59000
Variance     1.26190
StDev        1.12334

Row  X    Px
  1  0  0.18
  2  1  0.34
  3  2  0.23
  4  3  0.21
  5  4  0.04
```

The mean is 1.6 calls on hold with a standard deviation of 1.1. Most likely 1 will be on hold since that outcome has the highest probability (.34). The probability that X is 3 or less is .75 acquired by adding .18; .34; and .23, the cumulative probability. The probability that none are on hold is .18. It is not likely that 5 would be on hold as that probability is only .04, less than a 4 percent.

MINITAB should still be running ready to start a new project.

Step 11. Select **File>Save Project As,** and save the new project as **BINOM.mpj**.

Saving the project at the start allows you to click the Disk icon and save this project at any time as you work.

5-4 Binomial Distributions

Binomial experiments have two possible outcomes. The probability of "success" in each of n independent trials is p. This may be the known proportion of a population that has some characteristic. For example, it is thought that 15 percent of the population are left-handed. The probability that each individual is left-handed would be .15. the probability is the same for each individual. The trials are independent.

Suppose that a random sample of 20 is selected. The random variable X is the number out of 20 that are left-handed. X could be any number from none of them (X = 0) to all of them (X=20). These classical probabilities are calculated with the binomial formula.

$$P(X) = \frac{n!}{(n-X)!*X!} * p^X * q^{(n-X)}$$ MINITAB can do this for us.

Calculate a Binomial Probability

What is the probability that 3 in a random sample of 20 are left-handed?

Use these instructions to calculate one probability.

Step 1. Select

Calc>Probability Distributions>Binomial.

a. Click the button for Probability.

b. Click in the box for Number of trials:. Type **20**.

c. Click in the box for Probability of success:, then type **.15**.

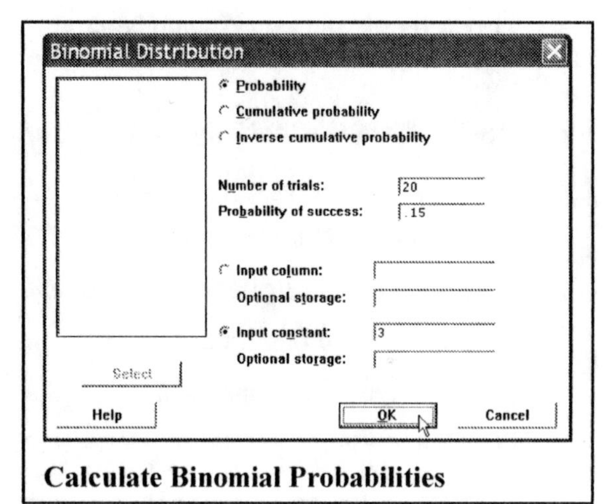

Calculate Binomial Probabilities

d. Click the button for Input constant: then press <Tab>. Type **3**, the value of X.

Step 2. Click **[OK]**.

The result will be displayed in the Session window.

The probability that 3 out of 20 will be left-handed is .2428, a fairly likely probability.

Step 3. Click the icon for Edit last dialog box, then click the button for Cumulative probability.

Step 4. Click **[OK]**. The probability that 3 *or less* will be left-handed is .6477. Very likely!

Probability Density Function

```
Binomial with n = 20 and p = 0.15
x  P( X = x )
3     0.242829
```

Cumulative Distribution Function

```
Binomial with n = 20 and p = 0.15
x  P( X <= x )
3     0.647725
```

Next, a complete distribution will be calculated.

Construct a Binomial Distribution

n=20 and p = .15 Use a shortcut for entering the sequence of integers from 0 to 20 into the first column.

Step 1. Select **Calc>Make Patterned Data>Simple Set of Numbers.** You must enter three items:

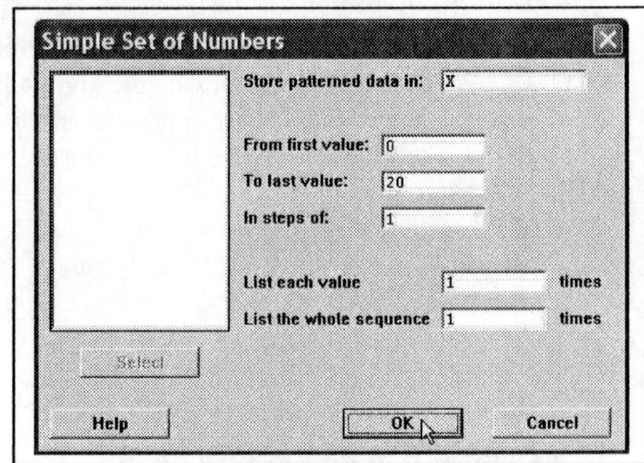

- **a.** Type **X** in the box for Store patterned data in:. MINITAB will use the first empty column of the active worksheet and name it X.
- **b.** Press <Tab>. Enter the value of **0** for From first value:. Press <Tab>.
- **c.** Type **20** for the last value. This value should be n. In steps of:, the value should be **1.**
- **d.** Click **[OK].**

Step 2. Select **Calc>Probability Distributions>Binomial.** In the dialog box you must enter 5 items.

- **a.** Click the button for Probability.
- **b.** In the box for Number of trials:, enter **20.**
- **c.** Enter **.15** for the Probability of success:.
- **d.** Check the button for Input columns:, then type the column name, **X**, in the text box.
- **e.** Click in the box for Optional storage:, then type **Px.**

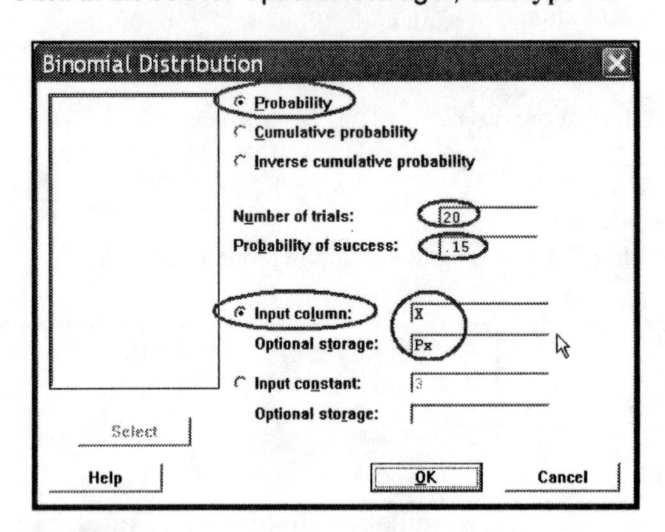

Step 3. Click **[OK]**.

The first available column will be named **Px,** and the calculated probabilities will be stored in it. To view the completed table, click the

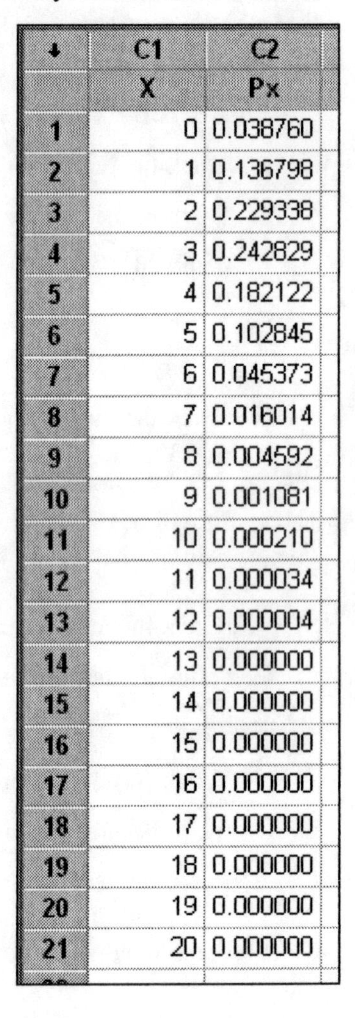

↓	C1	C2
	X	Px
1	0	0.038760
2	1	0.136798
3	2	0.229338
4	3	0.242829
5	4	0.182122
6	5	0.102845
7	6	0.045373
8	7	0.016014
9	8	0.004592
10	9	0.001081
11	10	0.000210
12	11	0.000034
13	12	0.000004
14	13	0.000000
15	14	0.000000
16	15	0.000000
17	16	0.000000
18	17	0.000000
19	18	0.000000
20	19	0.000000
21	20	0.000000

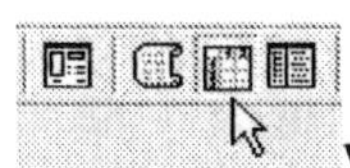 worksheet icon on the toolbar.

Graph a Binomial Distribution

Step 1. To graph this discrete distribution, select **Graph>Scatterplot> Simple.**

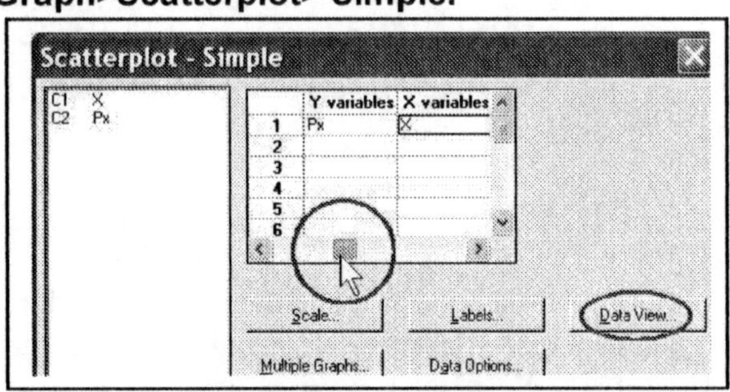

- **a.** Double click C2 Px for the Y variable and the C1 X for the X variable.
- **b.** Click **[Scale],** then the tab for Gridlines. Check all four.
- **c.** Click **[Labels].** Type an appropriate **title** and **your name** and **date** in the footnote.
- **d.** Click **[Data View],** then the Display tab. Change the Data display type to Project lines. No other types should be checked.
- **e.** Click **[OK].**

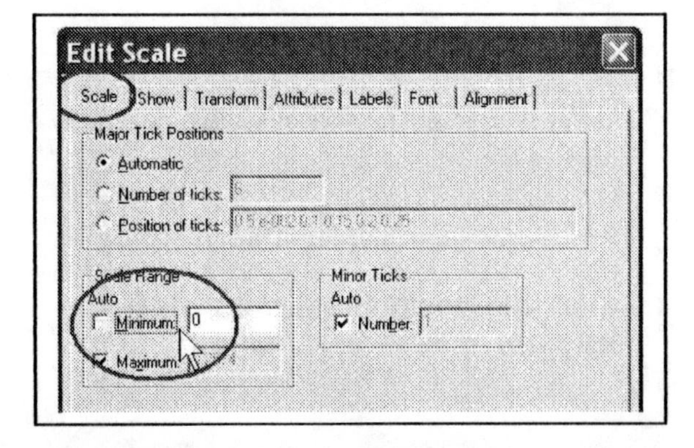

One more graph detail:

Step 2. Select **Editor>Select Item>Y-scale.**

- **a.** Select **Editor>Edit Y-scale.**
- **b.** Remove the check from automatic minimum in the Scale Range, then type **0** in the box.
- **c.** Click **[OK].**

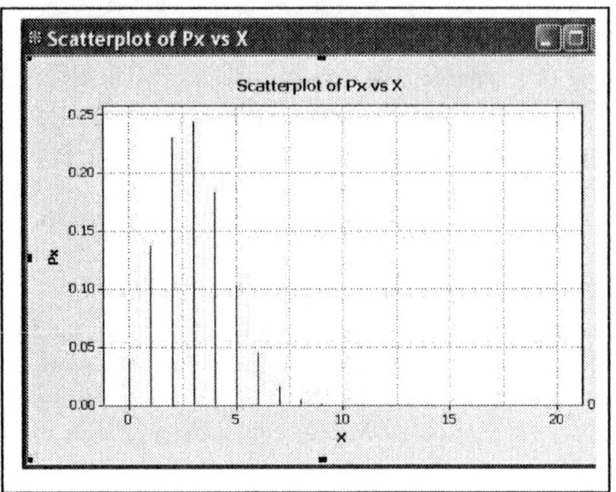

The graph shows the distribution of X is skewed right. The most likely outcomes surround the peak at the mean, X = 3.
P(3) = .243.

Note: The peak is always at the mean.
The most likely outcome is always the mean.

Mean, Variance, and Standard Deviation for a Binomial Distribution

The mean, variance, and standard deviation can be calculated using the formulas from the previous section or use the calculator and the formula for the mean and standard deviation of a binomial distribution.

$$\mu = np$$
$$\mu = 20 * .15 = 3$$

and

$$\sigma = \sqrt{npq}, \quad q = 1 - p$$
$$\sigma = \sqrt{20 * .15 * .85} \approx 1.6$$

Because we determined the probability distribution (table), we could use the instructions for finding the mean, variance, and standard deviation of a discrete probability distribution. However, the calculator can also be used with the formula to calculate them and store them in constants.

Step 1. Select **Calc>Calculator.**

  a. Type **K1** for Store result in:, then type the Expression: **20*.15.** Click **[OK].**

Step 2. Click the Edit last dialog box icon, then change the K1 to a **K2.**

  a. In the Expression: box type the formula **20*.15*.85** (the variance).

  b. Press <Enter> or click **[OK].**

Step 3. Select **Edit>Edit last dialog,** then change the K2 to a **K3.**

  a. Press <Tab> to move to the Expression: box. The variance formula should be highlighted.

  b. Scroll the function list, and then double click on square root function, SQRT(). That should enclose the variance formula inside the square root. The expression should be: SQRT(20*.15*.85).

  c. Click **[OK].**

Step 4. Click the Project Manager icon or select **Window>2 Project Manager.**

  a. Click the Constants folder in the left pane.

  b. In the right pane, right click each constant and then rename them **Mean, Variance**, and **Sigma.**

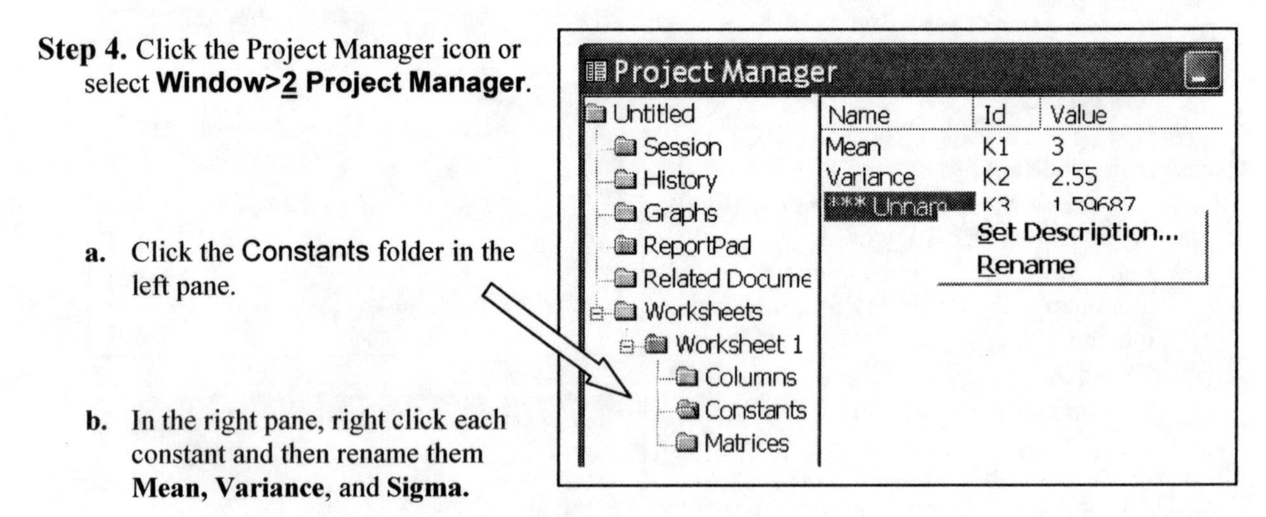

The mean is 3. The standard deviation is 1.6.

  c. Select **File>New>Minitab Worksheet** then continue to the next section.

Binomial Probabilities for Large Values of n.

Example 5-23: The "Statistical Bulletin" published by Metropolitan Life Insurance Co. reported that 2 percent of American births result in twins. If a random sample of 8,000 births is taken, find the mean, variance, and standard deviation of the number of births that would result in twins. The table is not practical to calculate. With a regular calculator or the Windows calculator:

$$\mu = np = 8000 * .02 = 160$$

$$\sigma^2 = 8000 * .02 * .98 = 156.8$$

$$\sigma = \sqrt{156.8} \approx 12.5$$

On average, 160 out of 8,000 births would be twins.

Using the empirical rule regarding bell-shaped curves, chapter 3 in textbook, almost all of the distribution is within three standard deviations of the mean.

$3*\sigma = 3*12.5 = 37.5 \approx 40.$

$\mu \pm 3\sigma$ → 160 ± 40 →120 to 200 is the range for X.

Most likely between 120 and 200 out of 8,000 births would be twins. The probabilities for this range can be calculated and graphed.

Construct the Binomial Distribution

Step 1. Enter the sequence of integers calculated above: X = 120 to X = 200 into C1. Use the shortcut.

The values of X from 0 to 8,000 could be used, however, the values from 120 and 200 are the ones that will have nonzero probabilities.

- **a.** Select **Calc>Make Patterned Data>Simple set of numbers.**
- **b.** Type **X** into the box for Store patterned data in:.
- **c.** Type **120** for first value and **200** for ending value.
- **d.** Click **[OK].**

Step 2. Select **Calc>Probability Distribution>Binomial.**

- **a.** Click Probability; enter **8000** for number of trials.
- **b.** Enter **.02** for Probability of success:.
- **c.** The Input column: is X and Store results in:, Px.
- **d.** Click **[OK].**

Step 3. Edit last dialog box and check Cumulative probabilities. Store them in cumPx. Click **[OK].**

Construct the Graph

Step 4. Select **Graph>Scatterplot > Simple.** The settings from the previous graph are ok. Check:

- **a.** The Y variable is Px and the X variable is X.
- **b.** Click **[Data View]** then the tab for Data Display. Select **project lines.**
- **c.** Click **[Labels],** and enter the title.

  Binomial Distribution

  n = 8000 p = .02

- **d.** Click **[Scale],** then click the tab for Reference lines and type a **0** (zero). Click **[OK]** twice.

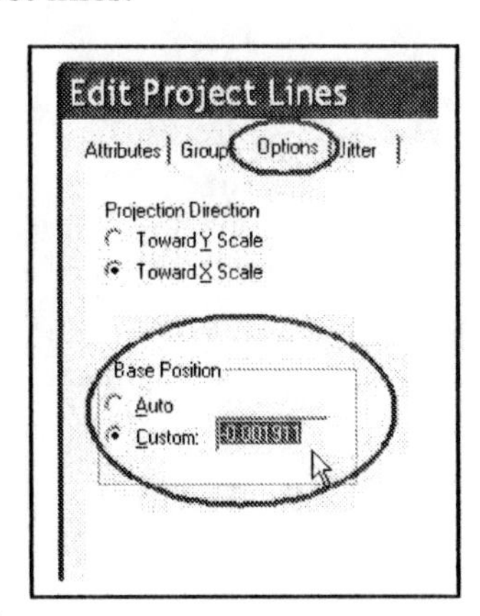

Step 5. Clean up the bottom of the scale.

- **a.** Select **Editor>Select Item>Project Lines.** Click **[OK].**
- **b.** Select **Editor>Edit Project Lines** then the tab for Options. Choose the Custom button and change whatever number is there to a **0.** Click **[OK].** The scales will look much better!

Step 6. Click the disk icon to save the project then select **File>New>Minitab Project** or **File>Exit** if ready to quit. The project should still be named **BINOM.mpj.**

The graph is in a Graph window. The table is in the worksheet. The table shows those values of X that have a probability of .005 or higher.

The bell-shaped distribution is symmetrical! Notice how small the probabilities are for each outcome. The scale for Px is .01, .02, and so on.

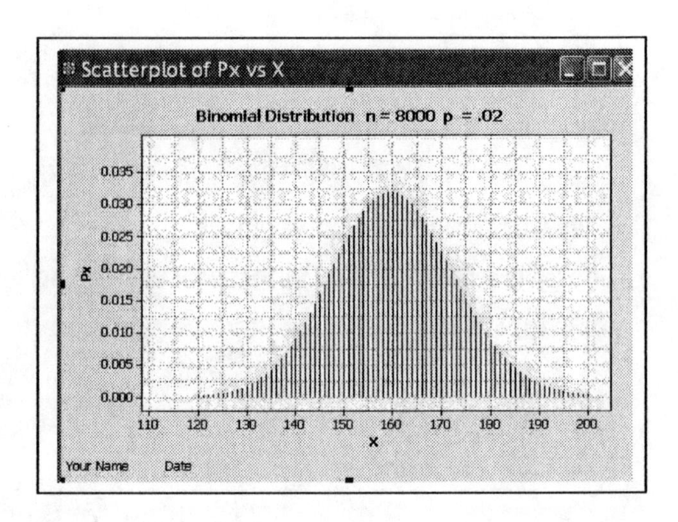

Data Display

X	Px	cumPx
137	0.0058391	0.033804
138	0.0067898	0.040594
139	0.0078376	0.048432
140	0.0089812	0.057413
141	0.0102174	0.067630
142	0.0115405	0.079171
143	0.0129421	0.092113
144	0.0144113	0.106524
145	0.0159346	0.122459
146	0.0174960	0.139955
147	0.0190772	0.159032
148	0.0206582	0.179690
149	0.0222173	0.201907
150	0.0237317	0.225639
151	0.0251782	0.250817
152	0.0265338	0.277351
153	0.0277761	0.305127
154	0.0288840	0.334011
155	0.0298386	0.363850
156	0.0306232	0.394473
157	0.0312243	0.425697
158	0.0316316	0.457329
159	0.0318387	0.489168
160	0.0318428	0.521010
161	0.0316450	0.552655
162	0.0312503	0.583906
163	0.0306673	0.614573
164	0.0299079	0.644481
165	0.0289868	0.673468
166	0.0279213	0.701389
167	0.0267304	0.728119
168	0.0254348	0.753554
169	0.0240557	0.777610
170	0.0226147	0.800225
171	0.0211330	0.821358
172	0.0196310	0.840989
173	0.0181280	0.859117
174	0.0166418	0.875759
175	0.0151882	0.890947
176	0.0137811	0.904728
177	0.0124320	0.917160
178	0.0111506	0.928310
179	0.0099442	0.938255
180	0.0088178	0.947072
181	0.0077749	0.954847
182	0.0068167	0.961664
183	0.0059433	0.967607
184	0.0051529	0.972760

5-5 Other Distributions

The binomial distribution lists the probability that X out of N successes occur in a sample if the population parameter, p, is known. The probability must be the same for each observation in the sample. That would be true if the population is large. It would be true if the population is finite, N is known, and the sampling is done with replacement. However, if the population is finite and the sampling is done without replacement, then the probability changes for each individual. The hypergeometric distribution would apply.

As an example, six numbers are selected, without replacement, from a population of fifty in a state lottery game. Players take a chance that their set of six numbers match those drawn by the state. It is possible that a player's set of numbers may match anywhere from none to all six.

The formula for computing the probabilities is:

$$\frac{{}_aC_x * {}_bC_{n-x}}{{}_NC_x}$$

a = 6, *possible* matches (successes) in N   b = 44, *possible* failures in N

n = sample size out of N   X = number of matches on a ticket

Calculate a Table of Hypergeometric Probabilities

Step 1. To enter the first column of integers from 0 to 6, select **Calc>Make Patterned Data>Simple Set of Numbers.** Name the variable **X**.

Calculate the probabilities.

Step 2. Select **Calc>Probability Distribution,** then **Hypergeometric Distribution**.

- **a.** Click the button for Probability.
- **b.** Enter **50** for Population size (N).
- **c.** Enter **6** for Successes in population. This value is referred to as "a" in the textbook.
- **d.** Enter a **6** for the Sample size.
- **e.** Click the button for Input column. and type in the column name, **X.**
- **f.** Type in **Px** for Optional storage.
- **g.** Click **[OK]**.
- **h.** Choose **Edit>Edit Last Dialog,** then change from Probability to Cumulative probabilities. Change Storage to **cumPx.**
- **i.** Click **[OK]**. The new table will have three columns.

Optional: Click any cell in C2 Px. Right click and select **Format column>numeric.** Make the set number of decimal places equal to nine.

C1 X	C2 Px	C3 cumPx
0	0.444225365	0.444225364521
1	0.410054183	0.854279547
2	0.128141932	0.982421479
3	0.016668869	0.999090348
4	0.000892975	0.999983324
5	0.000016613	0.999999937
6	0.000000063	1.000000000

Step 2 Hypergeometric Distribution

The worksheet contains the calculated values.

The probability of matching four or less (no prize) is .999983324. The probability of matching all six is .000000063. The highlighted cell shows all available decimal places.

The completed table is shown.

Step 7. Click the Disk icon .
Step 8. Save this project as **OtherDistr.MPJ**.
Continue.............Other Discrete Distributions

The Poisson Probability Distribution

The Poisson distribution can be used when n is large and p is small and the independent variable X occurs over time. Poisson distributions also apply when a density of items is distributed over a given area or volume, such as the number of plants growing per acre or the number of defects in a given length of videotape. The probability of X occurrences in an interval of time, volume, area, and so on for a variable where λ (lambda) is the mean number of occurrences per unit is given by the formula:

$$P(X:\lambda)=\frac{e^{x} * \lambda^{X}}{X!}$$

Calculate Poisson Probabilities

Example 5-27: If there are 200 typographical errors randomly distributed in a 500-page manuscript, find the probability that a given page contains exactly three errors.
The mean number of errors per page is less than one. $\lambda = \frac{200}{500} = .4$ The mean is .4 errors per page.
Find the probability that a page will contain three errors. X = 3
Step 1. Select **Calc>Probabilitity Distributions>Poisson.**

Step 2. Click Probability.

This option will calculate the probability of exactly three. Click on **[Help]** for more information including formulas used.

Step 3. Click in the box for Mean: and type **.4.**

Step 4. Click Input constant:.

Step 5. Press <Tab>, then type 3, the value of X.

Step 6. Click **[OK].**

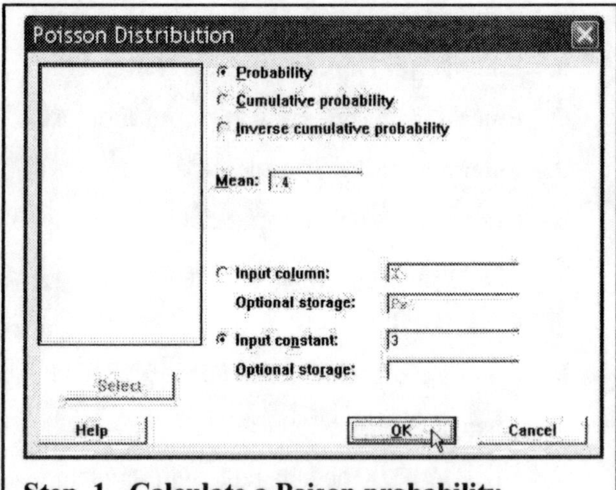

Step 1 Calculate a Poison probability

The probability will be displayed in the Session window.

The probability that three errors occur in a page is .0072.

Probability Density Function

```
Poisson with mu = 0.400000
     x          P( X = x )
  3.00             0.0072
```

Example 5-28: A sales firm receives, on average, three calls per hour on its toll-free number. For any given hour, find the probability that it will receive the following:
a) At most three calls b) At least three calls c) Five or more calls

Use MINITAB to calculate the cumulative probabilities for at most three calls.

Step 1. Select **Calc>Probability Distributions>Poisson.**

- **a.** Click Cumulative probability.
- **b.** Click in the box for Mean and type in **3**, the value of λ.
- **c.** Click on the button for Input constant: and type in **3,** the value of X.

Step 2. Click **[OK].**

As you can see in the Session window, the probability is .6472. The cumulative probability is the sum of the probabilities up to and including the value given. There isn't any data in the worksheet. None is required.

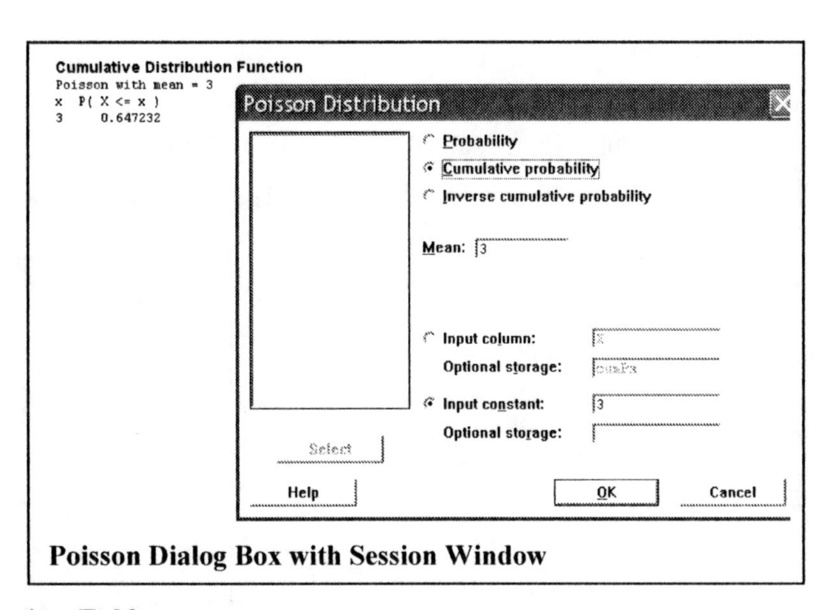

Poisson Dialog Box with Session Window

Construct a Poisson Distribution Table

For a Poisson distribution, X can be any number greater than or equal to zero. The upper limit of the values can be determined in the following way. The maximum value of X can be determined by finding the value for X such that 99.99 percent of the distribution is lower. Here is how.

Determine the Upper limit of X for a Poisson Distribution

Step 1. Select **Calc>Probability Distributions>Poisson.**

- **a.** Click Inverse cumulative probability.
- **b.** Click in the box for Mean and enter **3**.
- **c.** Check the box for Input constant: then type **.9999**, the cumulative probability. 99.99 percent of the distribution is lower than the value of X it calculates. If you leave the Optional storage box blank, the result will be displayed in the Session window.

Step 2. Click **[OK].**

The Session window may give two values. Use the largest or a number a little higher.

The probability that $X \leq 11$ = .999929.
The largest value you need for X is 11. This is a discrete distribution, so there may not be an exact value of X.

Inverse Cumulative Distribution Function

```
Poisson with mean = 3

 x   P( X <= x )      x   P( X <= x )
10      0.999708     11      0.999929
```

Continue to make the table.

Step 3. Select **File>New>Minitab Worksheet.** Enter the values of X from 0 to 11.

Step 4. Select **Calc>Make Patterned Data>Simple set of numbers.**

- **a.** Name the column **X.**
- **b.** The first value should be **0** and the last value should be **11**.
- **c.** The increment is **1**.

Step 5. Select **Calc>Probability Distributions>Poisson.**

- **a.** Click Probability.
- **b.** Click in the box for the Mean: and type **3**.
- **c.** Click on Input column: and double click the column named X.
- **d.** Click in Optional storage: and enter **Px**. The probabilities will be stored in this column.
- **e.** Click **[OK].**

Step 6. Click Edit last dialog box icon.

- **a.** Change to Cumulative probability and the storage to **cumPx.**
- **b.** Click **[OK].**

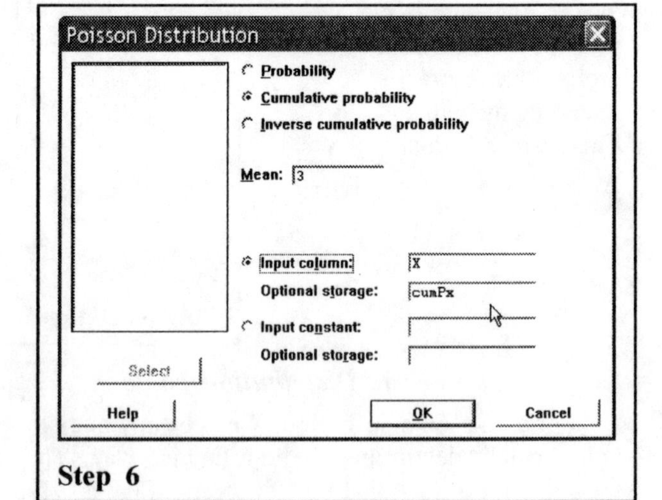

Step 6

The worksheet will contain the distribution (table).
Use the table to complete Example 5-28.

a) $P(X \leq 3) =$ .6472

b) $P(X \geq 3) = 0.5768$
1 - P(2 or less) = 1 - 0.423190 =

c) $P(X \geq 5) = 0.193945$
$1 - P(X \leq 4) = 1 - .8152 = .1848$.

X	Px	cumPx
0	0.049787	0.049787
1	0.149361	0.199148
2	0.224042	0.423190
3	0.224042	0.647232
4	0.168031	0.815263
5	0.100819	0.916082
6	0.050409	0.966491
7	0.021604	0.988095
8	0.008102	0.996197
9	0.002701	0.998898
10	0.000810	0.999708
11	0.000221	0.999929

Step 7. Close the program. Save the project as **OtherDistr.MPJ.**

Chapter 5 Textbook Problems Suitable for MINITAB

Page	230	5.2	19 - 36
Page	238	5.3	2 - 12
Page	247	5.4	2 – 15, 17 -27
Page	248	Review	4 – 18, 22
Page	246	Data Project	Simulation; rolling three dice 100 times See also EC problems 20, 21 and 23

Chapter 5 EndNotes:

Chapter 6 The Normal Distribution

6-1 Introduction

Continuous variables can assume any real value. Unlike a discrete distribution, it is not possible to list all values because there are infinite numbers of possibilities. A number line represents all the possible values. Rational numbers (fractions and integers such as $\frac{-15}{7}$) and irrational values (such as $\sqrt{11}$) are possible, not just integers used to mark the scale.

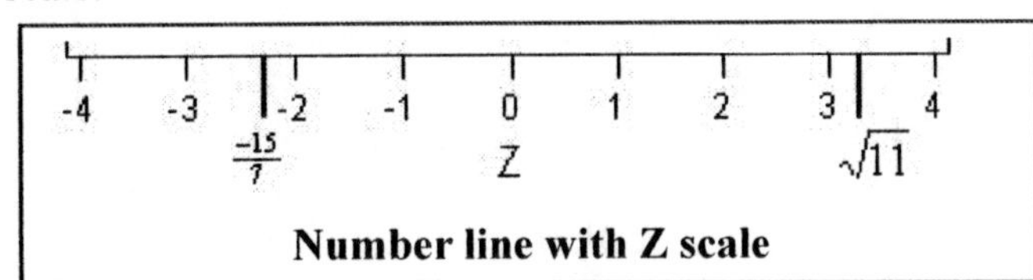

Number line with Z scale

For continuous distributions, the height of a bell-shaped curve does not represent the probability. The area under the curve represents the probability that an individual outcome, Z, is in this interval. The height of the curve is the value of a formula, the probability density function, or PDF.

The probability density function for a normal distribution is $y = \dfrac{e^{\frac{-(X-\mu)^2}{2\sigma^2}}}{\sigma\sqrt{2\pi}}$.

The curve depends on μ, the mean of the distribution, and σ, the standard deviation of the distribution. Many continuous variables have distributions that are bell-shaped and are approximately normal.

6-2 The Standard Normal Distribution

The standard normal distribution has a mean of 0 and a standard deviation of 1. It is customary to use Z as the variable for this distribution.

$$Y = \frac{e^{\frac{-Z^2}{2}}}{\sqrt{2\pi}}$$

To illustrate, we will calculate values of this function and plot them.

Calculate Probability Density Function Values for the Standard Normal Distribution

Step 1. Set up the table with values of Z from -3.0 to +3.0 by tenths.

a. Select **Calc>Make Patterned Data>Simple Set of Numbers.**
b. Type **Z** for the name of the storage column.
c. Type in **-3.0** for the starting value and **+3.0** for the last value.
d. Type **.1** for the increment. Make sure it is one tenth, not one.
e. Click **[OK].** There should be 61 values in the column named Z.

Step 2. Calculate the Y values of the probability density function.

a. Select **Calc>Probability Distributions>Normal.**
b. Click the button for **Probability Density.**
c. Do not change the mean or standard deviation. They should be zero and one.

d. Click the button for **Input Column** then enter the name of the column, **Z.**

e. In Optional storage: type **Y.**
The completed dialog box is shown.

f. Click **[OK]** twice.

The values in Y are calculated using the PDF. A column named Y will store the function values. Remember, these are *not probabilities.* They are the heights of the bell-shaped curve for the given values of Z. Continue to the next section.

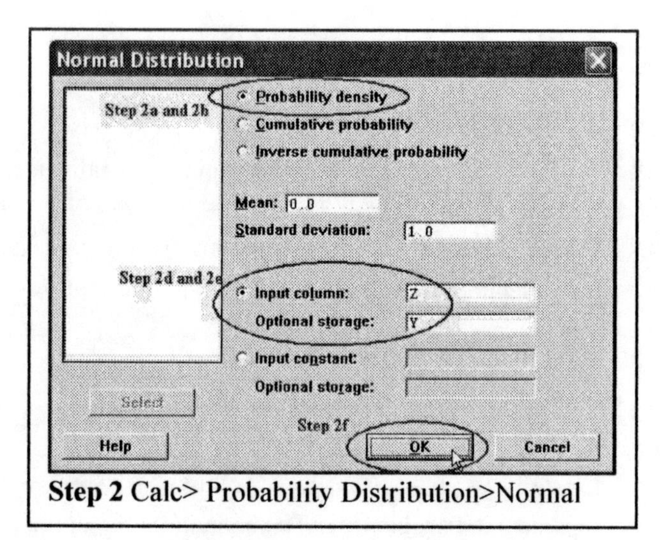

Step 2 Calc> Probability Distribution>Normal

6-3 The Standard Normal Distribution

Plot the PDF for the Standard Normal Distribution

Step 3. Select **Graph >Scatterplot >Simple.**

a. The graph variables should be Y and then Z.

b. Select **[Scale].**
   1. Click the tab for Gridlines, then check the options for Y major ticks and X major ticks.
   2. Click the tab for Reference Lines, then type in a **0** (zero) for both positions, Y and X. Click **[OK].**

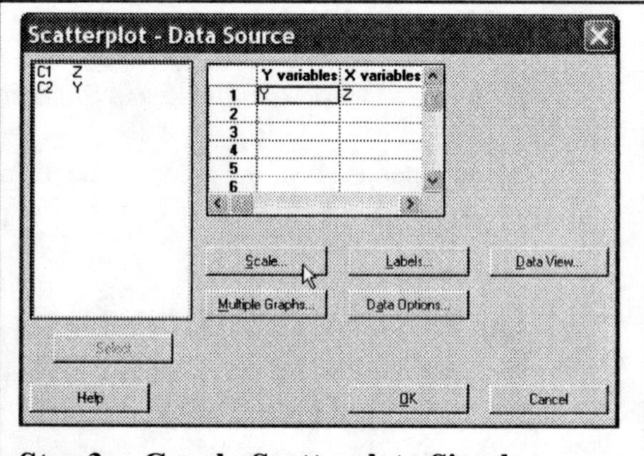

Step 3a. Graph>Scatterplot >Simple

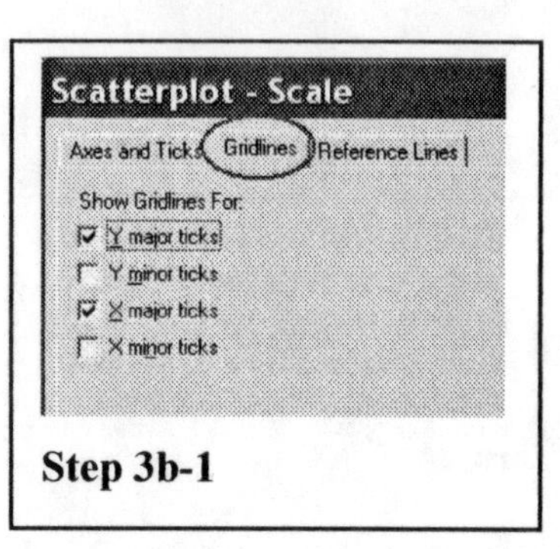

Step 3b-1

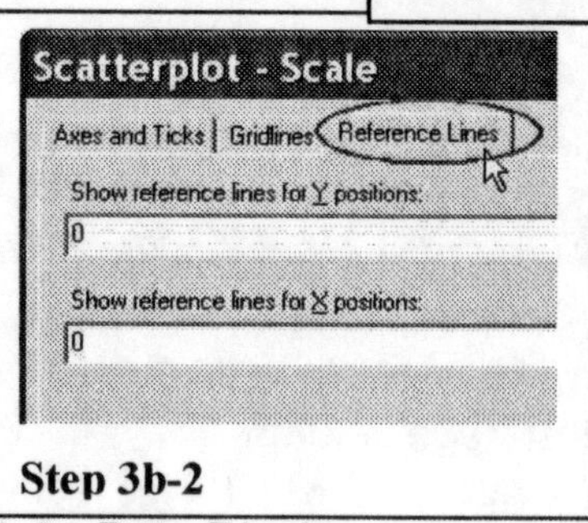
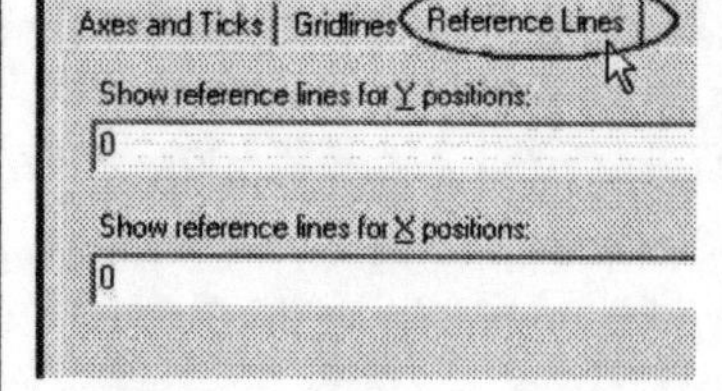

Step 3b-2

c. Select **[Data View],** then the tab for Data Display.
   1. Uncheck the option for Symbols.
   2. Check the option for Connect line.
   3. Click **[OK].**

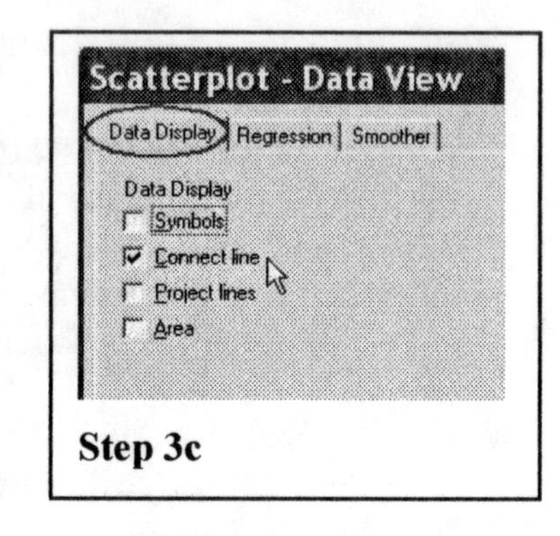

Step 3c

d. Select **[Labels],** then the tab for Titles and Footnotes.
   1. Type in the name of the graph for Title 1, **Standard Normal Distribution.**
   2. Type **Mean = 0 Standard Deviation = 1** for the subtitle.
   3. Type **your name** in Footnote1, then click **[OK]** twice.

In a separate window the graph will be displayed. The maximum height of the curve is less than .4. The total area is 1 square unit, one-half on each side of the mean.

To give a better perspective of these characteristics, graph the distribution again.

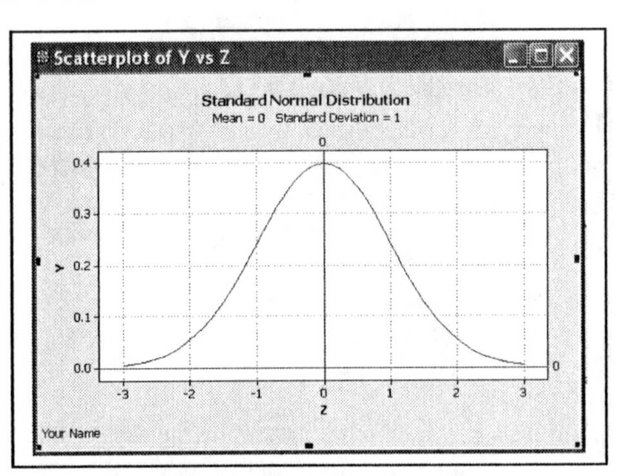

Change the type of graph to Area and change the scale for the vertical axis. Here is how.

Step 4. Click the Edit last dialog box icon on the toolbar.

a. Select **[Data View],** then the tab for Data Display.
   1. Deselect Connect.
   2. Select Area.
   3. Click **[OK]** twice.

The new graph will open.

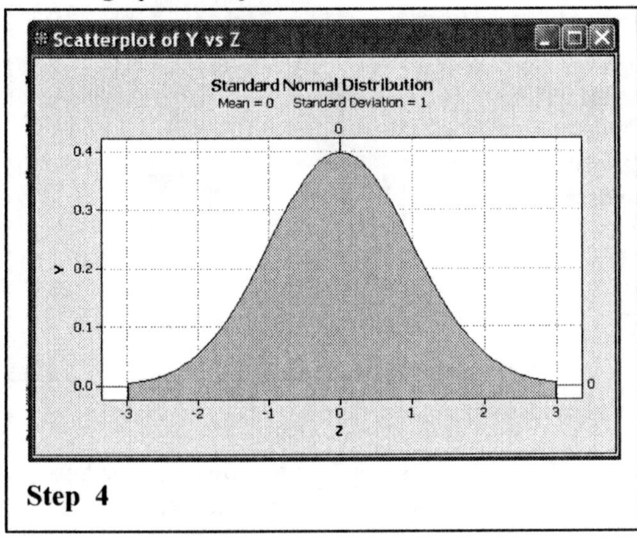

Step 4

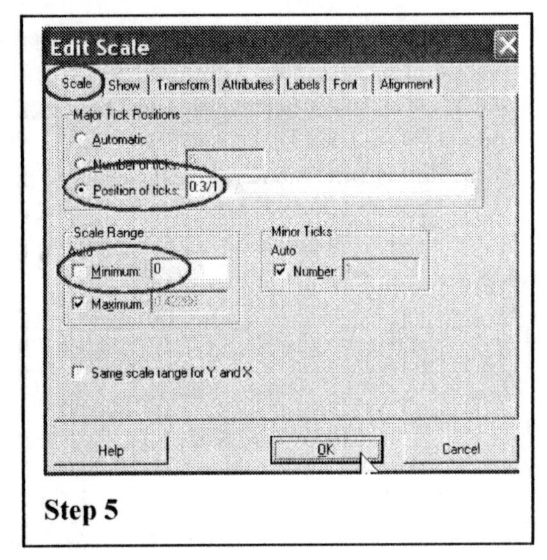

Step 5

Step 5. To change the scale, click the mouse over the Y-axis, then right click. A menu will appear.

a. Select Edit Y Scale, then the tab for Scale.
b. In the Position of ticks type **0:3/1.**
c. Uncheck Minimum in the Scale range.
d. Type **0**.
e. Click **[OK]**.

The new graph will show the standard normal distribution and the fact that the "curve" is slight when drawn with a larger scale. That the total area is only 1 square unit should seem more reasonable.

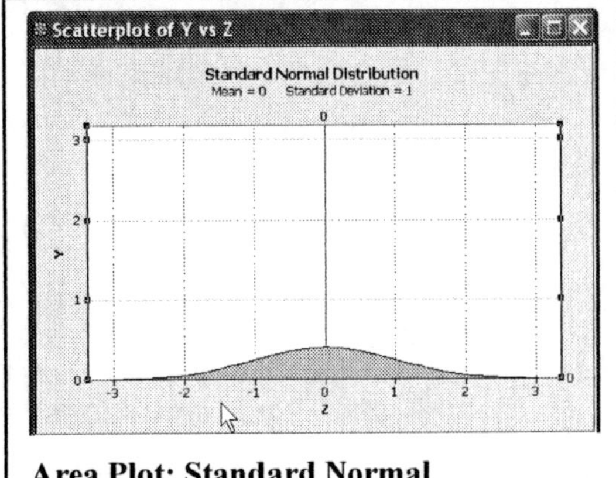

Area Plot: Standard Normal

Comparing the original plot to the plot with tick marks you can see what a difference the scaling makes. The bell shape appears very pronounced in the original graph. The second graph appears almost flat. Usually we draw a bell-shaped curve like the first with a pronounced peak. The height of the curve is determined by the function. Area represents probability. The total area under the curve is 1 square unit.

Step 6. Save the project as **Normal.MPJ.**

Find the Area Under the Curve

Finding areas (probabilities) using MINITAB uses the rules of appendix B-3 beginning on page 711 in the textbook. Cumulative areas are calculated by MINITAB. One of three rules will apply:

1. Find the area to the left of any Z value directly.
2. Find the area to the right of any Z value by finding the area to the left, then subtracting the value from 1.
3. Find the area between any two Z values by finding the area for each and then calculating the absolute value of the difference.

An example of each case follows.

Example 1. Example: Determine the area under the curve to the left of Z = +1.28. MINITAB calculates cumulative area (less than or equal to the given value of Z).
The area is highlighted.

Step 1. Select **Calc>Probability Distributions>Normal.**

Step 2. Click the button for Cumulative probability.

Step 3. Click Input constant: and enter **1.28,** the value of Z.
If a constant such as **K1** is entered in the Optional storage: dialog box, the result will be stored and not displayed in the Session window. Choose this *if* you need to use this value in another calculation. Leave it blank for this example.

Step 4. Click **[OK]**.

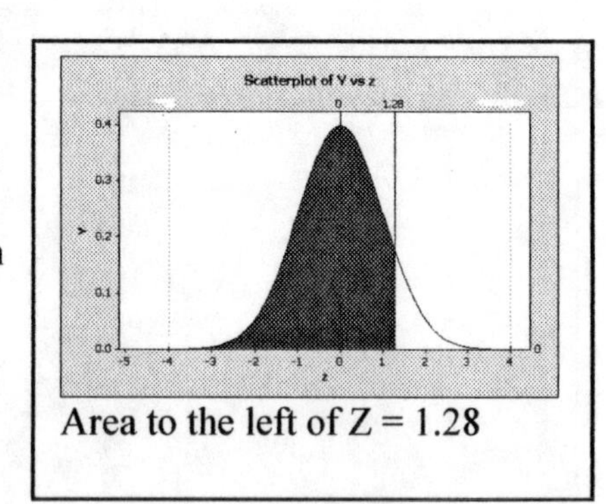

Area to the left of Z = 1.28

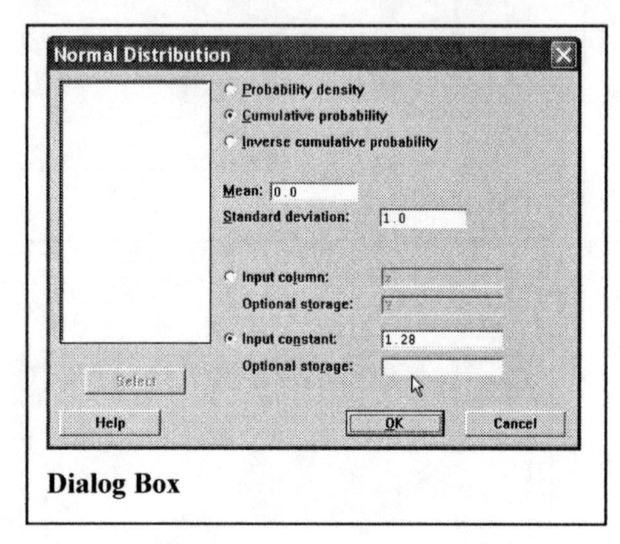

Dialog Box

The completed dialog box is shown. In the Session window, the cumulative area is .899727 .

Cumulative Distribution Function

```
Normal with mean = 0 and standard
deviation = 1
   x    P( X <= x )
1.28     0.899727
```

Example 2. This is Example 6-8, page 278 in the textbook.
Find the area between Z = - 1.37 and 1.68.

Step 1. On the menu bar, click the Edit Last Dialog Box icon as shown or select **Calc>Probability Distributions>Normal.**

Step 2. Change the Input constant: to **-1.37,** then type **K1** for the Optional storage:. Click **[OK].**

Step 3. Click the Edit Last Dialog Box icon.

   a. Change the **Input constant**: to **1.68.**
   b. Change the **Optional storage**: constant to **K2.**
   c. Click **[OK].**

Step 4. To calculate the difference between the two areas, select **Calc>Calculator.**

   a. Type **K3** in the **Store result in variable**: text box.
   b. In the **Expression**: box enter **K2 – K1.**
   Note: Names of constants are not case sensitive. It doesn't't matter if you type K1 or k1.
   c. Click **[OK].**

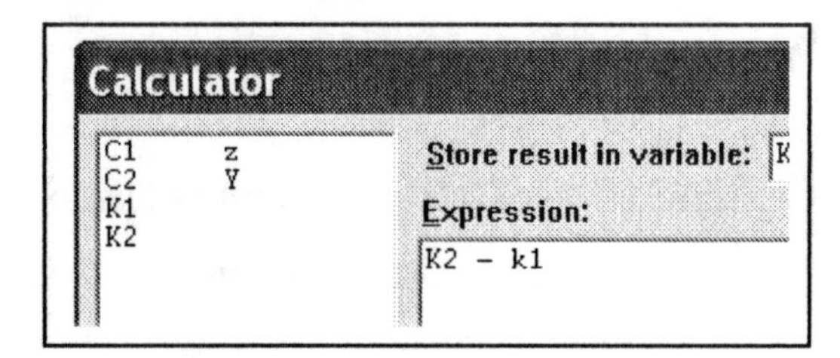

Step 5. Select **Data>Display Data.**

   a. Drag the mouse over the three constants.
   b. Click **[Select],** then **[OK].**

The area is displayed in K3: .868178.

Data Display

K1	0.0853435
K2	0.953521
K3	0.868178

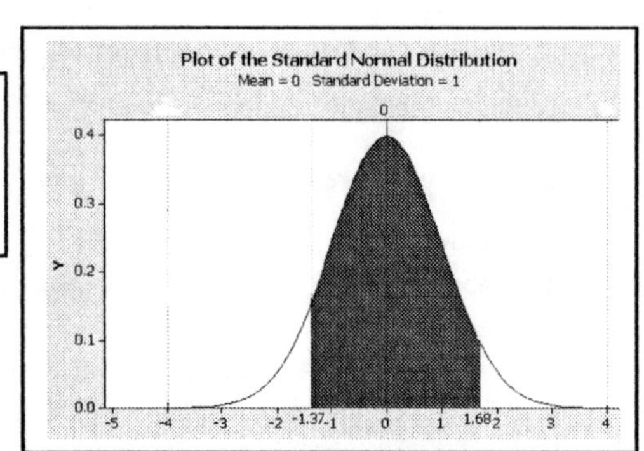

This value would be negative (incorrect) if the subtraction is done in the wrong order.

Example 3: **To find the area to the right of any Z value, find the area to the left of the Z value, then subtract from 1. Find the area to the right of +2.33.**

Step 1. To erase the constants K1-K3, select **Data>Erase variables** then drag the mouse over the constants K1-K3. Click **[Select]** then press <Enter>.

Step 2. Select **Calc>Probability Distributions>Normal.**

   a. Change the **Input constant**: to **2.33.**
   b. Type **K1** for **Optional storage**:.
   c. Click **[OK].**

Step 3. Select **Calc>Calculator.**

   a. Type **K2** for **Store result in**:.
   b. **Expression**: should be **1 – k1.**
   c. Click **[OK].**

Step 4. View the results in the Project Manager window. The area rounded to thousandths is .010.

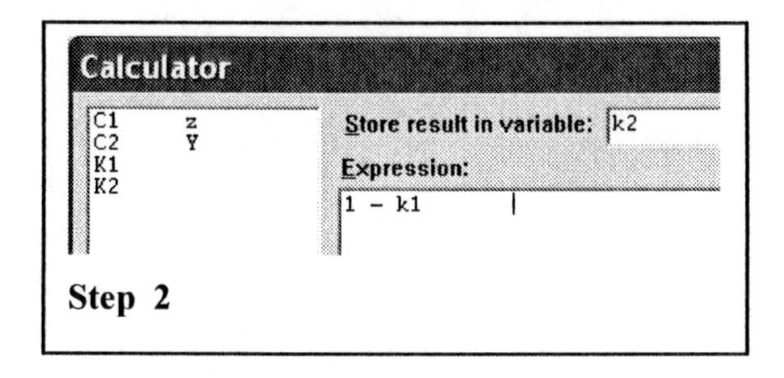

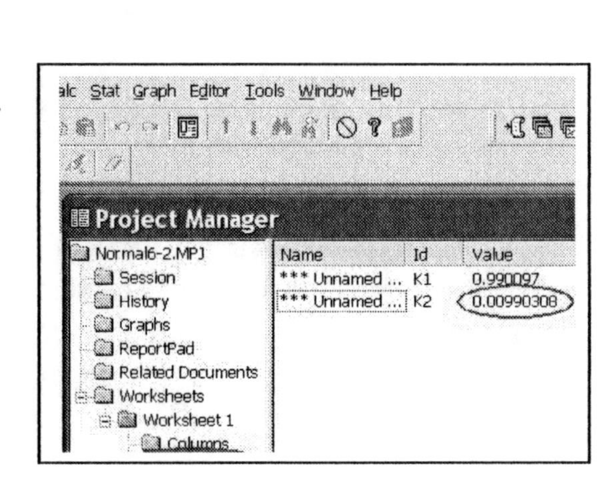

Example 4: This is textbook Example 6-13. Find the Z value such that the area under the normal distribution curve between 0 and the Z value is 0.2123. The area (.2123) is given.

Step 1. Select **File>New>Minitab Worksheet**. This is an alternative to erasing the constants. Each worksheet has its own constants.

Step 2. Determine the cumulative area to the left of the desired Z value.

For this example, mentally add: .5 + .2123 = .7123 because the area to the left of this Z value would be .7123.

Step 3. Select **Calc>Probability Distributions>Normal.**

a. Check the option for Inverse cumulative probability.

b. Change the Input constant: to **.7123.**

c. Clear the constant **K1** from Optional storage:.

d. Click **[OK].**

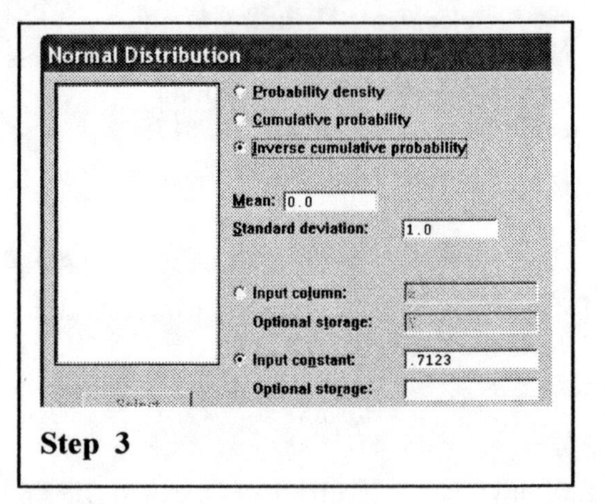

Step 3

Step 4. View the result in the Session window.

View the result in the Project Manager or use **Data>Display data** as in previous examples.

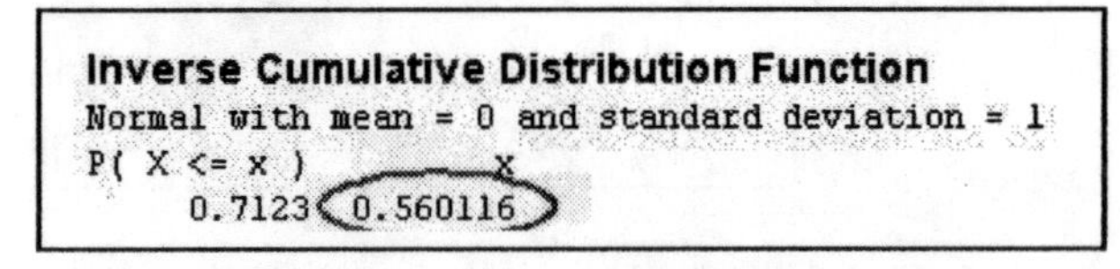

Inverse Cumulative Distribution Function

```
Normal with mean = 0 and standard deviation = 1
P( X <= x )       x
     0.7123  0.560116
```

Z = 0.56 (rounded to hundredths)

6-4 Applications of the Normal Distribution

So far the examples have been for the standard normal distribution and Z values. The same procedure works for other normal distributions. Remember, to use the tables of MINITAB you must use the area to the left of the Z value, the cumulative area. The next example will illustrate.

Textbook example 6-17: To qualify for a police academy, candidates must score in the top 10 percent on a general abilities test. The test has a mean of 200 and a standard deviation of 20. Find the lowest possible score to qualify. The test scores are normally distributed.

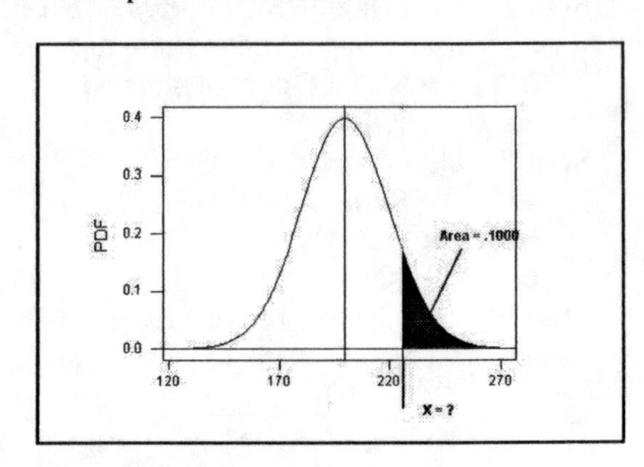

Step 1. Mentally determine that the area to the left of this value of X = 1 – 0.10 = 0.90.

Step 2. Select **Calc>Probability distributions>Normal.**

a. Click the button for Inverse cumulative probability.

b. Change the Mean: to **200.**

c. Change the Standard deviation to **20.**

d. Type **.90** in the text box for Input constant:.

Remember, this must be the cumulative area.

Step 3. Click **[OK].**
The result is viewed in the Session window. X = 225.631.

Step 4. Select **File>New>Minitab Project.**
Save your work as **Normal.MPJ.**

Determining Normality

There are several ways statisticians test a data set for normality.

Four are shown here.

Construct a Histogram

Step 1. Enter the data for Example 6-19 in the first column of a new worksheet.

a. Name the column **Inventory.**

b. Use **Stat>Basic Statistics>Graphical Summary** presented in Section 3-4 to create the histogram. Inspect the histogram for shape.

Step 2. Is it symmetrical?

Is there a single peak? Are there outliers? If there are two or more, reject normality.

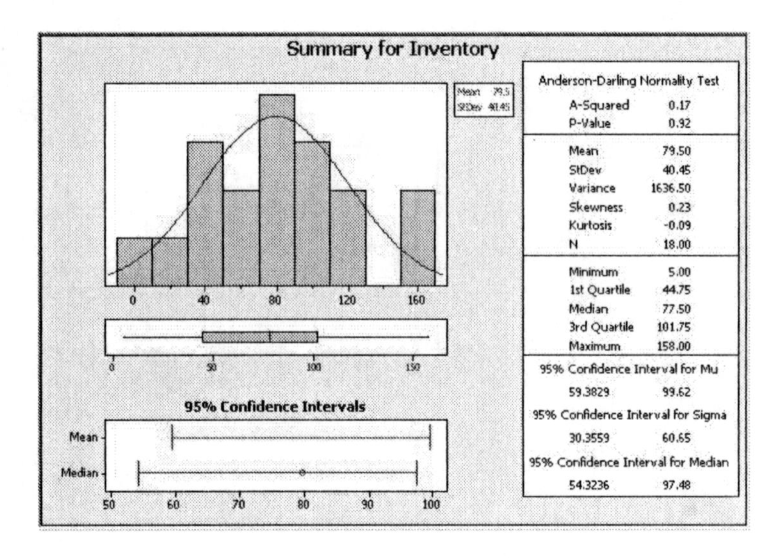

Check for Outliers

Inspect the boxplot for outliers. There are no outliers in this graph. Furthermore, the box is in the middle of the range and the median is in the middle of the box.

Calculate the Pearson's Index of Skewness

The measure of skewness in the graphical summary is not the same as Pearson's index. Use the MINITAB calculator and the formula $PI = \frac{3(\overline{X} - median)}{s}$.

Step 3. Select **Calc>Calculator,** then type in **PI** in the text box for Store result in:.

a. In Expression: type **3*(79.5-77.5)/40.5,** then click **[OK].** These statistics are from the frequency distribution, not the data.

b. Click **[OK].** The result, 0.148318, will be stored in the first row of C2 named PI. Because it is smaller than +1 the distribution is not skewed.

Note: If the data is used directly, enter in Expression: **3*(MEAN(C1)-MEDI(C1))/(STDEV(C1)).**

The result from the data is -0.0593982.

Construct a Normal Probability Plot

Step 4. Select **Graph>Probability Plot,** then Simple Single and click **[OK].**

a. Double click C1 Inventory to select the data to be graphed.

b. Click **[Distribution],** and make sure that Normal is selected. Click **[OK].**

c. Click **[Labels],** and enter the title for the graph: **Quantile Plot for Inventory**.

d. Type **your name** in the subtitle.

e. Click **[OK]** twice. Inspect the graph to see if the graph of the points is linear.

f. What do you look for in the plot?

1. An "S-curve" indicates a distribution too thick in the tails, a uniform distribution, for example.
2. Concave plots indicate a skewed distribution.
3. If one end has a point extremely high or low, there may be outliers.

This data set appears to be nearly normal by every one of the four criteria!

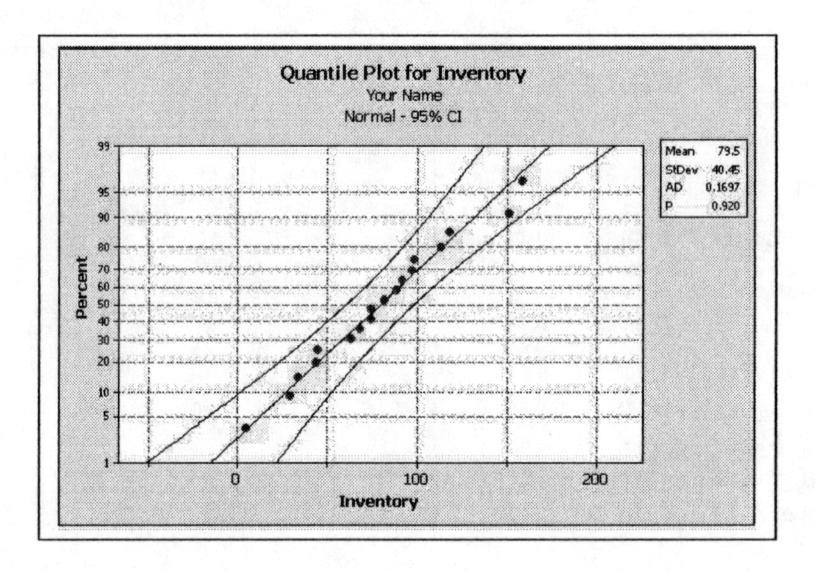

6-5 The Central Limit Theorem

This theorem enables us to find the probability that the mean of a sample is a specified interval.

Example 6-22: The average age of a vehicle registered in the United States is 8 years (96 months). Assume the standard deviation is 16 months. If a random sample of 36 vehicles is selected, find the probability that the *mean of a sample* is between 90 and 100 months.

Step 1. Plan the procedure! Find the area in the normal distribution between 90 and 100 if the mean is 96 and the standard error of the mean is $\frac{16}{\sqrt{36}}$.

Step 2. Calculate and store the standard error of the mean.

a. Select **Calc>Calculator.**

b. Type **K1** in the Store result in variable: text box.

c. For Expression:, enter **16 / SQRT(36),** then click **[OK].**

Note: You could replace the SQRT(36) with 6 because the result is an integer and no rounding error would occur.

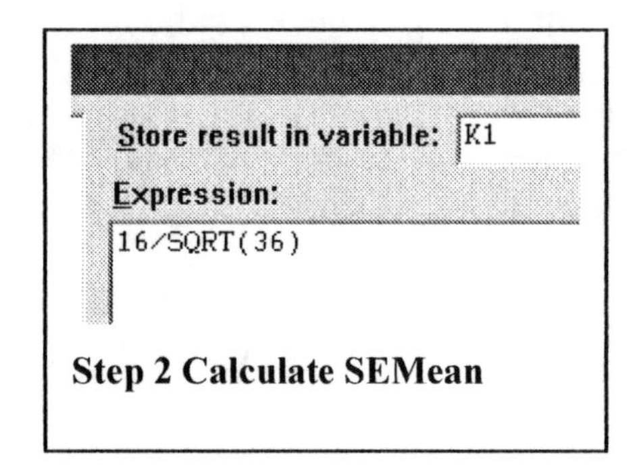

Step 2 Calculate SEMean

Step 3. Calculate the probability (find the area).

a. Select **Calc>Probability Distributions>Normal.**

1. Check option for Cumulative probability.
2. Change the Mean: to **96.**
3. Change the Standard deviation: to **K1.**
4. Enter **90** for the Input constant:.
5. Type **K2** for Optional storage:.

b. Click **[OK].** The completed dialog box is shown.

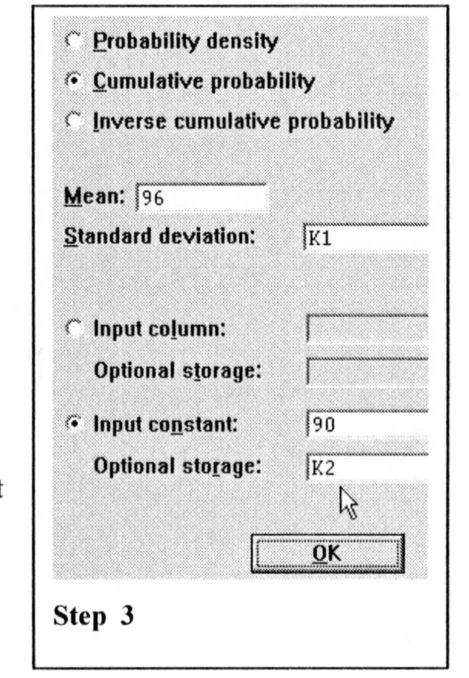

Step 3

Step 4. Repeat steps 4 and 5 changing the Input constant: to **100** and the Optional storage: to **K3**. This dialog box is not shown.

Step 5. Calculate the probability by subtracting K3 from K2.

a. Select **Calc>Calculator.**

b. Enter **K4** for the Variable.

c. Enter **K3 – K2** for the Expression:.

d. Click **[OK].**

Step 6. View the result in the Project Manager window.

The probability is 0.920968, which rounds to 0.921.

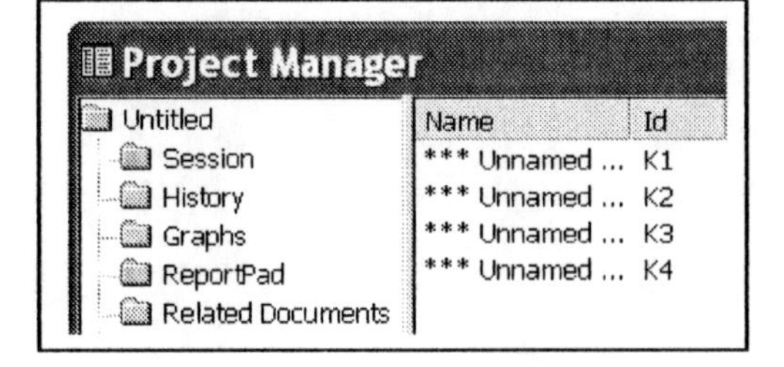

Step 7. Select **File>Exit,** then save the project as **Chapter6-5.MPJ.**

Chapter 6 Textbook Problems Suitable for MINITAB

Page 282	6.3	6 – 51
Page 296	6.4	1 – 30, 39 – 42
Page 309	6.5	8 – 25
Page 318	Review 1 – 11, 16, 17	

Chapter 6: Endnote

Chapter 7 Confidence Interval Estimates and Sample Size

7-1 Introduction

One aspect of inferential statistics is estimation, the process of estimating the value of a parameter from sample data. A population proportion is the preferred parameter for qualitative data. For quantitative data, the mean and standard deviation would be used to describe central tendency and variation. A parameter is a measurement using data from the population. Recall that a statistic is calculated from a sample. The symbols:

(population) Parameter		concept	(sample) Statistic
μ	mu	mean	$\overline{X}$
σ^2	sigma squared	variance	s^2
σ	sigma	standard deviation	s
p		proportion	$\hat{p}$

The statistic is the point estimate for the corresponding parameter. Thus, if we wish to know the mean of a population, the mean of a sample would be an estimate. A confidence interval is a range of values that probably contains the parameter being estimated. A confidence interval is an estimate.

7-2 Confidence Interval for the Mean, σ known

Use MINITAB to "Look Up" Critical Values of Z

For each critical value use the following instructions to find $Z_{\alpha/2}$,. No worksheet data is required. There may be a time when a table is needed and the paper variety is not available. It is the process we used in the previous chapter to find a Z value given the cumulative area.

Exercise 8-9 (a): Find the critical value of Z for a 99 percent confidence level.

Step 1. Select. **Calc>Probability Distribution>Normal**

Step 2. Click the button for Inverse cumulative probability.

Do not change the Mean: or Standard deviation:. The default of 0 and 1 are for the standard normal distribution.

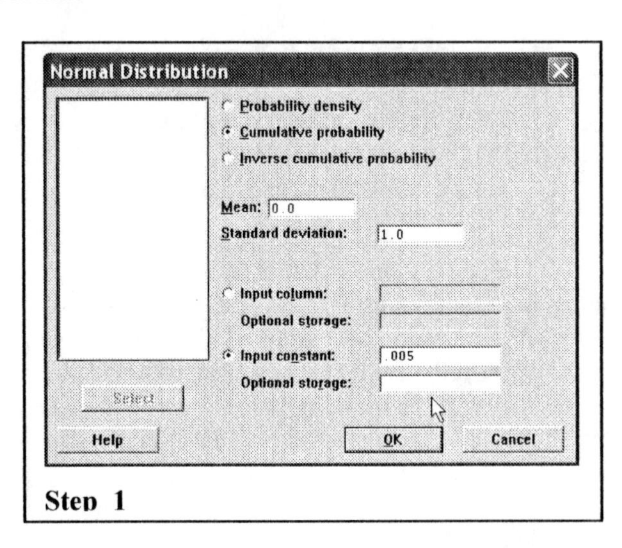

Step 1

Step 3. Mentally calculate the cumulative area in the left tail of the distribution.

(1-.99)÷2= .005

a. Click the button for Input constant:.
b. Click in the box and enter **.005.**

Step 4. Click **[OK].**
The Session window will display the left tail result. $Z_{\alpha/2} = 2.57583$

Inverse Cumulative Distribution Function

```
Normal with mean = 0 and standard deviation = 1
P(X <= x )          x
    0.005       -2.57583
```

Calculate the Confidence Interval for the Mean.

Example 7-3: The following data represent a sample of the assets (in millions of dollars) of thirty credit unions in southwestern Pennsylvania. Find the 90 percent confidence interval of the mean.

```
ASSETS
   12.23    2.89   13.19   73.25   11.59    8.74    7.92   40.22    5.01
    2.27   16.56    1.24    9.16    1.91    6.69    3.17    4.78    2.42
    1.47   12.77    4.39    2.17    1.42   14.64    1.06   18.13   16.85
   21.58   12.24    2.76
```

Step 1. Enter the data into a column of MINITAB. Name the column **Assets.**
Calculate the standard deviation of the sample. This value will be used as the point estimate for σ. Here is how.

Step 2. Select **Calc>Column Statistics.**

- **a.** Click the button for Standard deviation.
- **b.** Click in Input variable:.
- **c.** Double click Assets in the variable list.
- **d.** Type **s** in the box for Store results in:.
- **e.** Click **[OK].**

The standard deviation will be stored in the first empty constant, K1.

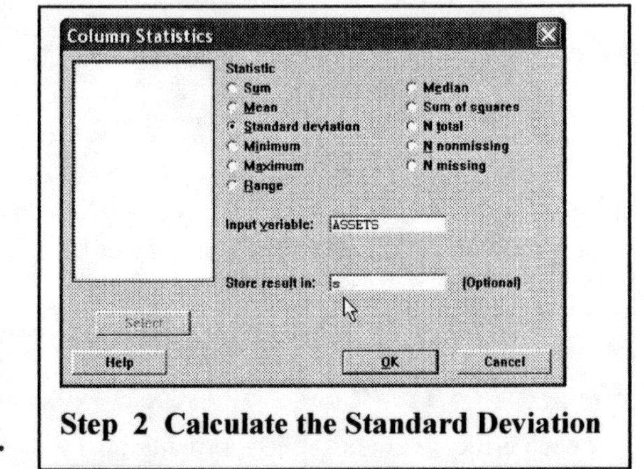

Step 2 Calculate the Standard Deviation

Step 3. Select **Stat>Basic Statistics>1-Sample Z.**

- **a.** Click in the box for Samples in columns then double click C1 Assets.
- **b.** Type in **s** for the standard deviation, an estimate sigma. σ is unknown.

Leave the box for Test mean: empty.
This item is used for the hypothesis test presented in the next chapter.

- **c.** Click on the **[Options]** button.
- **d.** Enter the confidence level, **90%** in the box.
- **e.** The Alternative should be not equal.
- **f.** Click **[OK]** twice.

The Session window will contain some of the summary statistics for Assets and the confidence interval. The mean asset value of all credit unions in southwestern Pennsylvania is estimated between 6.76 million dollars and 15.42 million dollars. The mean amount of assets for the credit unions in the sample is $11.1 million dollars.

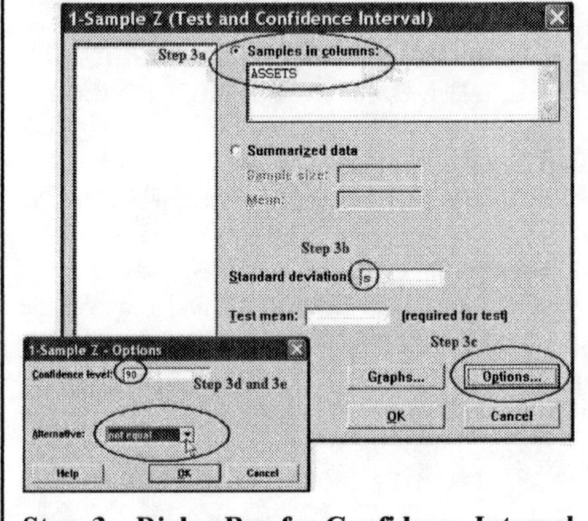

Step 3 Dialog Box for Confidence Interval

When the 1-sample Z is used, a critical value will be used from the standard normal distribution in the formula and the standard deviation supplied in the text box. The confidence interval calculated using the one-sample t command uses a critical value from the t-distribution and the standard deviation calculated from the sample, s.

One-Sample Z: ASSETS

```
The assumed standard deviation = 14.4054
Variable    N      Mean    StDev  SE Mean           90% CI
ASSETS     30   11.0907  14.4054   2.6301  (6.7646, 15.4167)
```

Calculate a Confidence Interval from the Summary Statistics

Many of the menus of MINITAB 14 now allow use of summary statistics in addition to data files.

Example 7-1: The president of a large university wishes to estimate the average age of students presently enrolled. From past studies, the standard deviation is known to be two years. A sample of fifty students is selected and the mean is found to be 23.2 years. Find the 95 percent confidence interval for the population mean. There is no data for this example. However, the statistics required for calculating a confidence interval are given:

$$\bar{x} = 23.2 \ years$$
$$\sigma = 2 \ \ years$$
$$n = 50$$

It doesn't matter what the active worksheet contains. No need to start a new project or worksheet. Continue from previous example.

Step 1. Select **Stat>Basic Statistics>1-Sample Z.**

The dialog box will contain settings from the previous example.

Press <F3> to clear the dialog box. This will reset all of the default values.

Step 2. Click the option for Summarized data.

a. Press <Tab> or click in the box for Sample size: and then type **50**.

b. Press <Tab>, then type **23.2**

c. Press <Tab> once more, then type **2** for the Standard deviation:.

d. Click **[Options]**. Check the confidence level (95%) and the Alternative should be "not equal to."

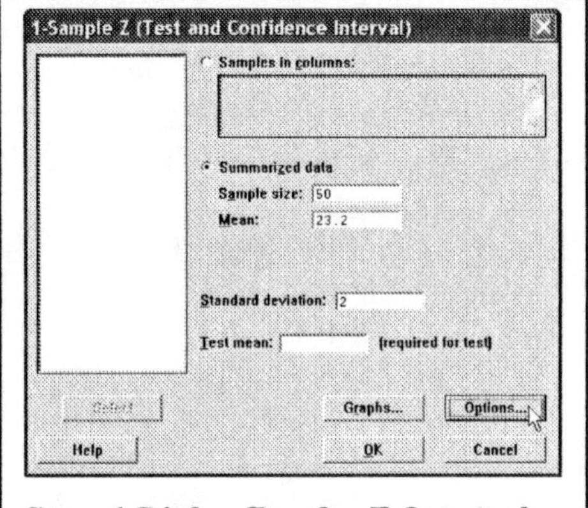

Step 1 Dialog Box for Z-Interval

Step 3. Click **[OK]** twice.

The Session window will contain the confidence interval. The average age of all students is probably between 22.6 and 23.8 years.

One-Sample Z

```
The assumed standard deviation = 2
 N     Mean   SE Mean           95% CI
50  23.2000    0.2828  (22.6456, 23.7544)
```

Calculate the Sample SizeRequired to Estimate a Mean

Use the calculator and the formula for sample size to determine the sample size.

The formula is n $\geq \left(\dfrac{z * \sigma}{E} \right)^2$

Example 7-4: A college president asks a statistics teacher to estimate the average age of the students at their college. How large a sample is necessary? The statistics teacher would like to be 99 percent confident that the estimate would be accurate to within one year. From a previous study, the standard deviation of the ages is known to be three years. No data is required. Keep going!

Step 1. Select **Calc>Calculator.**

a. In the box for **Store result in variable:** type **K2.** This will store the sample size as a constant. If a variable name such as "n" is used here, the result will be stored in the first row of a *column* named **n**.

Note: **K1** already contains the standard deviation from the first example. Using **K1** would erase the previous contents.

b. Look up the critical value of z for a 99 percent confidence interval. From section 7-2 $Z_{\alpha/2} = 2.57583$.

c. In the **Expression:** box enter the formula.

$$(2.57583*3/1)**2$$

View the results.

Step 1 Calculate Sample Size

Step 2. Click **Window>Project Manager** or click the Project Manager icon.

a. Click the **Constants** folder in the **Worksheets** section. The constant will be listed as *****Unnamed*****. The value for sample size, 59.7141 would be rounded up to 60.

b. *Optional:* To name the constant "n", right click *****Unnamed*****, then select **Rename.**

c. Type in the desired name for the constant, **n**.

Step 3. Using **File>Save Project As...,** save your work so far as **Chapter07.MPJ**.

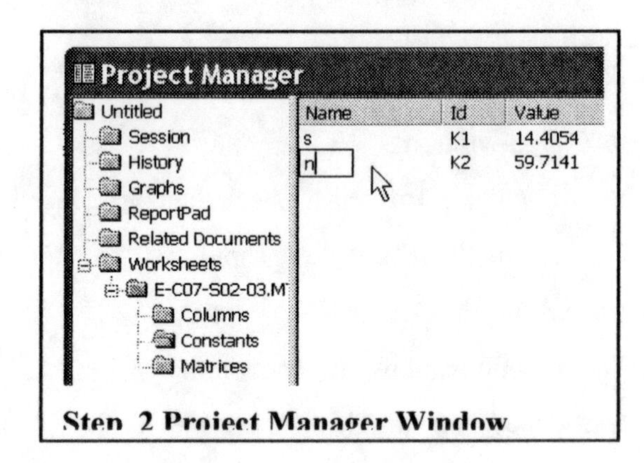

Step 2 Project Manager Window

7-3 Calculate a Confidence Interval for the Mean, Sigma Unknown, and n < 30

The confidence interval for the mean would be calculated with the formula $\overline{X} \pm t_{\alpha/2} * \frac{s}{\sqrt{n}}$. The critical value is from the student-t distribution. This is the formula MINITAB uses with the command, 1-Sample t. The standard deviation calculated from the sample data.

Example 7-7: The data represents a sample of the number of home fires started by candles for the past several years. Find the 99 percent confidence interval for the mean number of home fires started by candles each year.

Fires						
5460	5900	6090	6310	7160	8440	9930

Step 1. Click **File>News>New Worksheet.**

- **a.** Enter the data into a column of MINITAB.
- **b.** Name the column **Fires.**

Step 2. Select **Stat>Basic Statistics>1-Sample t.**

- **a.** Click the option for Samples in columns:.
- **b.** Press <Tab> then double click C1 Fires.
- **c.** Click **[Options]**, then type **99** into the text box for Confidence level.
- **d.** Click **[OK]** to close the Options dialog box.
- **e.** Click **[Graphs].**
  1. Check the box for Boxplot.
  2. Click **[OK]** twice.

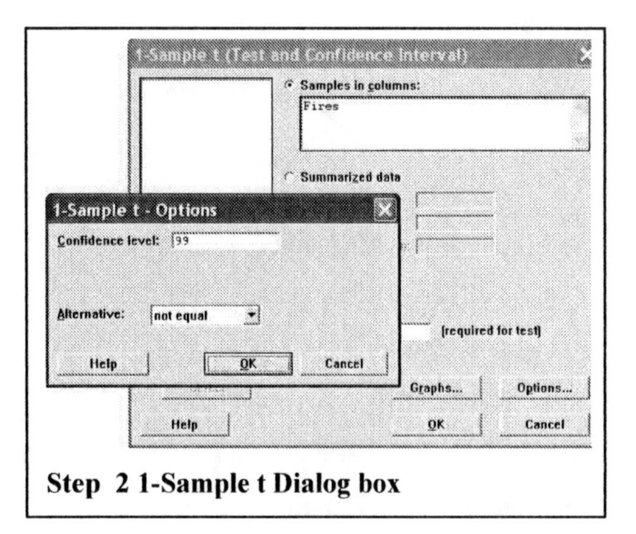

Step 2 1-Sample t Dialog box

Some of the summary statistics will be presented in the Session window and a boxplot will include a graphic representation of the confidence interval.

One-Sample T: Fires

```
Variable  N     Mean    StDev  SE Mean         99% CI
Fires     7  7041.43  1610.27   608.63  (4784.99, 9297.87)
```

The mean number of fires per year would be between 4785 and 9298. This is a wide range. The right whisker is longer than the left and the median is at the left end of the box. This suggests the distribution may be skewed right. Fires may not be a normal distribution, a necessary assumption for a t-interval.

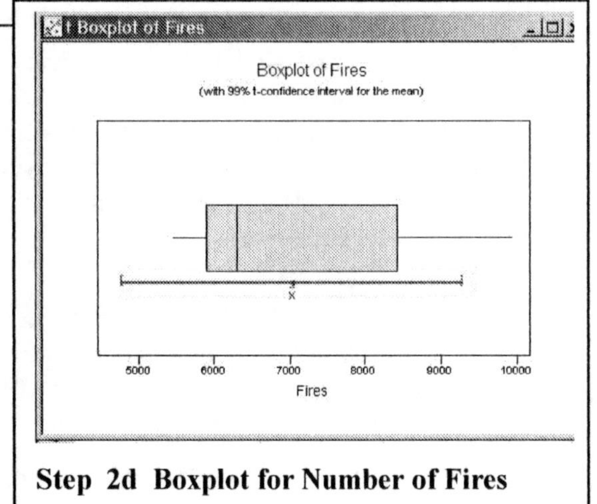

Step 2d Boxplot for Number of Fires

MINITAB will use the appropriate t-value when it calculates the confidence interval. Though it isn't necessary to find this value, it is possible to use MINITAB when a paper table is unavailable.

Use MINITAB to Find Critical Values of t

Exercise 4 (a) of section 7-3: There is no data. n = 18 and 99 percent is the confidence level.

Step 1. Select
Calc>Probability Distributions>t.

a. Click Inverse cumulative probability.

b. Click in the box for Degrees of freedom: and enter **17.** df = n - 1.

c. Mentally calculate the cumulative area in the left tail, (1-.99)/2=.005, then click Input constant: and type **.005.**

f. Click **[OK].** $t_{\alpha/2} = 2.89823$

Note: **In the Session window it is negative because we calculated the left-tailed value.**

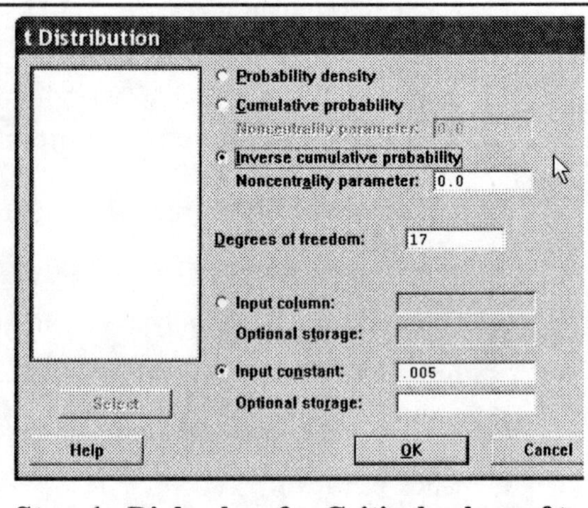

Step 1 Dialog box for Critical values of t

Next exercise.
Exercise 4 (b) n = 23, 95 percent level

d. Click the Edit last dialog icon or select **Edit>Edit Last Dialog.**

e. Change the Degrees of freedom: to **22** and the Input constant: to **.025.**

f. Click **[OK].** $t_{\alpha/2} = 2.07387.$

Step 2. Click the Disk icon to save the project as **Chapter07.MPJ** .

7-4 Confidence Intervals and Sample Size for Proportions

Proportions are desirable for qualitative data not means and standard deviations. What percent (proportion or fraction) of the population has some characteristic? Three fourths (¾) of doctors recommend aspirin. What proportion of adults is left-handed? What proportion of households has more than two cars? What proportion of the buyers purchase our product (market share)? What percent of television households are tuned into the Super Bowl (television rating)? What proportion of households has central air conditioning?

Calculate the Confidence Interval for p, the Population Proportion

Example 7-9: A sample of 500 nursing applications included sixty from men. Find the 90 percent confidence interval of the true proportion of men who applied to the nursing program. No data is required.

Step 1. Select **Stat>Basic Statistics>1-Proportion.**

a. Click the button for Summarized data.

b. Number of trials: n = **500**

c. Number of events: X = **60**

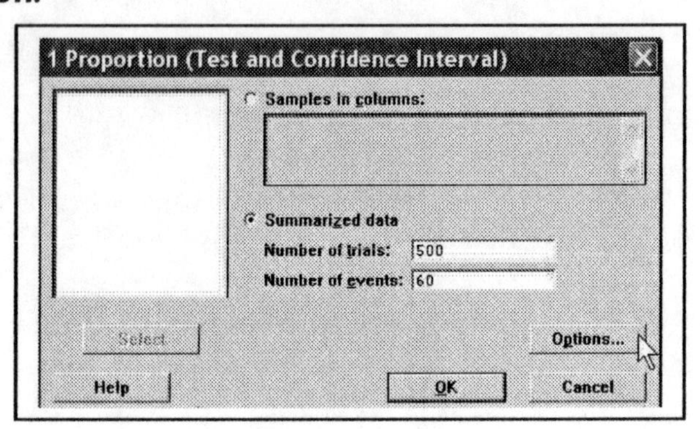

If the sample proportion were given, you would need to calculate X by multiplying n by the proportion.

Step 2. Click **[Options].** Type in the confidence level of **90.**

Note: It doesn't matter what Test proportion contains. This option will be used in the hypothesis test in the next chapter. Ignore it for now.

- **a.** The Alternative should be not equal.
- **b.** VERY IMPORTANT! Check the box for Use test and interval based on normal distribution.

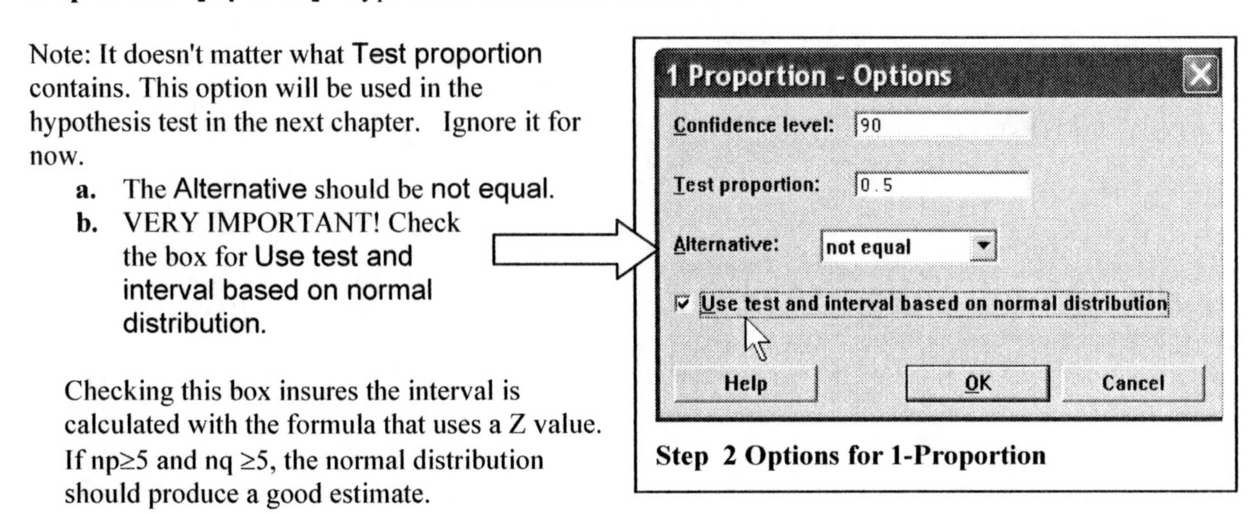

Step 2 Options for 1-Proportion

Checking this box insures the interval is calculated with the formula that uses a Z value. If np≥5 and nq ≥5, the normal distribution should produce a good estimate.

- **c.** Click **[OK]** twice.

In the Session window the sample proportion and confidence interval will be displayed. The percentage of male applicants for the nursing program is between 9.6 percent and 14.4 percent.

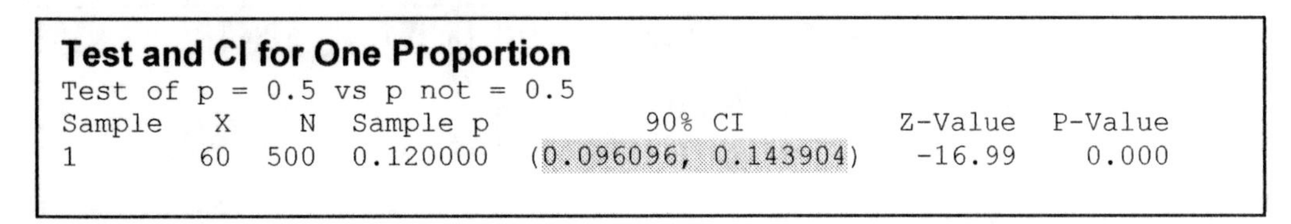

Test and CI for One Proportion

```
Test of p = 0.5 vs p not = 0.5
Sample   X    N  Sample p         90% CI          Z-Value  P-Value
1       60  500  0.120000  (0.096096, 0.143904)    -16.99    0.000
```

Note: Ignore the Z-value and P-value for now. They will be used in the next chapter.

Calculate the Sample Size Required to Estimate a Proportion

Example 7-11: A researcher wishes to estimate, with 95 percent confidence, the proportion of people who own a home computer. A previous study shows that 40 percent of those interviewed had a computer at home. The researcher wishes to be accurate within 2 percent of the true proportion. Find the minimum sample size required.

Known sample statistics: $\hat{p} = .4$ $\hat{q} = .6$ The calculator will be used with the formula. $\hat{p} * \hat{q} * \left(\frac{Z_{\alpha/2}}{E} \right)^2$

If there is no estimate for p, use a p-hat of .5.

Step 1. Look up the critical value of Z. It is 1.96.

Step 2. Select **Calc>Calculator.**

- **a.** Store the result in K1.
- **b.** Type the formula in the Expression: box.
- **c.** Click **[OK].**

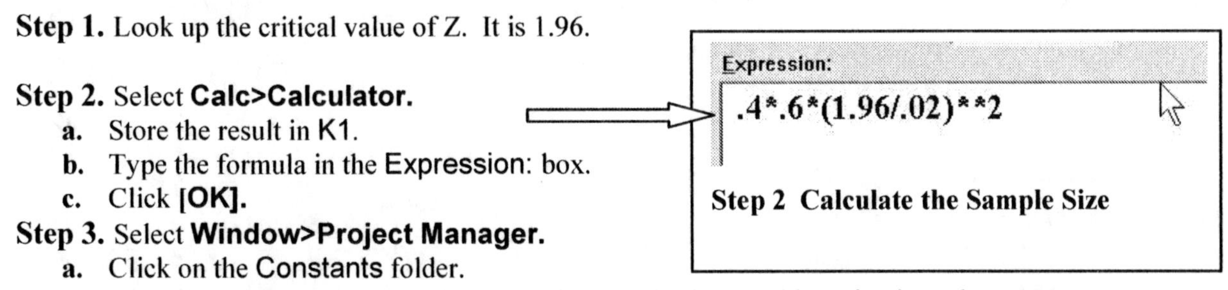

Step 2 Calculate the Sample Size

Step 3. Select **Window>Project Manager.**

- **a.** Click on the Constants folder.
- **b.** View the results. K1 = n = 2304.96. The sample size would need to be at least 2305.

Step 4. Type your name at the top of the session window then select **File>Print Session Window**.

Step 5. Select **File>New>Minitab Project**. Save your work as Chapter07.MPJ then continue.

7-5 Confidence Intervals for Variance and Standard Deviations

The chi-square distribution is used to estimate the variance of a population. Like the normal and student-t distributions, the total area is 1. Unlike the normal and student-t distribution, the chi-square variable cannot be less than 0, and it is skewed, not symmetrical. Chi-square values can be very large. There is no menu item to calculate this confidence interval. Use MINITAB and the formulas to calculate them. The plan is to find the chi-square values, then use the calculator with the formula to find the interval.

Use MINITAB to "Look Up" Critical Values of the Chi-square Distribution

The distribution is not symmetric, so the left and the right side critical values will be different values. Chi-square values cannot be negative.

Example: Look up the critical values of chi-square for a 90 percent confidence interval when n = 25.

df = n-1 = 17 $\quad\alpha/2 = \left(\frac{1-.90}{2}\right) = .05$ and $1 - \alpha/2 = .95$

The last two values are the cumulative area in the left tail and the cumulative area up to the right tail that will be needed for the chi-square distribution. No data is required. The result will be stored in a constant.

Get and store the right-tail critical value of chi-square.

Step 1. Select **Calc>Probability Distributions>Chi-Square.**

- **a.** Click Inverse cumulative probability.
- **b.** Click in the box for Degrees of freedom:.
- **c.** Type in **24.**
- **d.** Click the button for Input constant:.

The value entered is the cumulative probability. For the right-tail value, type **.95.**

- **e.** Click in the box for Optional storage:, then type **RightCHI.**
- **f.** Click **[OK].**

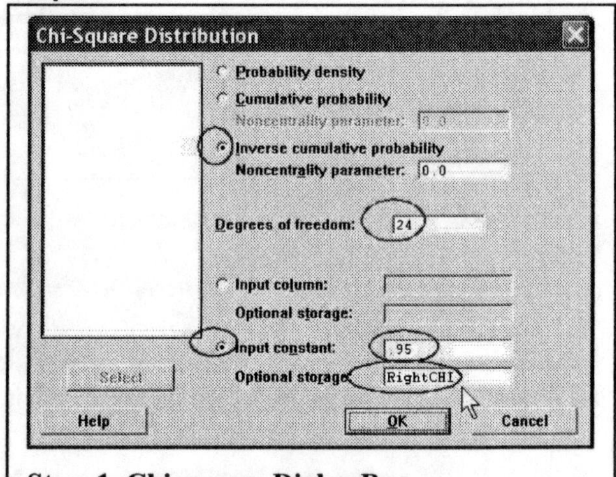

Step 1 Chi-square Dialog Box

Get and store the left-tail critical value of chi-square.

Step 2. Select **Edit>Edit Last Dialog box** and change the Input constant: to **.05** and the Optional storage: name to **LeftCHI.**

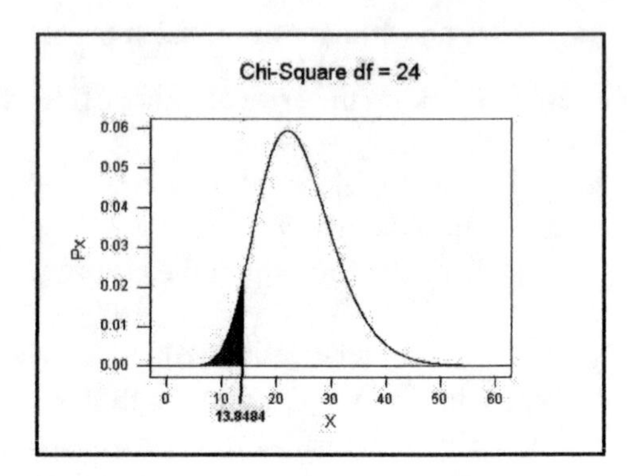

Display the constants in the Session window.

Step 3. Select **Data>Display Data.**

Step 4. Double click the two constants and then **[OK].**

The two constants will be displayed in the Session window.

Data Display

```
RightCHI    36.4150
LeftCHI     13.8484
```

Calculate a Confidence Interval for Variance using Data

Section 5 Exercise 10: A random sample of stock prices per share is shown. Find the 90 percent confidence interval for the variance and standard deviation for the prices. Assume the variable is normally distributed.

StockPrice

26.69	13.88	28.37	12.00	75.37	7.50	47.50	43.00	3.81	53.81
13.62	45.12	6.94	28.25	28.00	60.50	40.25	10.87	46.12	14.75

The plan is to use MINITAB to:

1. Calculate the standard deviation and size of the sample.
2. Look up and store the critical values for chi-square.
3. Calculate the lower and the upper confidence limits for variance using the formulas:

$$\text{Lvar} = \frac{(n-1)s^2}{\text{RightCHI}} \text{ for the lower and } \quad \text{Rvar} = \frac{(n-1)s^2}{\text{LeftCHI}} \text{ for the upper limit.}$$

4. The estimates for the standard deviation are the square roots of the variance.

Step 1. Enter this data into a column of MINITAB. Name the column **StockPrice.** Calculate the sample size and standard deviation.

Step 2. Select **Calc>Column Statistics.**

- **a.** C1 StockPrice is the Input variable:.
- **b.** Store the sample size in a constant, **n**.
- **c.** Click the button for N nonmissing.
- **d.** Click **[OK].**
- **e.** Click the Edit last dialog icon and repeat the column statistics with these changes.
- **f.** Click the button for Standard deviation.
- **g.** Store result in: **s.**
- **h.** Click **[OK].**

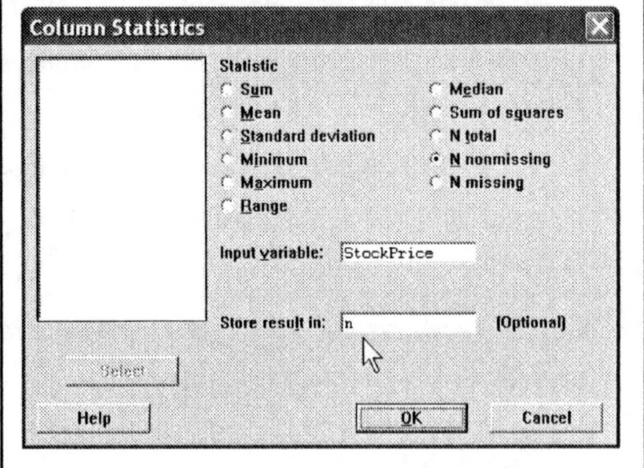

Step 2a-e Calculate n and s for the Sample.

K1 = n = 20
K2 = s = 20.2844

Use the instructions from the previous example to look up the critical values of chi-square.

df = 19 and $\alpha/2 = .05$ and $1 - \alpha/2 = .95$

Step 3. Select **Calc>Probability Distributions>Chi-Square.**

- **a.** Click Inverse cumulative probability.
- **b.** Click in the box for Degrees of freedom:, then type **19.**
- **c.** Click Input constant: and enter **.05.**
- **d.** Click in Optional storage: and type **LeftCHI.**
- **e.** Click **[OK].**
- **f.** Click the Edit last dialog icon and repeat using **.95** for the constant and **RightCHI** for storage.

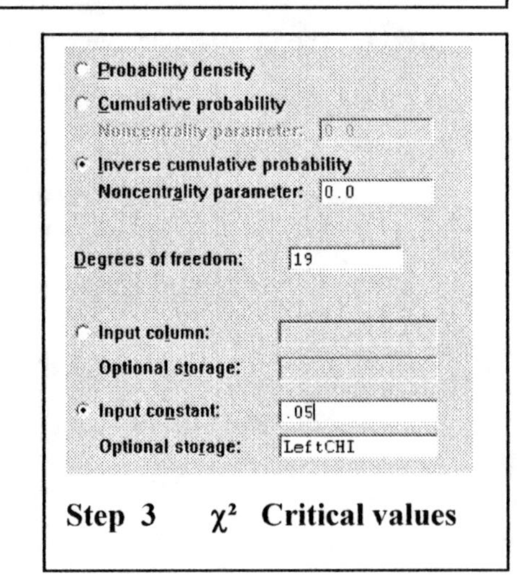

Step 3 χ^2 Critical values

g. In the Program Manager window, view **LeftCHI** and **RightCHI** in the constants folder.

K3 = LeftCHI = 10.117
K4 = RightCHI = 30.1435

Use the calculator to find the confidence interval. First the left and then the right variance limits. Then find the square root to determine the confidence interval for the standard deviation. It may seem odd, but the smaller variance is calculated from the larger χ^2. Variance is inversely proportional to chi-square.

Step 4. Select **Calc>Calculator.**

a. Type **K5** to Store result in variable:.

b. Click in the Expression: box and enter the formula.

(n-1)*'s'**2 / 'RightCHI'

c. Click **[OK].**

d. Click Edit last dialog icon and repeat step 7, changing the variable name to **K6** and in the formula replace RightCHI with **LeftCHI.**

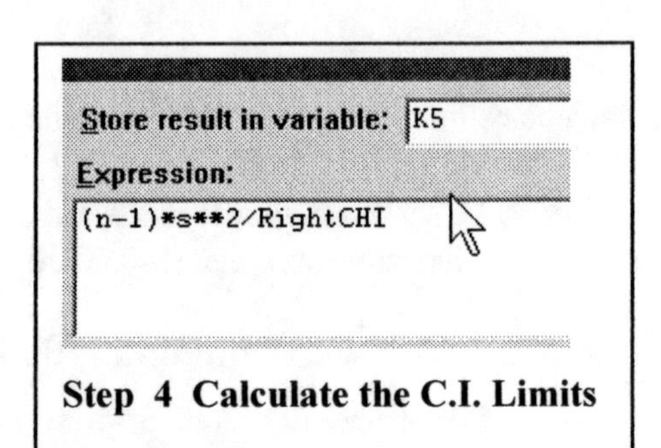

Step 4 Calculate the C.I. Limits

e. In the Project Manager window change the names to **Lvar** and **Rvar.**

K5 = Lvar = 259.348
K6 = Rvar = 772.723

Calculate the square root of the variance to get the standard deviations

Step 5. Select **Calc>Calculator.**

a. Type **K7** for Store result in variable: and **SQRT(K5)** for the Expression:.

b. Edit the last dialog box changing K7 to **K8** and the Expression: to **SQRT(K6).**

Step 6. The Program Manager window could be used to view the Constants folder.

Step 7. To display the new constants in the Session window , select **Data>Display data.**

Step 8. Drag the mouse over all of the constants and click **[Select]** then **[OK].**

In the Session window, the confidence interval for the variance is 259.348 to 772.723. The 90 percent confidence interval estimate for the population standard deviation (sigma) is between 16.1 and 27.8.

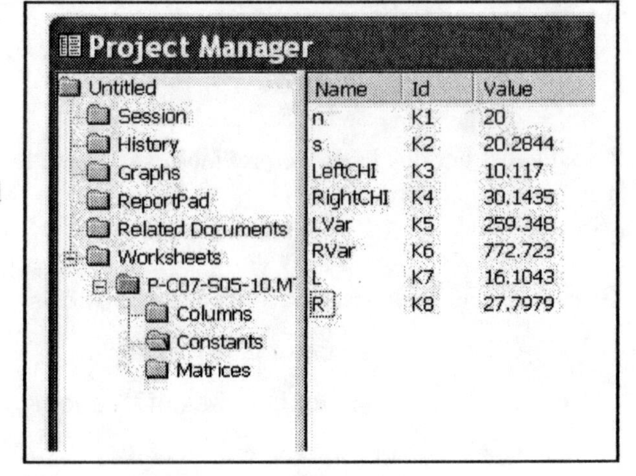

Step 9. Click the Disk icon then save your work as **ChiSq07.MPJ.**

Step 10. Select **File>New>Minitab Project.**

7-6 Data Analysis

Problem #3 is found on page 362: Construct a confidence interval for the proportion of individuals who did not complete high school. Use a sample of at least thirty subjects. Thirty-six was arbitrarily chosen by the author.

Select a sample of thirty-six from the Databank file.

Step 1. Select **File>Open Worksheet** to open the file, Databank.MTP.

Step 2. Select **Calc>Random Data>Sample from Columns.**

a. Enter **36** in the box for Number of rows.

b. Double click C3 ED-LEVEL, then press <Tab>.

c. Type **EDUC** for Store samples in:, then click **[OK].**

The 1-proportion test requires no more than two codes and the higher code numerically or alphabetically is considered the success. Code the data so that a 1 will represent No diploma and a zero will represent codes 1, 2 or 3 (HS Diploma or higher).

Step 3. Select **Data>Code>Numeric to Numeric.**

a. Double click the column with the new sample, EDUC. This is the From column. Press <Tab>.

b. Type **EduCode** for the Into columns:.

c. In the Original values:, first cell, type the number **1:3** then press <Tab>.

d. In New: values type in **0.**

e. Press <Tab>, then type **0** press <Tab> and type **1** for the New: value.

f. The window should look like the one shown here. Click **[OK].**

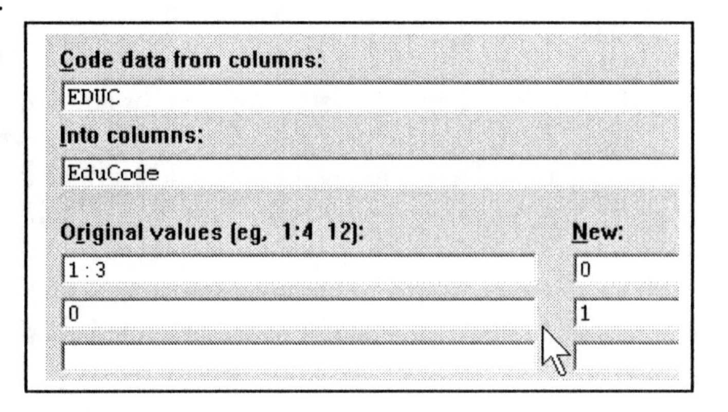

Calculate the confidence interval.

Step 4. Select **Stat>Basic Statistics>1 Proportion.**

a. Click the button for Samples in columns:.

b. Press <Tab> then double click the new column, EduCode.

c. Click **[Options].**

1. Type in the confidence level, **95%.**
2. Check the box for Use test and interval based on normal distribution. This check box results in the use of the Z-interval formula desired.
3. Click **[OK]** twice.

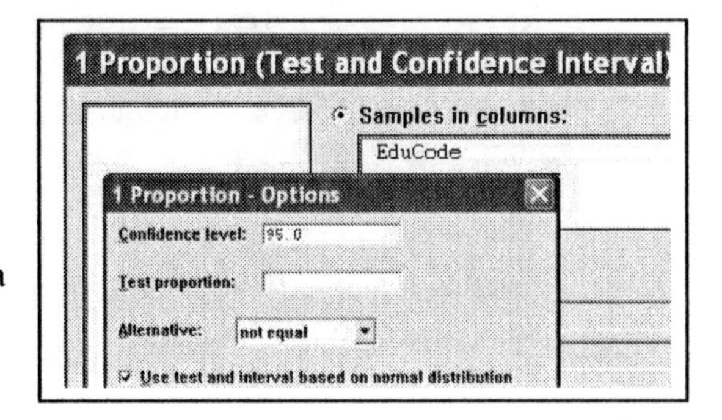

There may be a warning that reminds us that np and nq must be greater than five for the normal distribution to approximate a binomial distribution. In the session window we see the results. Random sample results will vary. Yours will be different.

Test and CI for One Proportion: EduCode

```
Test of p = 0.5 vs p not = 0.5    Event = 1

Variable  X   N  Sample p          95% CI            Z-Value  P-Value
EduCode   4  36  0.111111  (0.008452, 0.213771)      -4.67    0.000

NOTE * The normal approximation may be inaccurate for small samples.
```

The proportion who did not complete high school education is between 1 percent and 21 percent.

Step 5. Click the Disk icon, then save the project as **DataAnalysis07.MPJ**.
Step 6. Use **File>Exit** to close the program.

Note: The 1-Proportion command will calculate the confidence interval from raw data, however, there can only be two codes. Recode the data and then calculate the confidence interval. MINITAB considers the highest code, either numerically or alphabetically, a success. The help file for 1-Proportion explains:

Raw data

Enter each sample in a numeric, text, or date/time column in your worksheet. Columns must be all of the same type. Each column contains both the success and failure data for that sample. Successes and failures are determined by numeric or alphabetical order. Minitab defines the lowest value as the failure; the highest value as the success. For example:

- For the numeric column entries of "20" and "40," observations of 20 are considered failures; observations of 40 are considered successes.

- For the text column entries of "alpha" and "omega," observations of alpha are considered failures; observations of omega are considered successes. If the data entries are "red" and "yellow," observations of red are considered failures; observations of yellow are considered successes.

You can reverse the definition of success and failure in a text column by applying a value order (see Ordering Text Categories)

With raw data, you can generate a hypothesis test or confidence interval for more than one column at a time. When you enter more than one column, Minitab performs a separate analysis for each column.

Minitab automatically omits missing data from the calculations.

Chapter 7 Textbook Problems Suitable for MINITAB

Page	Section	Exercises
Page 336	7-2	9 – 20, 21 – 25
Page 343	7-3	4a-e, 5 – 20
Page 350	7-4	3 – 14, 15-17, 19, 20
Page 358	7-5	3 – 12
Page 361	Review	1 - 16

Chapter 7 Endnotes:

Chapter 8 One-Sample Hypothesis Tests

8-1 Introduction

Hypothesis testing is a decision-making process for evaluating claims about a population, in particular a parameter such as the population mean. The data, however, is a sample not a census.

8-2 The Hypothesis Testing Procedure

When using statistical software such as MINITAB, the P-value approach works well because MINITAB calculates the test statistic and the P-value, steps 2 and 3 in the following table. The researcher must decide on the appropriate hypotheses and the correct command to use to calculate the test statistic. MINITAB will not choose the command nor will it tell the user to reject or not reject H_o.

Step	P-value approach to hypothesis tests	Procedure
1	State the null and alternate hypotheses. Identify which is the claim. The alternate hypothesis determines the critical region.	H_0: $\mu = k$ H_1: $\mu \neq k$ or $\mu > k$ or $\mu < k$
2	Calculate the test statistic.	Choose the appropriate MINITAB command.
3	Find the P-value.	Find the P-value in the Session Window.
4	Make the decision. **If P-value < significance level, reject H_o.**	Reject H_o or Can't Reject H_o ?
5	Summarize the result.	Use statistics and evidence from sample explained in paragraph form.

Most of the hypothesis tests are found in the **Stat>Basic Statistics** menu.

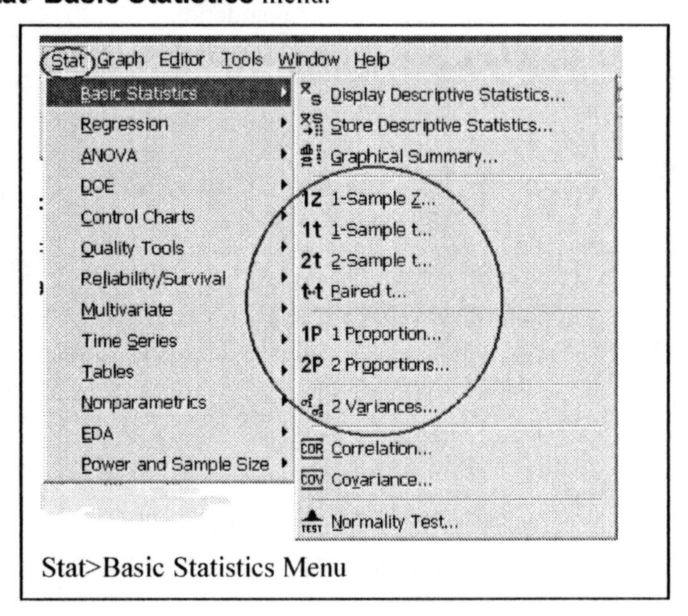

Stat>Basic Statistics Menu

Though all five steps of the test are necessary for the researcher to complete, MINITAB is used for step 2, calculating the test statistic. A few examples will illustrate.

8-3 Testing a Hypothesis About a Mean (Quantitative variable)

Z-test for Large Samples, σ is known

Section 3 Example 5: Test the claim that the average cost of rehabilitation for stroke victims is $24,672. Use a significance level of 0.01. The standard deviation of the population is $3,251.
The summary statistics are given not the data.

$\bar{x} = 25226$

$n = 35$

$\alpha = .01$

$\sigma = 3251$

Step 1. State the hypotheses:

- **a.** H_o: μ = $24672
- **b.** H_1: $\mu \neq$ $24672

There is no data. It doesn't matter what is in the worksheet. New to MINITAB 14 is an increased ability to use summarized data. Calculate the test statistic from the summarized data.

Step 2. To calculate the test statistic, select **Stat>Basic Statistics>1 Sample Z**

- **a.** Click the option button for Summarized data.
- **b.** Click in the text box for Sample size:, then type **35**.
- **c.** Press <Tab>, then type the sample mean, **25226**, no commas or dollar signs.
- **d.** Click in the text box for Standard deviation:, then type the population standard deviation, **3251**.

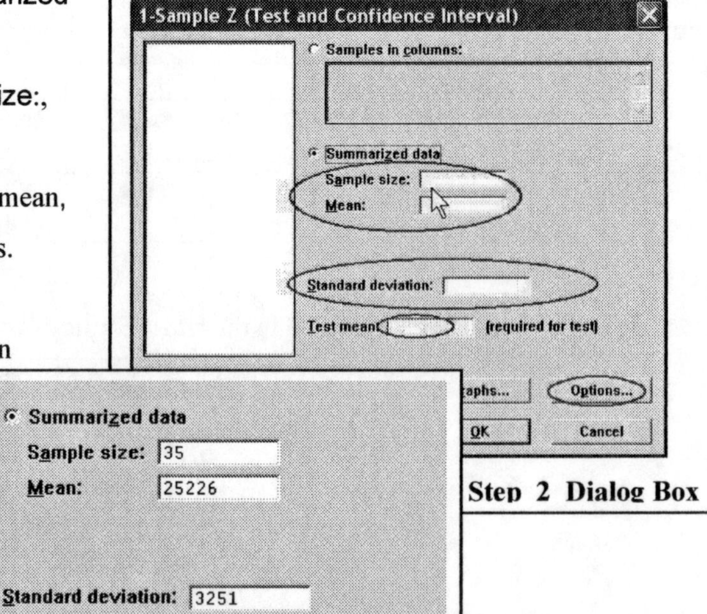

Step 2 Dialog Box

- **e.** Click in the text box for Test mean: and enter the hypothesized value of **24672**.

Step 2 Completed Dialog Box

- **f.** Click on **[Options].**
  1. Change the Confidence level: to **99,** the complement of alpha.
  2. The Alternative: should be not equal.
- **g.** Click **[OK]** twice.

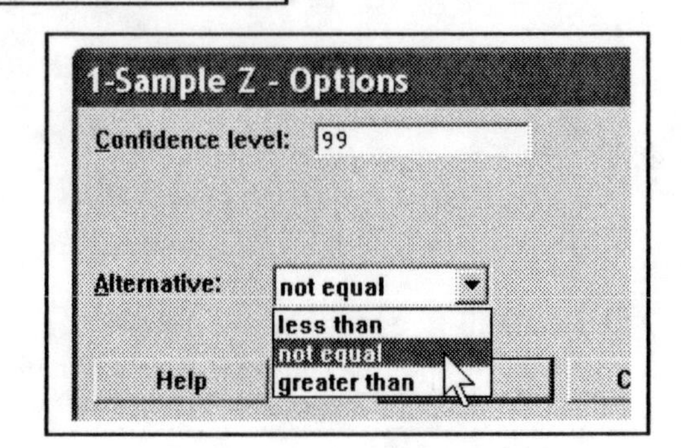

The results will be displayed in the Session window.

One-Sample Z

```
Test of mu = 24672 vs not = 24672
The assumed standard deviation = 3251
 N      Mean   SE Mean          99% CI              Z      P
35   25226.0     549.5   (23810.5, 26641.5)   1.01  0.313
```

Step 3. The test statistic is 1.01, just as the textbook example shows.

Step 4. The P-value for the test is 0.313. The P-value is greater than the significance level, .313 > .01, so the null hypothesis cannot be rejected.

Step 5. There is not enough evidence in the sample to reject the null hypothesis in favor of the alternative.

Z-test for Large Samples, σ is unknown

Section 3 Example 4: Test the claim that the mean shoe cost is less than \$80. $\alpha = 0.10$.
Instructions show how to calculate the test statistic and P-value. All other steps in the procedure are done by the researcher.

Step 1. State the hypotheses:

H_o: $\mu \geq 80$

H_1: $\mu < 80$

Calculate the test statistic.

a. Enter the data into a column of MINITAB. Do not try to type in the dollar signs. Name the column "**ShoeCost.**" *Note:* Type the values so the first row is entered into the first column of MINITAB. followed by the second row, etc. until all 36 are in C1 ShoeCost.

↓	C1
	ShoeCost
1	60
2	70
3	75
4	55
5	80
6	55
7	50
8	40
9	80
10	70
11	50

60	70	75	55	80	55
50	40	80	70	50	95
120	90	75	85	80	60
110	65	80	85	85	45
75	60	90	90	60	95
110	85	45	90	70	70

Note: If you use the file from the CD ROM, you will need to stack the data using **Data>Stack>Columns** as described in chapter 3 on page 66.

b. Calculate the standard deviation of the sample, s. Use this as point estimate for sigma. Select **Calc>Column Statistics.**

1. Check the button for Standard deviation:.
2. Select ShoeCost for the Input variable:.
3. Type s in the text box for Store the result in:.
4. Click **[OK].**

Step 2. Calculate the test statistic and P-value.

a. Select **Stat>Basic Statistics>1 Sample Z,** then press <F3> to reset the defaults.

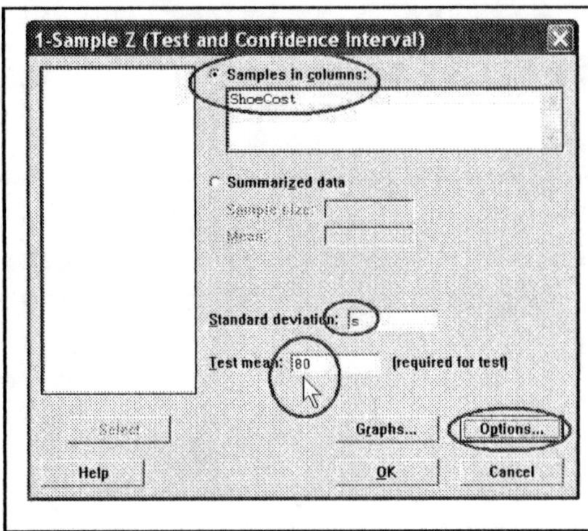

b. Press <Tab>.

c. Double click ShoeCost to select it for Samples in columns:.

d. Click in the text box for Standard deviation: then type s, the sample standard deviation previously stored.

e. Click in the text box for Test mean:, and enter the hypothesized value of **80**.

f. Click on **[Options].**

1. Change the Confidence level: to **90**.
2. Change the Alternative: to less than.

This setting is crucial for calculating the P-value.

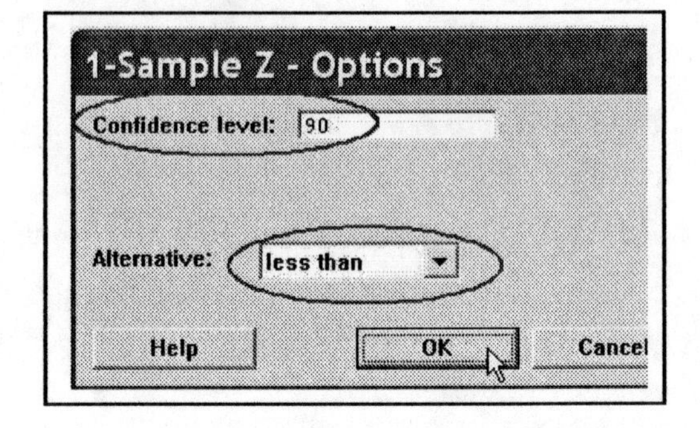

g. Click **[OK]** twice. The Session window shows the result. The test statistics is -1.57.

One-Sample Z: ShoeCost

```
Test of mu = 80 vs < 80
The assumed sigma = 19.161

                                                  90%
                                                 Upper
Variable    N      Mean     StDev   SE Mean      Bound      Z      P
ShoeCost   36   75.0000   19.1610    3.1935    79.0926  -1.57  0.059
```

Step 3. The P-value of .059 is less than $\alpha = 0.10$.

Step 4. Reject the null hypothesis.

Step 5. There is enough evidence in the sample to conclude the mean cost is less than \$80. The point estimate, the sample mean, would be \$75.

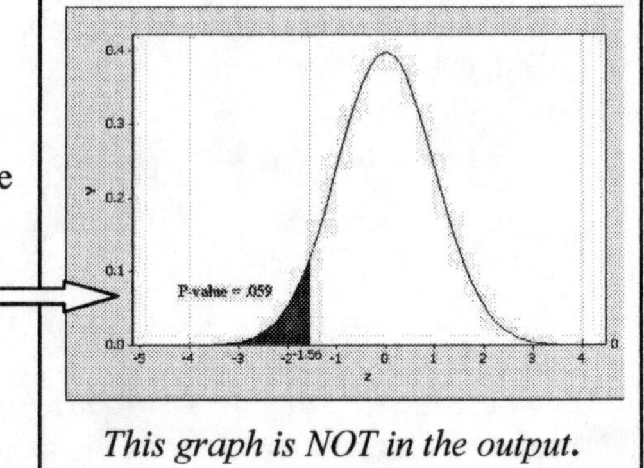

This graph is NOT in the output.

Note: In a left-tailed test, the P-value represents the area to the left of the test statistic. It can only be less than α if the value is to the left of the critical value, -1.28. It is not necessary to look-up the critical value when using the P-value approach.

Step 6. Click the Disk icon then save this project as **Chapter08.MPJ.** Continue.

8-4 One-Sample t-test; σ is Unknown and n < 30

Problem 17 Section 4 From past experience, a teacher believes that the average score on a real estate exam is 75. Use the P-value method to test the claim that the students' mean score is still 75. A sample of twenty students' exam scores follow. Use $\alpha = .01$.

Scores:	80	68	72	73	76	81	71	71	65	50
	63	71	70	70	76	75	69	70	72	74

The five steps of the hypothesis test are shown.

Step 1. State the hypotheses:

H_o: $\mu = 75$

H_1: $\mu \neq 75$

Step 2. Calculate the test statistic.

a. Select **File>New>Minitab Worksheet,** then enter the data in C1 of a new worksheet. Name the column **Scores.**

b. Select **Stat>Basic Statistics>1-Sample t.**

1. Click in the text box for Samples in columns:, then double click C1 Scores.

2. Type **75**, the hypothesized value in the box for Test mean:.

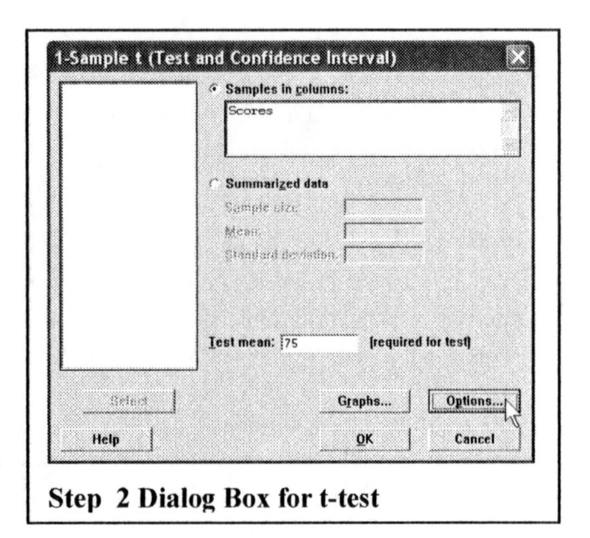

Step 2 Dialog Box for t-test

3. Click **[Options].** Change the Confidence level to **99**. The Alternative should be not equal.
4. Click **[OK].**

c. Click on **[Graphs],** then select Boxplot. If there is a checkmark for Histogram, click it to deselect it. Click **[OK]** twice.

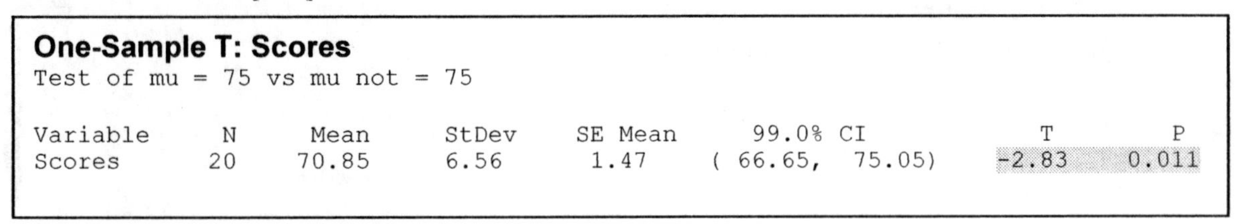

One-Sample T: Scores

Test of mu = 75 vs mu not = 75

Variable	N	Mean	StDev	SE Mean	99.0% CI	T	P
Scores	20	70.85	6.56	1.47	(66.65, 75.05)	-2.83	0.011

d. The test statistic is t = -2.83.

Step 3. The P-value is .011. It is slightly larger than the significance level of .01.

Step 4. Decision: Can't reject H_0.

Step 5. There isn't enough evidence in the sample to conclude that the mean number of pages is different from 75. The sample mean, 70.85 is not statistically significant.

The boxplot shows this data has an outlier of 50. One individual earned an extremely low score compared to the rest. This makes the distribution negatively skewed and suggests the data is not symmetric. Perhaps this score was typed in error? If not, the data may not be from a normal distribution and a non-parametric test should be used.

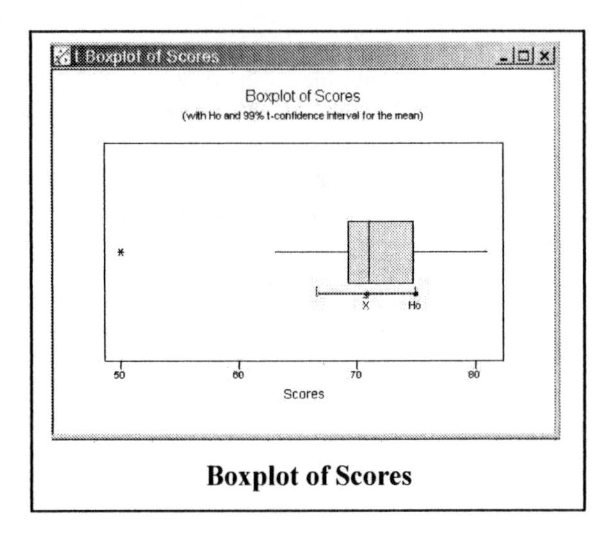

Boxplot of Scores

Step 6. Click the Disk icon then save this project as **Chapter08.MPJ.**

8-5 One-Sample Z test for a Proportion

For qualitative data, the population proportion would be the parameter of interest. If $np \geq 5$ and $nq \geq 5$ the normal distribution can be used to approximate the sampling distribution. MINITAB will calculate the test statistic and P-value.

Section 5 Example 17: There is no raw data igiven. The summarized statistics are given. An educator estimates that the dropout rate for seniors at high schools in Ohio is 15 percent. Last year, thirty-eight seniors in a random sample of 200 Ohio seniors withdrew. At $\alpha = .05$, is there enough evidence to reject the educator's claim? Use the P-value method.

Step 1. State the hypotheses:

H_0: $p = .15$ (claim)

H_1: $p \neq .15$

$np = .15*200 = 30$

$nq = .85*200 = 170$

Both are larger than 5. The normal distribution should be a good approximation.

Step 2. Calculate the test statistic.

a. Select **Stat>Basic Statistics>1 Proportion.**

1. Click the button for Summarized data.
2. The Number of trials: is **200.**
3. The Number of events: is **38.**

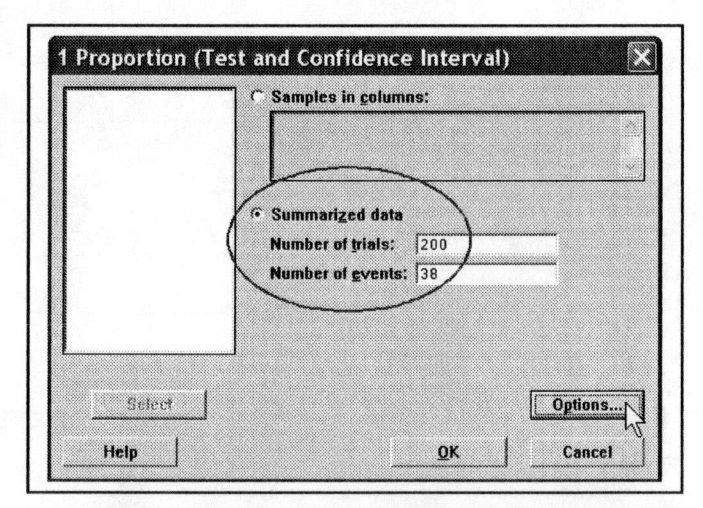

b. Click **[Options].**

1. Change the Test proportion: to the hypothesized value of **.15.**
2. Be sure the Alternative: is not equal. This should match the hypotheses in step 1.
3. Check the box to Use test and interval based on normal distribution. The test statistic will be calculated with the formulas given in the text.

$$Z = \frac{\hat{p} - p}{\sqrt{\frac{p \cdot q}{n}}}$$

4. Click **[OK]** twice.

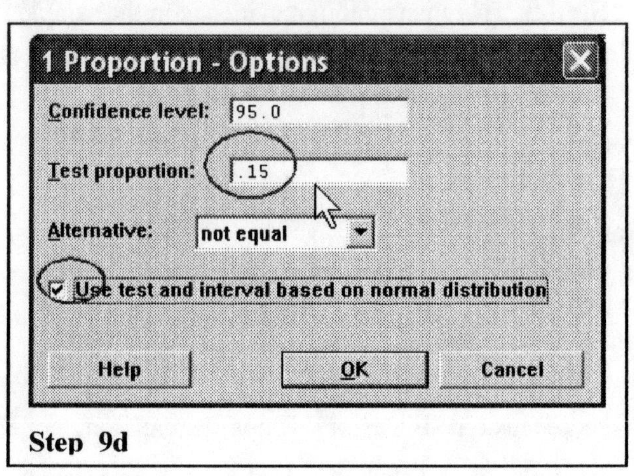

Step 9d

The Session window will contain the following output.

```
Test and CI for One Proportion
Test of p = 0.15 vs p not = 0.15

Sample       X     N  Sample p           95.0% CI          Z-Value  P-Value
1           38   200  0.190000  (0.135631, 0.244369)         1.58    0.113
```

c. The test statistic is +1.58.

Step 3. The P-value is .113.

Step 4. Because .113 > .05, do not reject H_0.

Step 5. There is not enough evidence in the sample to reject the claim. Nineteen percent of the sample dropped out. Even though the sample proportion is higher than 15 percent, it is not enough to conclude that the proportion for all is different than 15 percent.

Step 6. Click the Disk icon to save this project as **Chapter08.MPJ** .

If the proportion 19 percent were given instead of the number of successes, thirty-eight, you would need to calculate 19 percent of 200. The value of x is required in the dialog box, not the proportion.

One-sample Z test for a Proportion; Data

A sample of students is surveyed to determine whether they smoke and other items. The data is in a file named, PulseA.MTW. C4-T contains text that is either Smoke or Nonsmoker. MINITAB will use Smoke, the value that is "highest" alphabetically as a success. It is thought that more than 20 percent of college students smoke. Use $\alpha = .05$ and the P-value method to test the hypothesis that the proportion of college students who smoke is more than 20 percent.

Step 1. Open the data file.

a. Select **File>Open Worksheet**.

1. Navigate to the folder that contains this file, C:\Program Files\MINITAB 14\Studnt14.
2. The file type should be Minitab Worksheet [*.mtw, *.mpj] .
3. Double click PulseA.MTW.

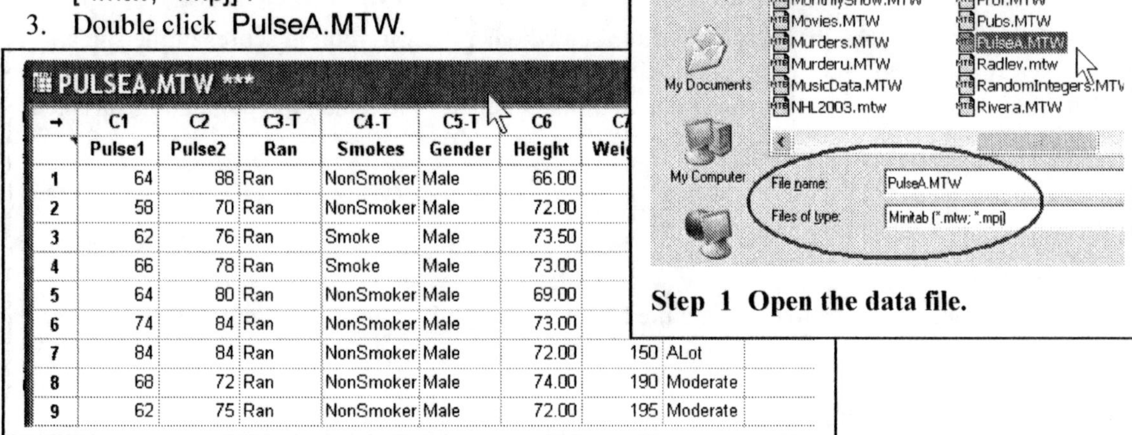

PULSEA.MTW ***

→	C1	C2	C3-T	C4-T	C5-T	C6	C7	
	Pulse1	Pulse2	Ran	Smokes	Gender	Height	Wei	
1	64	88	Ran	NonSmoker	Male	66.00		
2	58	70	Ran	NonSmoker	Male	72.00		
3	62	76	Ran	Smoke	Male	73.50		
4	66	78	Ran	Smoke	Male	73.00		
5	64	80	Ran	NonSmoker	Male	69.00		
6	74	84	Ran	NonSmoker	Male	73.00		
7	84	84	Ran	NonSmoker	Male	72.00	150	ALot
8	68	72	Ran	NonSmoker	Male	74.00	190	Moderate
9	62	75	Ran	NonSmoker	Male	72.00	195	Moderate

Step 1 Open the data file.

4. If a warning message appears, click [OK]. The first nine rows of data are shown.

b. State the hypotheses:

H_0: $p \leq .20$

H_1: $p > .20$ (claim)

Step 2. Calculate the test statistic.

a. Select **Stat>Basic Statistics>1 Proportion,** then press <F3> to reset the defaults.

b. Click the option for Samples in columns:.

c. Click in the box then double click C4 Smokes in the list.

d. Click **[Options].**

1. Enter **.2** for the Test proportion:.
2. Change the Alternative: to greater than.
3. Check the box to Use test and interval based on normal distribution.
4. Click **[OK]** twice.

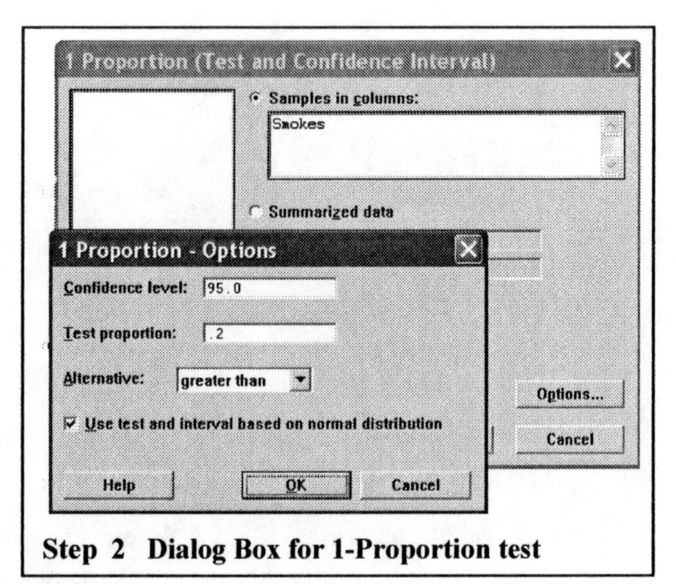

Step 2 Dialog Box for 1-Proportion test

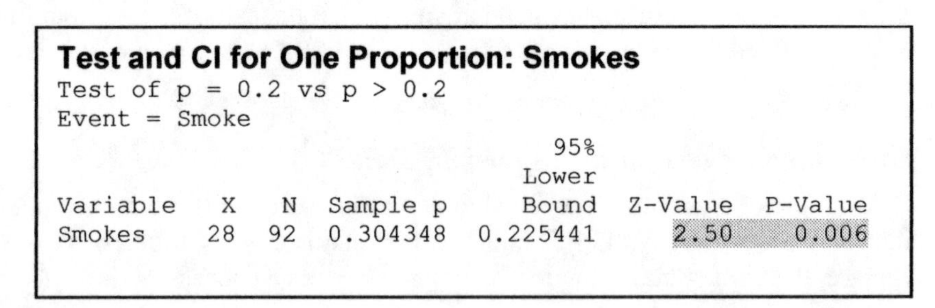

Test and CI for One Proportion: Smokes

Test of p = 0.2 vs p > 0.2

Event = Smoke

Variable	X	N	Sample p	95% Lower Bound	Z-Value	P-Value
Smokes	28	92	0.304348	0.225441	2.50	0.006

Step 3. The test statistic $Z = 2.50$.

Step 4. The P-value is .006. Reject H_0 .

Step 5. There is enough evidence in the sample to support the claim $p > .20$. Thirty percent of the students in the sample smoke. The sample proportion is the point estimate for p, the proportion of all who smoke.

Step 6. Select **File>New>Minitab Project** saving the project as **Chapter08.MPJ.** Continue.

8-6 Chi-square Test for a Variance or Standard Deviation; No Raw Data

Section 6 Example 24: An instructor wishes to see whether the variation in scores of the students in her class is less than the variance of the population. The variance in the sample of 23 is 198. Is there enough evidence to support the claim that the variance of her students is less than the population variance? Use a significance level of .05.

Step 1. State the hypotheses.

H_0: $\sigma^2 \geq 225$

H_1: $\sigma^2 < 225$

There isn't a menu command to calculate the test statistic and P-value. Calculate the test statistic with a formula and another menu command to calculate the P-value.

Step 2. Calculate the test statistic.

a. Name C1 **n,** C2 **s2**, and C3 **sigma2.**

b. In row 1 of the worksheet enter the values for this example.
$n = 23$, sample size
$s^2 = 198$, variance in the sample
$\sigma^2 = 225$, the hypothesized value

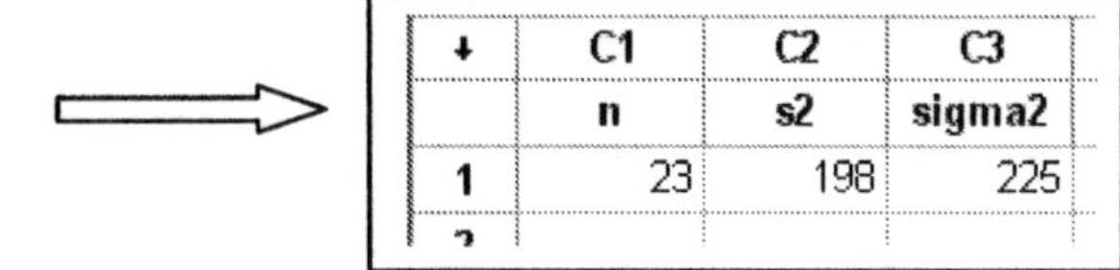

C1 C2 C3
n s2 sigma2
1 23 198 225

That sets up the worksheet with the values needed in the formula:

$$K1 = \text{chi-square} = \frac{(n-1)\,s^2}{\sigma^2}$$

c. Select **Calc>Calculator.**
1. Type **K1** in the box for Store result in variable:.
2. Type in the formula **(n-1)*s2/sigma2.**
3. The names used in the formula must match the names typed for the column names.
4. Click **[OK].**
   The chi-squared test statistic is 19.36.

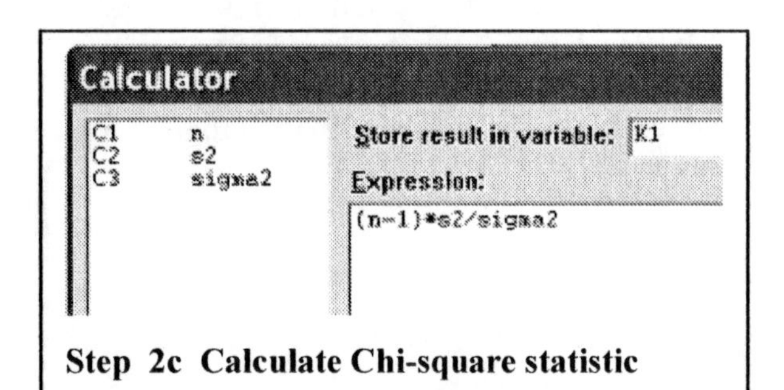

Step 2c Calculate Chi-square statistic

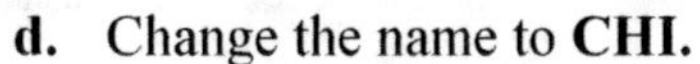

d. Change the name to **CHI.**
1. Select **Window>Project Manager.**
2. Select the Constants folder.
3. Right click the unnamed constant, K1.
4. Choose Rename, then type in **CHI.**
5. Press <Enter>.

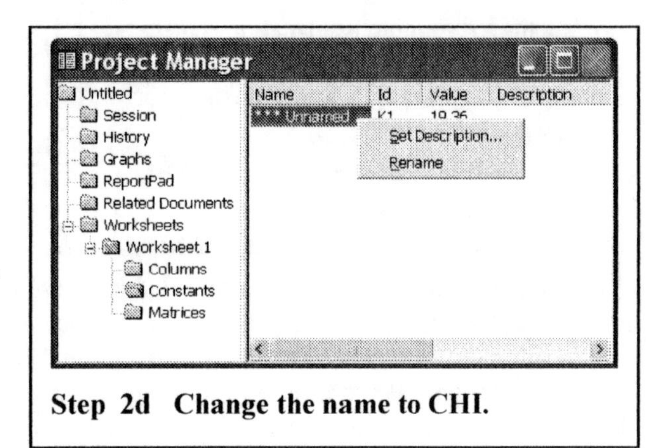

Step 2d Change the name to CHI.

Step 3. To calculate the P-value for a left-tailed test, select **Calc>Probability Distribution>Chi-square.**

a. In the dialog box you must set four items.
1. Check the button for Cumulative probability.
2. Type the value of **22** for the Degrees of freedom:.
3. Click on Input constant, then type **CHI.**
4. In the Optional storage: box type in **cdf**, then click **[OK].**

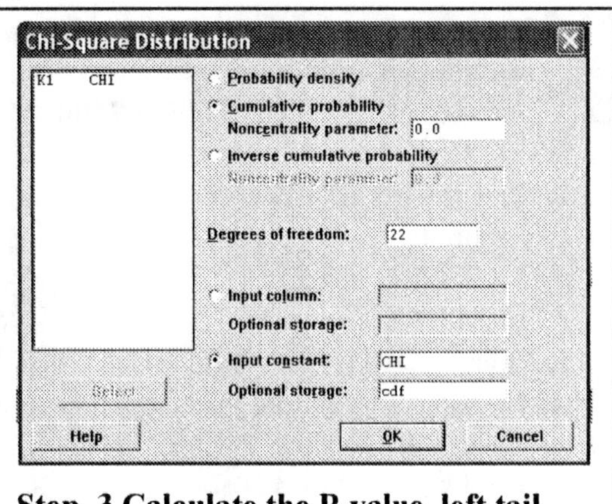

Step 3 Calculate the P-value, left tail

b. If this were a right-tailed test, subtract CDF from 1 to get the right-tailed P-value.

1. Select **Calc>Calculator.**
2. Type **K3** for the variable.
3. Type **1–cdf** for the **Expression:**.
4. Click **[OK].**

For a two-tailed test, double the smaller of cdf and 1-cdf.

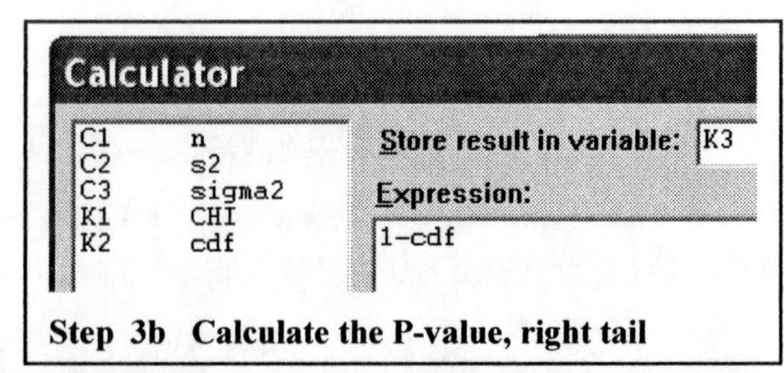

Step 3b Calculate the P-value, right tail

c. Display the constants in the Program Manager window.

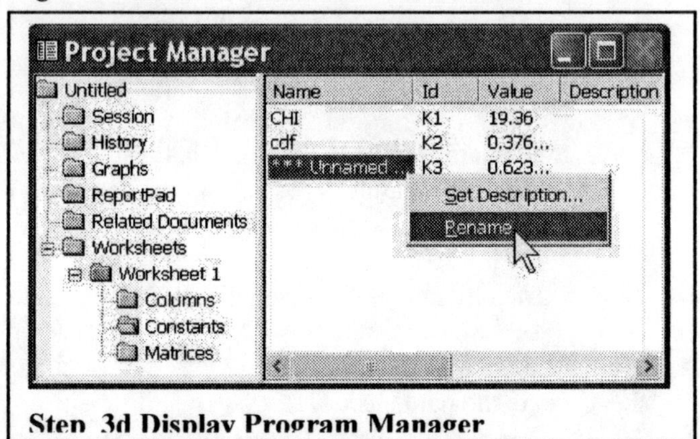

Step 3d Display Program Manager

The P-value for this left-tailed test is K2, the cumulative probability, .376994.

Step 4. Cannot reject H0.

Step 5. There is enough evidence in the sample to support the claim. The variance in the sample of 198 is not significantly lower than 225.

Step 6. Save this project as **Variance.MPJ.**

Chi-square Test for a Variance or Standard Deviation; Data

For raw data the procedure is the same. Calculate the sample variance using **Calc>Column Statistics** and proceed. Exercise 3 in chapter 8 section 6 is our practice problem.

A nutritionist claims that the standard deviation of the number of calories in one tablespoon of the major brands of pancake syrup is sixty. A sample of major brands of syrup is selected and the number of calories is shown. At a significance level of 0.10, can the claim be rejected?

Calories

53	210	100	200	100	220	210	100	240
200	100	210	100	210	100	210	100	60

The test for variance must be used even though the statement is about the standard deviation. The sampling distribution for variance is chi-square. Therefore, if the standard deviation is sixty, the variance, σ^2, is equal to 60^2 or 3600.

Step 1. State the hypotheses.

H_0: $\sigma^2 = 3600$

H_1: $\sigma^2 \neq 3600$

Step 2. Calculate the test statistic.

Use the calculator to store the sample size in n and the standard deviation, s. Both n and s are constants. Enter the data.

a. Select **File>New>Minitab Worksheet.**

b. Enter the data into **C1** of the new worksheet. Name the column **Calories.**

c. Select **Calc>Column Statistics.**
   1. Click the option for N non-missing. Press <Tab>.
   2. Double click C1 Calories.
   3. Click in the box for Storage and type the letter **n**.
      The sample size will be stored in a constant, n.
   4. Click **[OK].**

d. Click the Edit last dialog icon on the toolbar.
   1. Change the option to Standard deviation and the storage to **s**.
   2. Click **[OK].**

n = 18 and s = 64.616.

Calculate the test statistic.

e. Select Calc>Calculator.
f. Store the result in K3.
g. Enter the Expression: (formula) then click [OK].
   (n-1)*s2 / 3600**

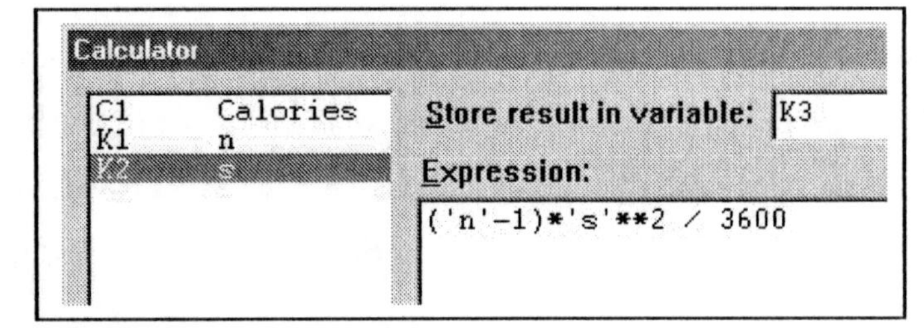

The test statistic will be stored in K3. The value should be 19.7166.

Step 3. Determine the P-value.

a. Select **Calc>Probability Distribution>Chi-square.**
   In the dialog box you must set four areas.
   1. Check the radio button for Cumulative probability.
   2. Type the value of **17** for the Degrees of freedom:.
   3. Click on Input constant:. Select K3 then type **cdf** for storage.
      This is the probability in the left-tail.
   4. Click **[OK].**

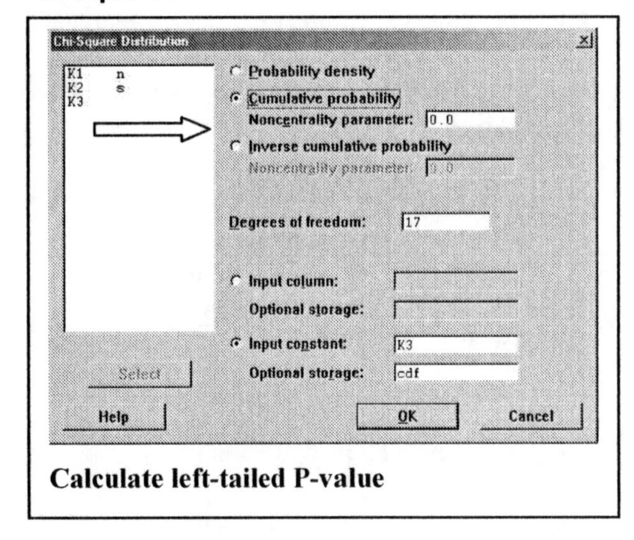

Calculate left-tailed P-value

The P-value will be the right tail probability, the complement of cdf.

b. Select **Calc>Calculator.**
   1. Store the result in K5.
   2. The Expression: is 1-cdf .
   3. Click **[OK].**

   K3 and K5 were renamed CHIS and 1-cdf, respectively.
   The P-value for this two-tailed test would be 2* 0.288983 = .577966.

View the Program Manager window.

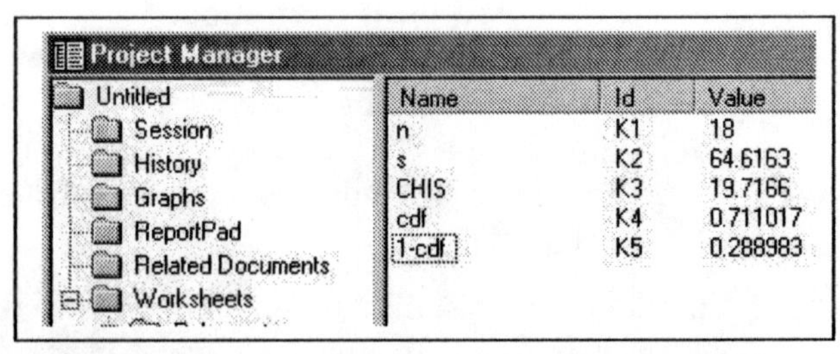

Step 4. Decision: Because .577966 > .1, do not reject Ho.

Step 5. There is not enough evidence in the sample to reject the null hypothesis, the claim that the standard deviation is sixty.

Step 6. Select **Data>Data Display,** then drag the mouse over C1 and K1–K5, then click **[Select].**

a. Print the Session window.

b. Click the disk icon to save the project as **Variance.MPJ.**

c. Select File>Exit to close or File>New>Minitab Project to continue.

Because MINITAB uses more decimal places in its calculations, some results like the previous exercise may differ from textbook solutions because of rounding. The difficulty in statistics is deciding the appropriate technique to use. First decide the objective of the research.
1) *Estimation* or *testing a hypothesis?* 2) Which *parameter?* 3) Which *procedure?*
The following table is a summary of the inferential techniques so far.

Chapter 7 Estimating parameters with point estimates or confidence intervals			
Parameter	Details /Condition	Procedure	MINITAB Command
Mean	σ known σ normal population if $n < 30$	Z interval	1-Sample Z
	σ unknown or $n \geq 30$		1-Sample Z
	σ unknown and $n < 30$ normal population	t interval, df = n-1	1-Sample t
Standard Deviation	$n \geq 30$ normal population if $n < 30$	χ^2 interval, df = n-1	none available
Proportion	$np \geq 5$ and $nq \geq 5$	Z interval	1-Proportion
Chapter 8 Testing a Hypothesis about a parameter: mean, proportion, or standard deviation			
Mean	σ known normal population if $n < 30$	critical Z	1-Sample Z
	σ unknown or $n \geq 30$	critical Z	1-Sample Z
	σ unknown and $n < 30$ normal population	critical t , df = n-1	1-Sample t
Variance	$n \geq 30$, $s \approx \sigma$ normal population	critical χ^2, df = n-1	not available
Proportion	$np \geq 5$ and $nq \geq 5$	Z interval	1-Proportion

Use this table to help you decide which process is needed. The next data analysis project will give you the idea.

8-7 Data Analysis

Problem 2b on page 427: Select a random sample of at least ten from the Databank file then test the hypothesis that the mean systolic blood pressure is 120 mmHg. Use a significance level of .05. Choose the appropriate technique:

i) The hypothesis is about the mean, σ unknown and $n < 30$.

H_0: $\mu = 120$

ii) If the population can be assumed symmetric, the critical value will be from the Student t distribution with 14 degrees of freedom.

	σ unknown and $n < 30$ normal population	t interval, df = n-1	1-Sample t

iii) Select a sample then proceed with the t- test. Use the P-value method.

- Open the Databank file using **File>Open Worksheet.**
- Select a sample of ten or more using **Calc>Random Data>Sample from columns.**
- Show the steps of the hypothesis test.

Step 1. State the hypotheses:

H_0: $\mu = 120$
H_1: $\mu \neq 120$

Step 2. Calculate the test statistic.

a. Select **Stat>Basic Statistics>1- Sample t**

- Variable is C1 Pressure15.
- Test mean is 120.

b. Click **[Options].** The level should be 95 percent and the alternative hypothesis is not equal.
c. Click **[OK],** then select **[Graph]** and check the option for the boxplot.
d. Click **[OK]** twice.

Your results will not match these.

What is the test statistic for your sample? t = ________ Mine was t = 5.10 .

Step 3. Find the P-value. Look in the Session window.
What is the P-value for your sample? P-value = ________ My P-value was 0.000.

Step 4. Decision: Reject H_0 or don't reject? ________ My conclusion. Reject H_0.

Step 5. Summary: There is enough evidence in my sample to conclude the mean systolic pressure for all is different than 120. The boxplot indicates a symmetric distribution. The mean of my sample, 135.7 is a point estimate for mu. It would appear the mean is higher than 120 mmHg.

Problem 3c on page 427. Select a sample of at least thirty individuals' exercise level. Test the hypothesis that 10% of the population exercises slightly. Level 1 is slight exercise. Both np and nq are more than five, so the test based on the normal distribution is appropriate.

Step 1. State the hypotheses:

H_0: $p = .1$
H_1: $p \neq .1$

Step 2. Select the sample and calculate the test statistics.

a. Open the Databank file.

1. Select **Calc>Random Data>Select from columns.**

2. Type **30** for the number of Rows to sample:.
3. The variable will be C5 EXERCISE.
4. Store the sample in Exercise30.
5. Click **[OK].**

There are more than two codes in this column so the 1 Proportion test will not work, yet. A solution to the dilemma is to recode the sample so there is 1 value for level 1 and all other levels are a zero. Change codes 2 and 3 to zeros. Leave the ones alone. That will make the highest code a 1, a "success."

b. Select **Data>Code>Numeric to Numeric.**
1. Code data From columns: Exercise30.
2. Into columns: Exercise30.

The new codes will replace the originals. If you chose a different name, a new column will be created with the new codes.

3. The Original values: 2 and 3 will be changed to a zero. The original codes 0 and 1 will stay the same.

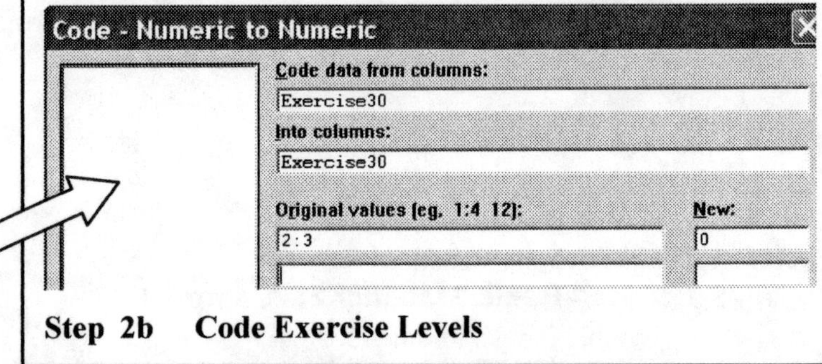

Step 2b Code Exercise Levels

Notice that there are no columns appearing in the variables list since the cursor is not in a location that can use them.

4. Click **[OK].**

Each row of the column of data should now have a 0 or a 1. Proceed with the hypothesis test.

c. Calculate the test statistic. **Select Stat>Basic Statistics>1-Proportion.**
1. Click the option for Samples in columns and enter Exercise30.
2. [Options] Test proportion is .1
3. [Options] Confidence level is 95 percent and Alternative is not equal.
4. [Options] Be sure the check the box for Normal distribution.
5. Click **[OK]** twice.

Test and CI for One Proportion: Exercise30

```
Test of p = 0.1 vs p not = 0.1
Event = 1
Variable      X   N  Sample p         95% CI            Z-Value  P-Value
Exercise30   10  30  0.333333  (0.164646, 0.502020)      4.26    0.000
```

What is the test statistic for your sample? ________ 4.26

Step 3. Find the P-value. Look in the Session window.
What is the P-value for your sample? ________ 0.000

Step 4. Decision: Reject H_o or don't reject? ________ Reject H_o

Step 5. Summary: There was enough evidence in my sample to conclude the proportion of students who get slight exercise is not 10 percent. It appears to be higher. In this sample it is 33 percent who exercise slightly.

Step 6. Select **File>Exit,** then save the project as **DataAnalysis09.MPJ.**

Chapter 8 Textbook Problems Suitable for MINITAB

Page	Section	Exercise
Page 387	8.3	1 – 13, 16 - 25
Page 398	8.4	5 - 20
Page 405	8.5	5 – 20
Page 416	8.6	3 - 14
Page 425	Review	1 – 19
Page 427	Data Analysis	1 - 4 Databank
		5 Dataset X
		6 - 7 Dataset XVI
Page 430	Data Projects	1 - 2 Original data

Chapter 8 Endnotes:

Chapter 9 Testing the Difference Between Two Parameters

9-1 Introduction

Basic concepts of hypothesis testing and the P-value approach were introduced in Chapter 8. So far, all of the tests involved a hypothesis with a parameter that had a specific value, for example, $\mu = 10$ or $p = .25$. Very often the researcher would like to compare the means, proportions, or variances for two groups. The focus isn't on the actual value of the means, variances, or proportions but whether they are the same or different. If the data was collected from the population, the parameter for each group would be calculated and compared. When the data is from a sample, the hypothesis test is evidence to support the claim that the means are the same or they are different for the populations from which the samples were drawn.

9-2 Testing the Difference Between Two Means: Large Samples

Section 2 Example 1: A survey found that the average hotel room rate in New Orleans is $88.42 and the average room rate in Phoenix is $80.61. The data were obtained from two random samples of 50 hotels in each city and the standard deviations were $5.62 and $4.83, respectively. Is there a significant difference in the room rates? Use $\alpha = .05$.

There is no data for this exercise. The summarized statistics will be used.

Step 1. State the hypotheses.

H_0: $\mu_{NewOrleans} - \mu_{Phoenix} = 0$

H_1: $\mu_{NewOrleans} - \mu_{Phoenix} \neq 0$ (claim)

Step 2. To calculate the test statistic, select **Stat>Basic Statistics>2-Sample t.**

a. Click the option for Summarized data, then fill in the statistics for each sample.

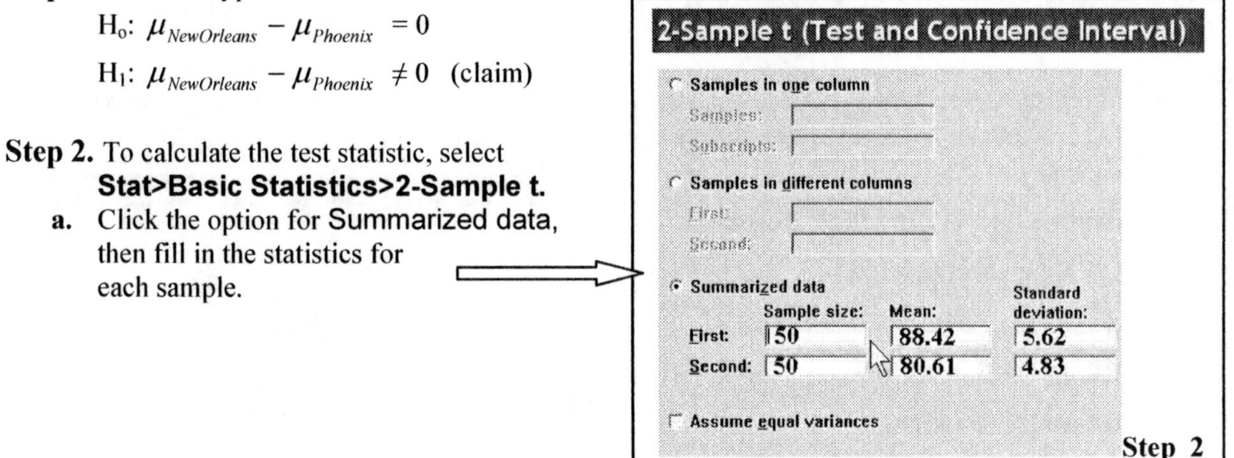

b. Click **[Options].**

The defaults of 95.0, 0, and not equal are correct.

c. Click **[OK]** twice.

The results will be displayed in the Session window.

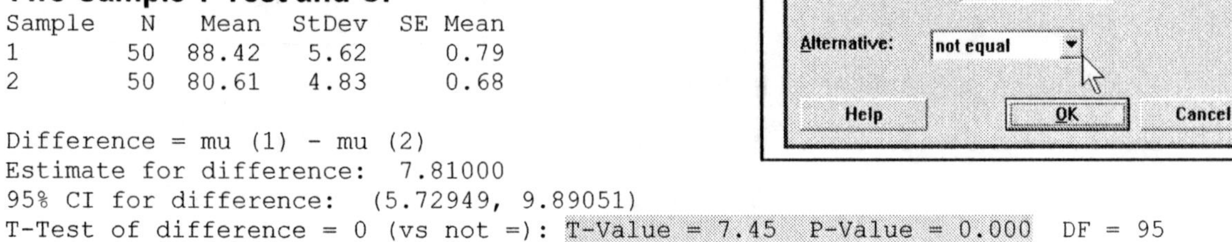

Two-Sample T-Test and CI

```
Sample    N    Mean   StDev   SE Mean
1        50   88.42    5.62      0.79
2        50   80.61    4.83      0.68

Difference = mu (1) - mu (2)
Estimate for difference:  7.81000
95% CI for difference:  (5.72949, 9.89051)
T-Test of difference = 0 (vs not =): T-Value = 7.45  P-Value = 0.000  DF = 95
```

d. The test statistic is +7.45.

Step 3. The P-value is 0.000.
Step 4. The P-value is less than $\alpha = .05$, therefore, reject the null hypothesis.
Step 5. There is a significant difference. The difference in the sample groups is $7.81, on average. Continue with the next example.

Section 2 Example 2: A researcher hypothesizes that the mean number of sports offered at colleges for men and women is different. Independent samples of the number of sports offered by colleges are shown. At $\alpha = .10$, is there enough evidence to support the claim?

Males					Females				
6	11	11	8	15	6	8	11	13	8
6	14	8	12	18	7	5	13	14	6
6	9	5	6	9	6	5	5	7	6
6	9	18	7	6	10	7	6	5	5
15	6	11	5	5	16	10	7	8	5
9	9	5	5	8	7	5	5	6	5
8	9	6	11	6	9	18	13	7	10
9	5	11	5	8	7	8	5	7	6
7	7	5	10	7	11	4	6	8	7
10	7	10	8	11	14	12	5	8	5

Step 1. State the hypotheses:
H_o: $\mu_{Male} - \mu_{Female} \leq 0$
H_1: $\mu_{Male} - \mu_{Female} > 0$ (claim)

a. Use **File>Open Worksheet** to open the file E-C09-S02-02.MTP or type the data into C1 and C2. Name the columns **MaleS** and **FemaleS.** There is one sample in each column.

Step 2. Select

Stat>Basic Statistics>2-Sample t.

a. Click the button for Samples in different columns.
b. Click in the box for First: then double click C1 MaleS in the list.
c. Double click C2 FemaleS in the list.
d. <u>Do not</u> check the box for Assume equal variances. If this box is checked, MINITAB uses the formula with pooled standard deviations for small samples.
e. Click **[Options].**
   1. Type in **90** for the Confidence level: and **0.0** for the Test difference:.
   2. Select greater than for the Alternative:. This option affects the P-value.
f. Click **[OK]** twice. The completed dialog boxes are shown.

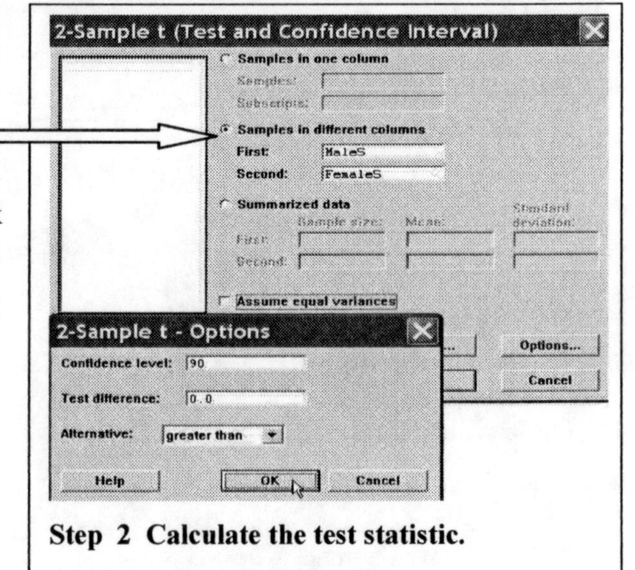

Step 2 Calculate the test statistic.

Step 3. The test statistic is 0.95.

The textbook solutions says t = 1.06. Please note the difference is due to rounding. All of the values in the text were rounded. MINITAB calculates with many more decimal places.

Step 4. The P-value is .172. The P-value is greater than the significance level, **.**172 > **.**1, so do not reject the null hypothesis.

Step 5. There is not enough evidence in the samples to support the claim that the means are different.

Two-Sample T-Test and CI: MaleS, FemaleS

```
Two-sample T for MaleS vs FemaleS
         N   Mean   StDev   SE Mean
MaleS    50  8.56    3.26      0.46
FemaleS  50  7.94    3.27      0.46
Difference = mu (MaleS) - mu (FemaleS)
Estimate for difference:0.620000
90% lower bound for difference:   -0.221962
T-Test of difference = 0 (vs >): T-Value = 0.95  P-Value = 0.172  DF = 97
```

Note: For large samples, the textbook calls this a Z-test. However, MINITAB will calculate the test statistic using the correct formula but the label will be a "t." It is a matter of semantics. This t-value is the test statistic we need. The two-sample test for means where the population standard deviations are known is so rare that there is not an option for that in MINITAB. All of the tests use the sample standard deviations. Therefore, MINITAB considers them a t-test.

Step 6. Save the project as **Chapter09.MPJ.** Continue.

Calculate the Test Statistic and P-value for Stacked Data

In the previous example, the data were separated into two columns. Each column contained the same measure, number of sports, one column for men and one column for women. The data is "unstacked."

Step 7. Select **File>OpenWorksheet.** Locate and open PulseA.MTW in the Studnt14 directory.

The resting pulse rates are in C1. The pulse rates are not separated for those who ran in place and those who did not run in place. This is stacked data. Stacked data are recorded in one column. Another column, the subscript column, determines the group. Our research question: Is the mean resting pulse rate of those who ran different than the mean resting pulse rate of those who didn't run? Test the claim using a significance level of .05.

PULSEA.MTW ***

→	C1	C2	C3-T	C
	Pulse1	Pulse2	Ran	Sm
1	64	88	Ran	NonS
2	58	70	Ran	NonS
3	62	76	Ran	Smo
4	66	78	Ran	Smo
5	64	80	Ran	NonS
6	74	84	Ran	NonS
7	84	84	Ran	NonS

Step 1. State the hypotheses.

H_0: $\mu_{ran} - \mu_{still} = 0$ (claim)

H_1: $\mu_{ran} - \mu_{still} \neq 0$

a. To see the size of each sample group, select **Stat>Tables>Tally Individual Variables.**

b. Double click C3-T **Ran.**

There are 35 who ran in place and 57 who were still. The large sample test is appropriate.

Calculate the test statistic.

Tally for Discrete Variables: Ran

```
  Ran  Count
  Ran     35
Still     57
   N=     92
```

Step 1b Frequency table for "Ran"

C1 **Pulse1** contains all the pulse rates. C3-T **Ran** contains two codes or subscripts, Ran or Still. Only the first seven rows of the data are shown. There are 92 rows of data.

Step 2. Select **Stat>Basic Statistics>2-sample t.**

a. Check **Samples in one column.**
b. Double click C1 Pulse1 in the box for Samples:.
c. Double click C3-T Ran in the box for Subscripts:.
d. Do not check Assume equal variances.
e. Click **[Options].**
   1. The Confidence level is 95 percent.
   2. The Alternative is not equal.
f. Click **[OK]** twice.

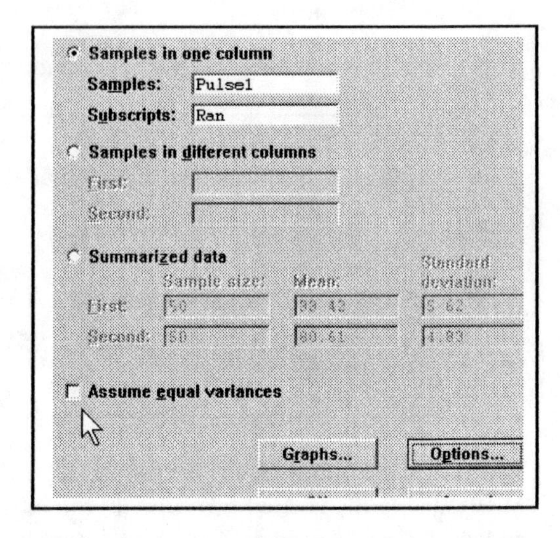

The test statistic is .49.

Step 3. The P-value is .626 for a two-tailed test.

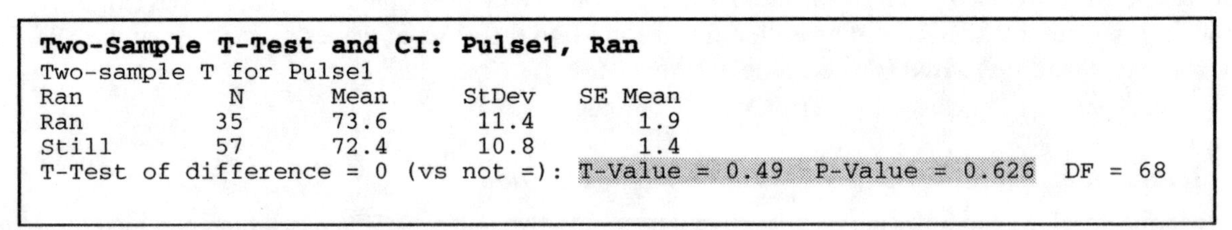

```
Two-Sample T-Test and CI: Pulse1, Ran
Two-sample T for Pulse1
Ran          N      Mean     StDev   SE Mean
Ran         35      73.6      11.4       1.9
Still       57      72.4      10.8       1.4
T-Test of difference = 0 (vs not =): T-Value = 0.49  P-Value = 0.626  DF = 68
```

Note: MINITAB uses alphabetical order to determine group 1 and group 2. The codes in C3-T Ran are Ran and Still, Ran is group 1. The pulse rates are separated by MINITAB, and the statistics for each group are calculated.

Step 4. The P-value is larger than alpha. Do not reject H_0.
Step 5. There is no significant difference in the mean resting pulse rates for those who ran in place and those who remained still.

Step 6. Click the disk icon on the Standard toolbar to save the project as **Chapter09.MPJ.**

9-3 Testing the Difference Between Two Variances: the F Distribution

Variance and standard deviation measure the variation in a set of quantitative data. The larger the variance, the more variation there is in the raw data. If two independent samples are selected from normal populations in which the variances are equal, the ratio of the sample variances, $F = \frac{s_1^2}{s_2^2}$.

The F distribution has a degree of freedom for the numerator (d.f.N) and a degree of freedom for the denominator (d.f.D). It is possible for the test to be right, left, or two-tailed. It is customary when using the traditional method of hypothesis testing for a one-tailed test to designate the larger variance as group 1, then a right-tailed critical value is determined from the table. We are using the P-value method so we are not limited to that custom.

Use MINITAB to Find Critical Values from the F Table

This table is not necessary if using the P-value method for hypothesis tests. These instructions are provided for your information or if you only need the table value.

Find the critical value for a two-tailed F test with $\alpha = .05$ and d.f.N = 25 and d.f.D. is 17.

Step 1. Select **Calc>Probability Distributions>F.**

a. Click Inverse cumulative probability.

b. Enter the d.f.N and d.f.D, **25** and **17**.

c. Click the button for Input constant:, then enter the value of $1-\alpha/2$, **.975**. This is the cumulative area in the F-distribution up to the critical value.

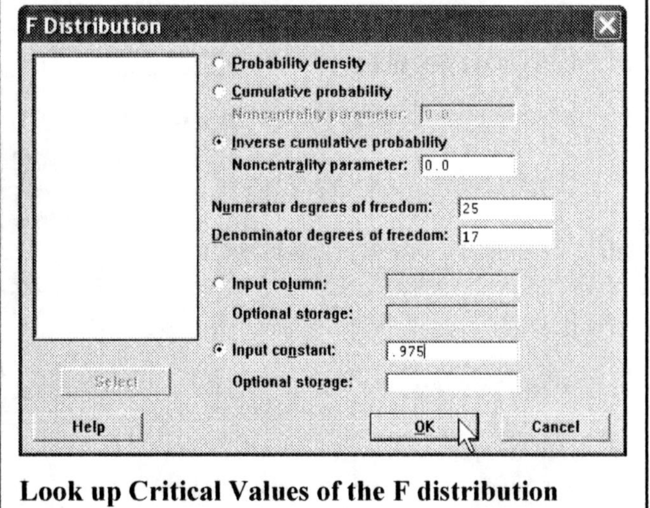

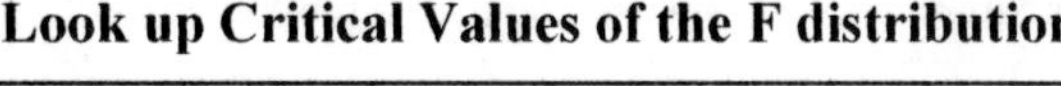
Look up Critical Values of the F distribution

For a right tailed F-test use $1-\alpha$ for the input constant. To find the critical value for a two-tailed F-test, it is customary to use $1-\alpha/2$ for the input constant and select the largest variance as s_1.

Step 2. Click **[OK]**. Look in the Session window for the critical value of F, + 2.548.

Inverse Cumulative Distribution Function

```
F distribution with 25 DF in numerator and 17 DF in denominator
P( X <= x )          x
    0.9750         2.54842
```

Calculate the Test Statistic and P-value for the Test of Variance

Section 3 Example 8: Test the hyothesis that the variance in the number of passengers for American and foreign airports is different. Use the P-value approach, $\alpha = .10$. Is there enough evidence to support the hypothesis that there is more variation at the U.S. airports?

Step 1. Enter the data into two columns of MINITAB. Name the columns **American** and **Foreign.**

↓	C1	C2
	American	Foreign
1	36.8	60.7
2	72.4	42.7
3	60.5	51.2
4	73.5	38.6
5	61.2	
6	40.1	

Step 2. Select **Stat>Basic Statistics>2-Variances.**

a. Click the button for Samples in different columns.

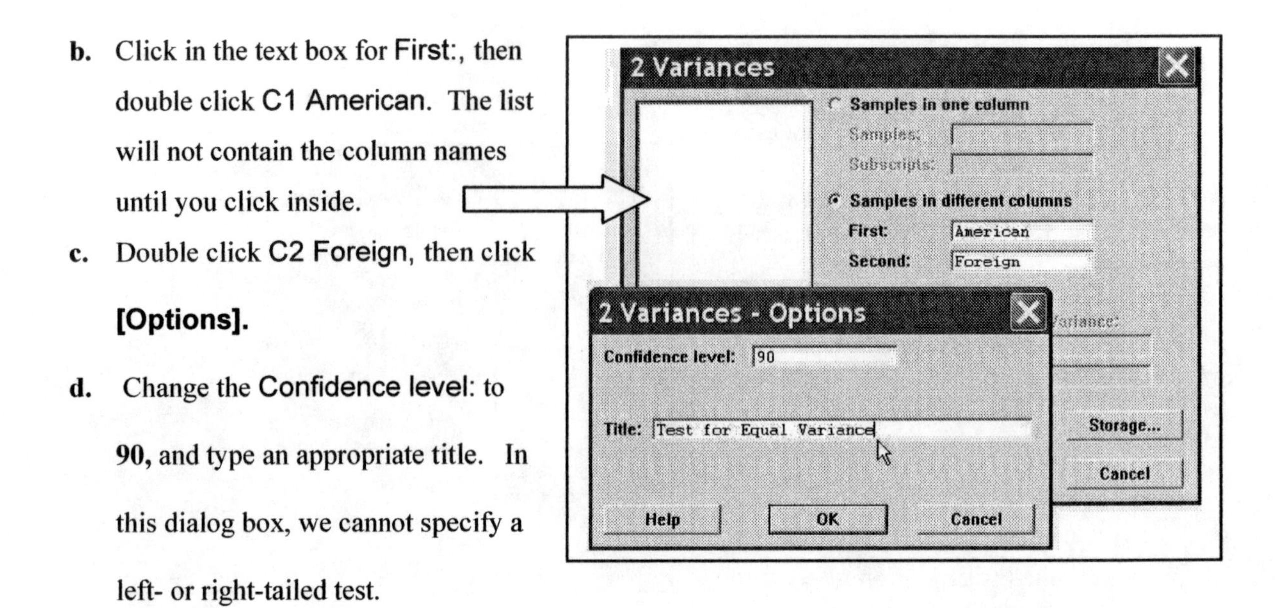

b. Click in the text box for **First:**, then double click **C1 American**. The list will not contain the column names until you click inside.

c. Double click **C2 Foreign**, then click **[Options]**.

d. Change the **Confidence level**: to **90,** and type an appropriate title. In this dialog box, we cannot specify a left- or right-tailed test.

e. Click **[OK]** twice.

Step 3. A Graph window will open that includes a small window that says F = 2.57.

Step 4. The P-value is .467.

F-Test	
Test Statistic	2.57
P-Value	0.467

Step 5. There is not enough evidence in the sample to conclude there is more variance in the number of passengers in American airports compared to foreign airports. Continue with the next section, which also uses this data.

9-4 Testing the Difference Between Two Means: Small Independent Samples

Suppose in the last example the hypothesis was that the mean number of passengers is different for American and foreign airports. Because one or both sample sizes is small, the first step would be to test the variances. We concluded that the variances were probably equal, so the appropriate critical value would be a t value with $n_1 + n_2 - 2$ degrees of freedom. The standard error for the test statistic and P-value should be calculated using the pooled estimate of the variance.

The hypothesis test for the means of two small samples is continued after the test for variances is complete.

Step 1. State the hypotheses:

H_o: $\mu_{American} - \mu_{Foreign} = 0$

H_1: $\mu_{American} - \mu_{Foreign} \neq 0$

Calculate the test statistic.

Step 2. Select **Stat>Basic Statistics>2-Sample t.**

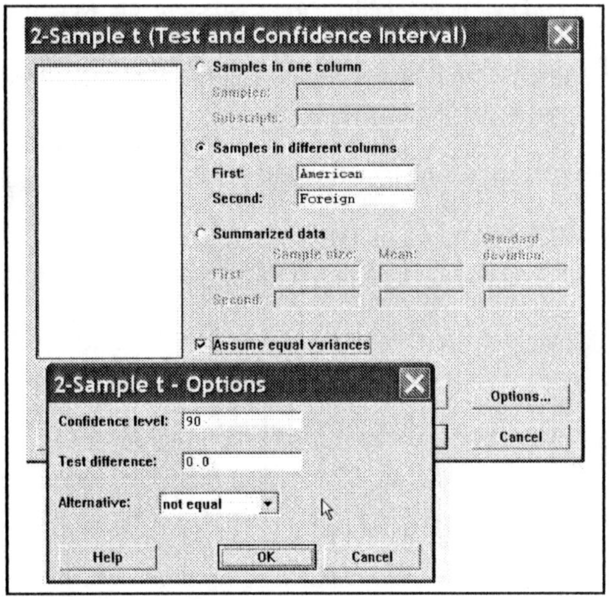

a. Press <F3> to reset the defaults.
b. Click on **Samples in different columns**.
c. Click in the box for **First:**, then double click **C1 American**.
d. Double click **C2 Foreign** for **Second:**.
e. Check the box for **Assume equal variances**, because the F test concluded we can make that assumption. The pooled standard deviation formula will be used to calculate the test statistic and P-value.
f. Click **[Options]**. Change the **Confidence: level** to **90**. The **Alternative:** should be **not equal**.
g. Click **[OK]** twice.

Two-Sample T-Test and CI: American, Foreign

```
Two-sample T for American vs Foreign
          N   Mean   StDev   SE Mean
American  6   57.4    15.7       6.4
Foreign   4   48.30    9.79      4.9

Difference = mu (American) - mu (Foreign)
Estimate for difference:  9.11667
90% CI for difference:  (-7.42622, 25.65955)
T-Test of difference = 0 (vs not =): T-Value = 1.02 P-Value = 0.335
Both use Pooled StDev = 13.7819
```

Step 3. The test statistic is 1.02.

Step 4. The P-value for the difference is .335. Do not reject the null hypothesis.

Step 5. There is no significant difference in the mean number of passengers at American airports compared to foreign airports.

Step 6. Save the project and then open a new worksheet. You know how!

9-5 Testing the Difference Between Two Means: Dependent Samples

Two dependent samples are related, and the data is recorded in "matched pairs." Before and after treatment plans are typical but not the only examples of dependent samples. The same individual is observed before and after some treatment. The change in the measurement is calculated for each individual by subtracting the values in each pair. The hypothesis test reduces to a one-sample test on the differences that is more sensitive than the two-sample test for independent samples. If the mean difference is 0, we can conclude there has been no significant change.

Section 5 Example 13: A dietician wishes to see if a person's cholesterol level will change if a certain mineral supplements their diet. Six subjects were pre-tested, and they took the mineral supplement for a six-week period. The results are shown in the table. Cholesterol level is measured in milligrams per deciliter. Can it be concluded that the cholesterol level has been changed at $\alpha = .10$? The variables are approximately normal.

Step 1. State the hypotheses. $H_o : \mu_D = 0$
$H_1 : \mu_D \neq 0$

Enter the **Before** and **After** data into two columns of a worksheet. No need to enter the number of the athlete. The first patient's cholesterol was 210 before and 190 after.

Calculate the test statistic.

Step 2. Select **Stat>Basic Statistics>Paired-t.**

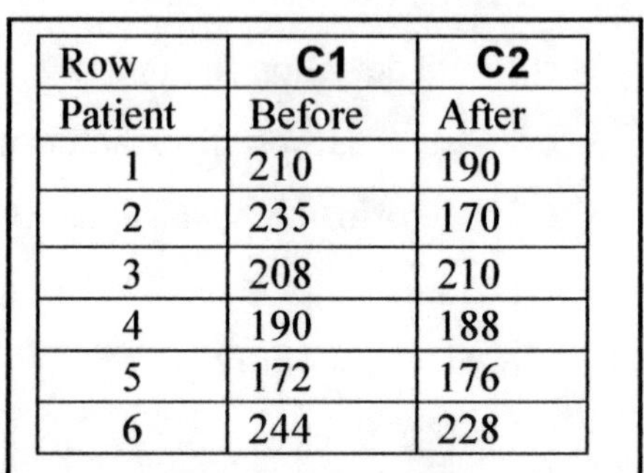

Row	C1	C2
Patient	Before	After
1	210	190
2	235	170
3	208	210
4	190	188
5	172	176
6	244	228

a. Double click C1 to select C1 for the First sample:.

b. Double click C2 to select it for the Second sample:.

c. Click **[Options].**

1. The confidence level should be **90.**
2. The Test mean: should be **0.0.**
3. The Alternative: should be less than.

d. Click **[OK]** twice.

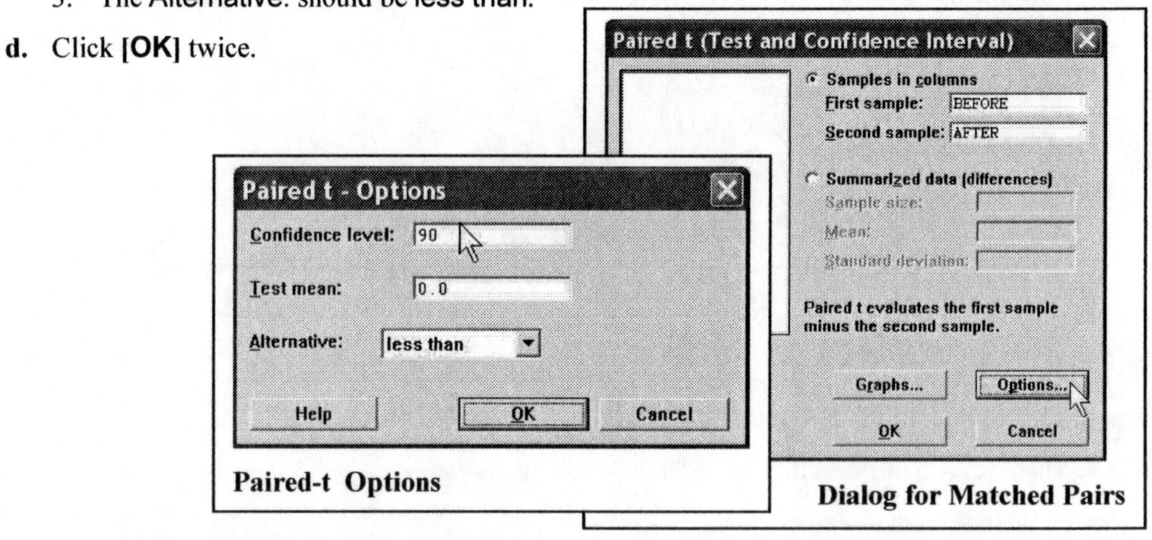

Paired-t Options **Dialog for Matched Pairs**

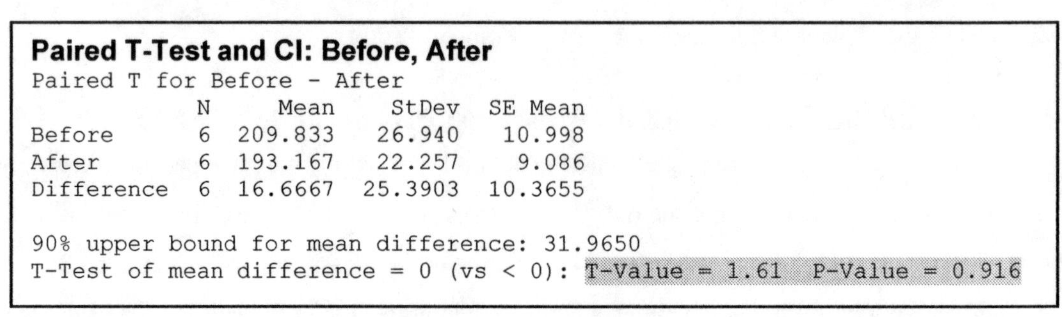

Paired T-Test and CI: Before, After

```
Paired T for Before - After
              N      Mean     StDev  SE Mean
Before        6   209.833    26.940   10.998
After         6   193.167    22.257    9.086
Difference    6   16.6667   25.3903  10.3655

90% upper bound for mean difference: 31.9650
T-Test of mean difference = 0 (vs < 0): T-Value = 1.61  P-Value = 0.916
```

e. From the Session window, the test statistic is 1.61.

Step 3. The P-value is .916.

Step 4. Do not reject the null hypothesis.
Step 5. On average, the difference is not statistically significant.
Step 6. Save the project as **Chapter09.MPJ** by clicking the Disk icon.

Hypothesis tests for quantitative data involve comparisons of measures such as the mean, variance, and standard deviation. For qualitative variables, hypothesis tests involve proportions. MINITAB can make comparisons for raw data or summary statistics.

9-6 Testing the Difference Between Proportions

Section 6 Example 16: A sample of 50 randomly selected men with high triglyceride levels consumed 2 tablespoons of oat bran daily for 6 weeks. After 6 weeks, 60 percent of the men had lowered their triglyceride level. A sample of 80 men consumed 2 tablespoons of wheat bran for 6 weeks. After 6 weeks 25 percent had lower triglyceride levels. Is there a significant difference in the two proportions at the .01 significance level?

Step 1. State the hypotheses.

$$H_o : \hat{p}_{Oat} - \hat{p}_{Wheat} = 0$$

$$H_1 : \hat{p}_{Oat} - \hat{p}_{Wheat} \neq 0$$

Calculate the test statistic. There is no data. It doesn't matter what is in the worksheet. Keep going.

Step 2. Select **Stat>Basic Statistics>2-Proportions.**

a. Click the button for Summarized data. The dialog box requires the count for each sample. How many in each group lowered their triglyceride level?

Calculate mentally: X1 = 60% of 50 = 30 and
X2 = 25% of 80 = 20.

Samples in one column
Samples:
Subscripts:
Samples in different columns
First:
Second:
Summarized data
Trials: Events:
First: 50 30
Second: 80 20
Options...

Two Proportion Test

Type in the sample sizes and number of successes for each group as shown.

b. Click **[Options].**

1. Change the Confidence level: to **99.**
2. The Test difference: should be **0.0** and
3. The Alternative: is not equal.
4. Check the option to

   Use pooled estimate of p for test.

   This option determines the formula used to calculate the test statistic.

5. Click **[OK]** twice.

The symbols may not be the same as those used in the textbook but the results are!

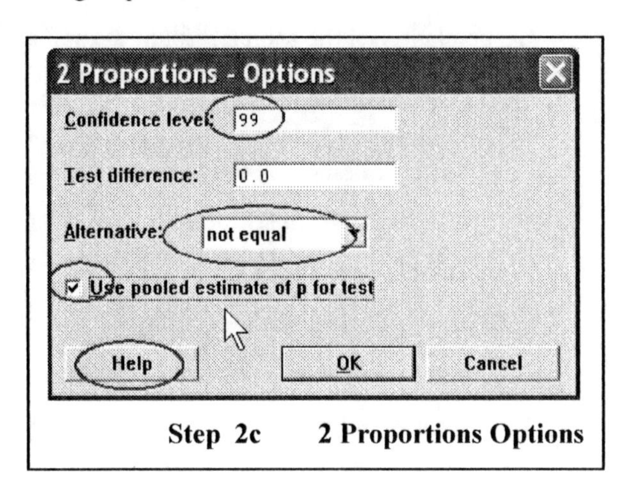

Step 2c 2 Proportions Options

```
Test and CI for Two Proportions
Sample    X   N  Sample p
1        30  50  0.600000
2        20  80  0.250000

Difference = p (1) - p (2)
Estimate for difference:  0.35
99% CI for difference:  (0.132289, 0.567711)
Test for difference = 0 (vs not = 0):  Z = 3.99  P-Value = 0.000
```

Step 3. The test statistic, Z = 3.99.

Step 4. The P-value is 0.000. Reject H_o.

Step 5. There is enough evidence in the sample to indicate the proportions are different, 60 percent of the oat bran sample compared to 25 percent of the wheat bran showed improvement. The inference is the oat bran is more effective at reducing triglyceride levels.

Step 6. Select **File>New>Minitab Project.** Save this project again as **Chapter09.MPJ.**

9-7 Data Analysis

The two-proportions test allows the use of raw data or summarized statistics. The following exercise compares the proportion of women who smoke to the proportion of men who smoke. In the Databank file, one column contains all the data for smoking and another column contains the gender codes, stacked data.

Exercise 3: Compare the proportion of men who are smokers with the proportion of women who are smokers. Choose random samples of size thirty or more. We are comparing two proportions, but there are three codes for smoking not two.

0	Doesn't smoke
1	Smokes less than 1 pack
2	Smokes 1 pack or more

Step 1. State the hypotheses:
$$H_o : \hat{p}_{men} - \hat{p}_{women} = 0$$
$$H_1 : \hat{p}_{men} - \hat{p}_{women} \neq 0$$

Step 2. The plan: Get the sample. Count the smokers and nonsmokers then use the summarized statistics.

a. Select **Calc>Random Data>Sample From Columns.**

1. Type **42** in the box for number of rows. Use a number that is 30 or higher.
2. Double click the columns for **Smoking Status**, then **Gender**.
3. Type in the names for two new columns, **Smoke** and **Sex.**
4. Click **[OK].**

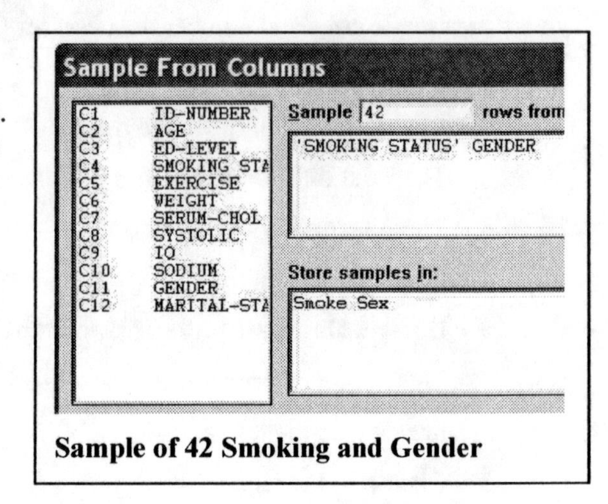

Sample of 42 Smoking and Gender

b. Select **Stat>Tables>Crosstabulation.**

1. Double click the two columns for **Smoke** and **Sex.**
2. Check the box for Counts.
3. Click **[OK].**

Expect yours to be different.

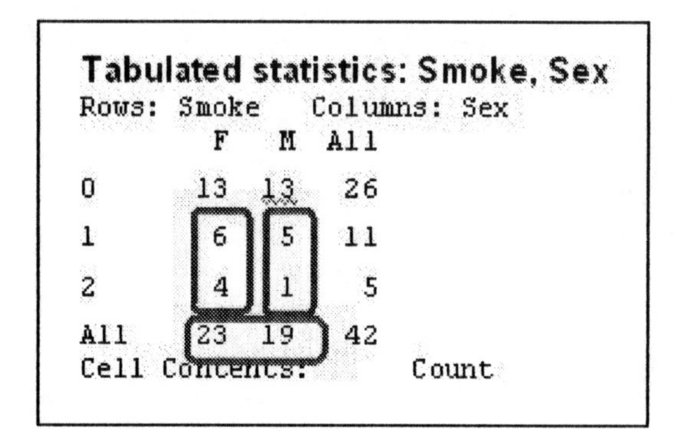

Tabulated statistics: Smoke, Sex

Rows: Smoke Columns: Sex

	F	M	All
0	13	13	26
1	6	5	11
2	4	1	5
All	23	19	42

Cell Contents: Count

In the Session window you will see the cross-tabulation. These are the statistics you will need to enter in the next dialog box. Your numbers will be different!

First	Trials = 23	Events = 10
Second	Trials = 19	Events = 6

Step 3. Select **Stat>Basic Statistics>2 Proportions.**

a. Click the button for Summarized data. Type in the statistics.

b. The **[Options]** should be

1. Confidence level = **95.0**.
2. Test mean = **0.0**.
3. Alternative = not equal.
4. **Very Important!** Check **Use pooled estimate of p for test.**

c. Click **[OK]** twice.

Test and CI for Two Proportions

```
Sample    X   N  Sample p
1        10  23  0.434783
2         6  19  0.315789

Difference = p (1) - p (2)
Estimate for difference:  0.118993
95% CI for difference:  (-0.172090, 0.410076)
Test for difference = 0 (vs not = 0):  Z = 0.79  P-Value = 0.429
```

Step 4. The test statistic is Z = .79

Step 5. The P-value is .429. Can't reject the null hypothesis.

Step 6. There is not enough evidence in this sample to conclude the proportions of men and women who smoke are different.

Step 7. Close MINITAB saving the project as **DataAnalysisCh09.MPJ.**

Chapter 9 Textbook Problems Suitable for MINITAB

Page		Section	Exercises
Page	439	9.2	5 - 20
Page	450	9.3	5, 7 - 20
Page	459	9.4	1 - 13
Page	472	9.5	2 - 10
Page	481	9.6	3 - 19
Page	485	Review	1 - 15
Page	487	Data Analysis	1 - 6
Page	491	Data Project	1- 3

Chapter 9: Endnotes

Chapters 8 and 9: Hypothesis Testing Worksheet

The following is an exercise in hypothesis testing that uses the data in PulseA.MTW data. Open the file PulseA.MTW in the Studnt14 subdirectory. Consider this to be a sample of ninety-two. Resting pulse rates are in C1. The pulse rates in C2 are pulse rates taken again after some of the students walked briskly for ten minutes. A coin was tossed to decide randomly whether they ran or remained at rest. Consider this file to be the sample of 92.

Hypothesis Testing Worksheet

For each problem, show all five steps of the hypothesis test using the P-value approach. Use a significance level of 0.05. Use MINITAB to calculate the test statistic and the P-value. At the end of the exercise, print the session window. Show all five steps of each test on this worksheet.

Hypothesis Testing Summary: Comparing Two Parameters
In the next exercise, several hypotheses will be proposed. You must choose the appropriate technique from chapters 8 and 9 in the textbook.
The following tables summarizes the items from the Stat>Basic Statistics menu are used to calculate the test statistic and P-values for the confidence intervals and hypothesis tests of Chapter 9. For your convenience, the tests from Chapter 8 are repeated.

Chapter 8 Testing a Hypothesis about a parameter: mean, proportion, or standard deviation			
Mean	σ known normal population if $n < 30$	critical Z	1-Sample Z
	σ unknown or $n \geq 30$	critical Z	1-Sample Z
	σ unknown and $n < 30$ normal population	critical t , df = n-1	1-Sample t
Variance	$n \geq 30$, $s \approx \sigma$ normal population	critical χ^2, df = n-1	not available
Proportion	$np \geq 5$ and $nq \geq 5$	Z interval	1-Proportion
Chapter 9 Testing the Difference Between Two Parameters			
Means independent samples	large samples, σ_1 and σ_2 known	critical Z	not available
	large samples σ_1 and σ_2 are unknown	critical Z	2-Sample t
	small samples equal variance $df = n_1 + n_2 - 2$	critical t	2-Sample t, check Assume equal variance
Variances	normal populations	critical F	2-Variances
Proportions		critical Z	2-Proportions

WORKSHEET

1) A medical journal claims that the mean resting pulse rate is 70 with a standard deviation of 9.8 beats per minute. Is there enough evidence in this sample to support the claim?

	P-value approach to hypothesis tests	Fill in with Details of Problem
1	State the null and alternate hypotheses. Identify which is the claim.	H_o: H_1:
2	Calculate the test statistic. Put in the space to the right.	
3	Find the P-value. Put in the space to the right.	
4	Make the decision. Reject H_o or Can't Reject H_o ? **If P-value < significance level, reject H_o.**	
5	Summarize then interpret the result.	

2) The same medical journal claims that the pulse rates of smokers are higher than the pulse rates of nonsmokers. Is there evidence in this sample to support that claim?

	P-value approach to hypothesis tests	Fill in with Details of Problem
1	State the null and alternate hypotheses. Identify which is the claim.	H_o: H_1:
2	Calculate the test statistic. Put in the space to the right.	
3	Find the P-value. Put in the space to the right.	
4	Make the decision. Reject H_o or Can't Reject H_o ? **If P-value < significance level, reject H_o.**	
5	Summarize then interpret the result.	

3) Is there enough evidence in this sample to support the claim that there is more variation in the resting pulse rates of women than the resting pulse rates for men? Compare the variance for the two groups.

	P-value approach to hypothesis tests	Fill in with Details of Problem
1	State the null and alternate hypotheses. Identify which is the claim.	H_o: H_1:
2	Calculate the test statistic. Put in the space to the right.	
3	Find the P-value. Put in the space to the right.	
4	Make the decision. Reject H_o or Can't Reject H_o ? **If P-value < significance level, reject H_o.**	
5	Summarize then interpret the result.	

4) Is there enough evidence in this sample to support the claim that the mean resting pulse rates for women is different than the resting pulse rates for men?

	P-value approach to hypothesis tests	Fill in with Details of Problem
1	State the null and alternate hypotheses. Identify which is the claim.	H_o: H_1:
2	Calculate the test statistic. Put in the space to the right.	
3	Find the P-value. Put in the space to the right.	
4	Make the decision. Reject H_o or Can't Reject H_o ? **If P-value < significance level, reject H_o.**	
5	Summarize then interpret the result.	

5) A textbook example claimed that more than 25 percent of students smoke. Is there evidence in this sample of students to support this claim?

	P-value approach to hypothesis tests	Fill in with Details of Problem
1	State the null and alternate hypotheses. Identify which is the claim.	H_o: H_1:
2	Calculate the test statistic. Put in the space to the right.	
3	Find the P-value. Put in the space to the right.	
4	Make the decision. Reject H_o or Can't Reject H_o ? **If P-value < significance level, reject H_o.**	
5	Summarize then interpret the result.	

6) Compare the mean resting pulse rate (pulse1) with the mean pulse rate after running in place (pulse2). How much higher, on average, is the pulse rate after running in place for those who did not run in place?

	P-value approach to hypothesis tests	Fill in with Details of Problem
1	State the null and alternate hypotheses. Identify which is the claim.	H_o: H_1:
2	Calculate the test statistic. Put in the space to the right.	
3	Find the P-value. Put in the space to the right.	
4	Make the decision. Reject H_o or Can't Reject H_o ? **If P-value < significance level, reject H_o.**	
5	Summarize then interpret the result.	

Chapter 10 Linear Correlation and Regression

10-1 Introduction

Correlation and regression are used to analyze a relationship between two quantitative variables. Correlation coefficients and the hypothesis test help to determine if there is a relationship and the strength of that relationship. Regression analysis results in a formula that tells us how the variables are related. This regression equation can then be used to estimate, forecast or predict values of the dependent variable. Simple linear regression detects a linear correlation between the dependent variable, Y, and one independent variable, X.

10-2 Scatter Plots

Section 2 Example 1: Construct a scatterplot for the data obtained from a study of age and systolic blood pressure of six randomly selected subjects.

Step 1. Type the data into three columns of a MINITAB worksheet as shown.

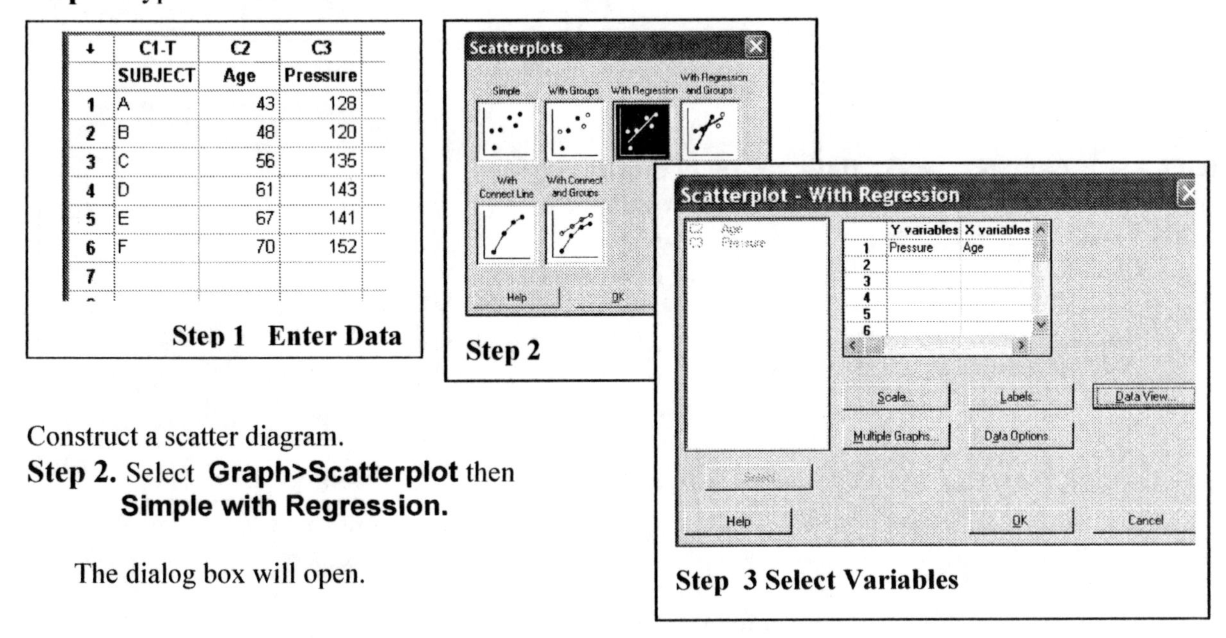

↓	C1-T	C2	C3
	SUBJECT	Age	Pressure
1	A	43	128
2	B	48	120
3	C	56	135
4	D	61	143
5	E	67	141
6	F	70	152
7			

Step 1 Enter Data

Step 2

Step 3 Select Variables

Construct a scatter diagram.

Step 2. Select **Graph>Scatterplot** then **Simple with Regression.**

The dialog box will open.

Step 3. In the Scatterplot – With Regression dialog box, double click on Pressure for the Y variables, then Age as the X variables. Always use the dependent variable as the Y variable.

Step 4. Click **[Data View].**

Click the tab for Data Display and make sure the option checked is Symbols.

a. Click the tab for Regression, selecting the options for Linear and Fit intercept.

b. Click **[OK].**

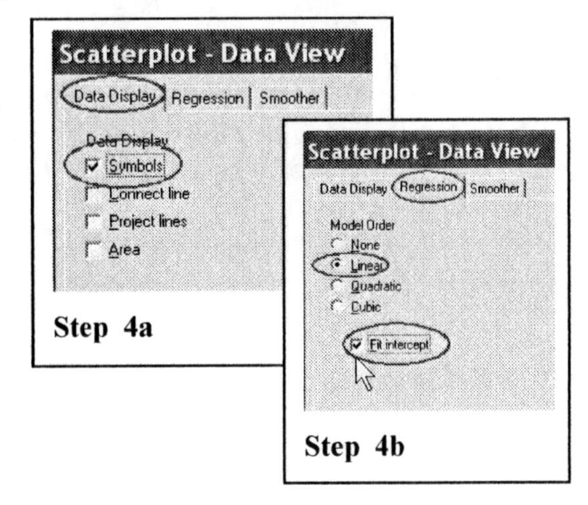

Step 4a

Step 4b

Step 5. Click **[Labels]** (not shown).

- **a.** Type in a title such as **Pressure vs Age.**
- **b.** In the footnote, type **Your Name** and **Date**.
- **c.** Click **[OK]** twice.

The Graph window will be created as shown.

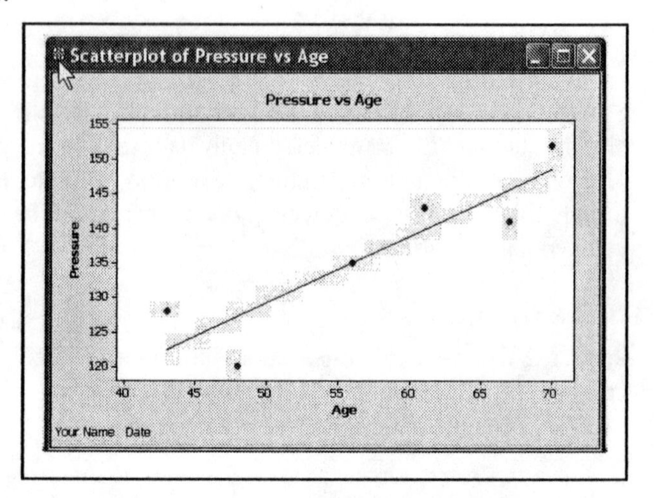

10-3 Calculate the Correlation Coefficient, r.

Step 6. Select **Stat>Basic Statistics>Correlation.**

- **a.** Double click on Pressure, the dependent variable first.
- **b.** Double click C1 Age.
- **c.** Click **[OK].**

Correlations: Pressure, Age

```
Pearson correlation of Pressure and Age = 0.897
P-Value = 0.015
```

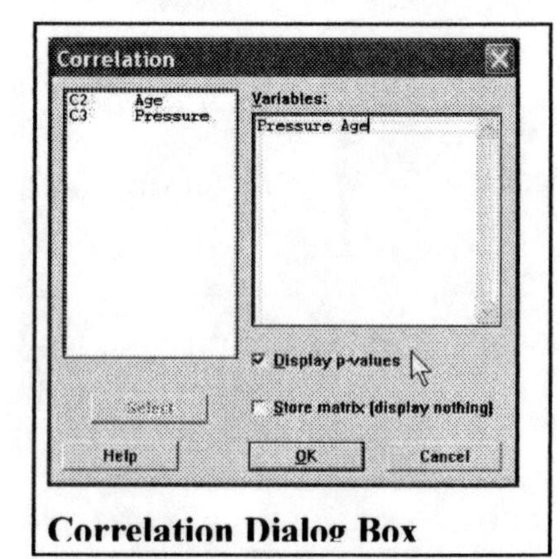

Correlation Dialog Box

Test the Significance or the Correlation Coefficient

The Session window will show r = + .897 has a P-value of 0.015. If the significance level were .05, we would reject the null hypothesis. There is a significant correlation between pressure and age.

The P-value for this test is correct for a two-tailed test.
If a one-tailed test is desired, divide this P-value by 2.

10-4 Regression Analysis

Determine the Equation of the Regression Line

Step 7. Select **Stat>Regression>Regression.**

- **a.** Double click Pressure in the list to select it for the Response: variable.
- **b.** Double click Age to select it as the only Predictors variable.
- **c.** Click on **[Storage].**

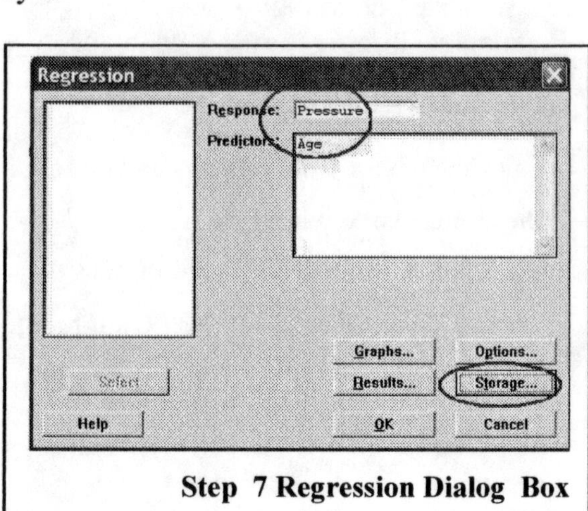

Step 7 Regression Dialog Box

1. Check the box for **Residuals**.
2. Check the box for **Fits**.
3. Click **[OK]** twice.

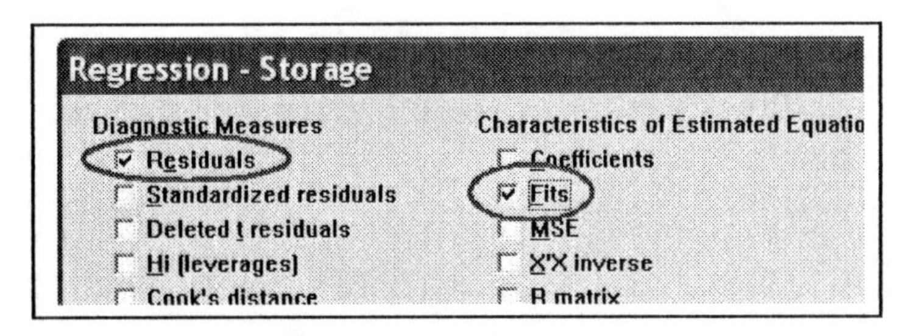

The regression equation, t- statistics, and an Analysis of Variance (ANOVA) table will be presented in the Session window. These will all be printed in step 10 then shown on the following page.

Step 8. Select **Stat>Basic Statistics>Display Descriptive Statistics.**

Choose both Pressure and Age as variables, then click **[OK].**

Step 9. Select **Data>Display.**

a. Drag the mouse over all five columns of data.

b. Click **[Select]**, then **[OK]**. The contents of the worksheet will be displayed in the session window. This is only recommended for small data sets.

c. Click the disk icon to save the project as **Chapter10.MPJ.**

Create and Print a Report

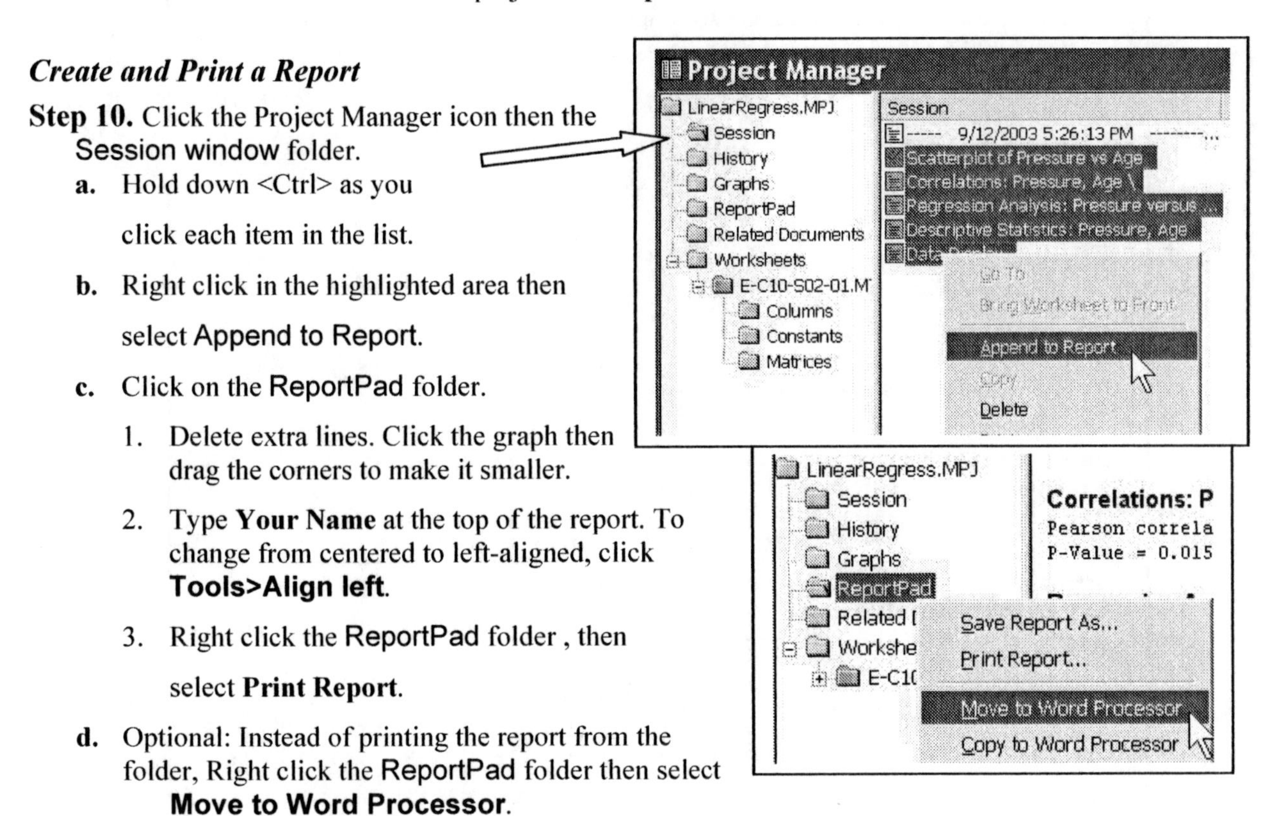

Step 10. Click the Project Manager icon then the Session window folder.

a. Hold down <Ctrl> as you click each item in the list.

b. Right click in the highlighted area then select Append to Report.

c. Click on the ReportPad folder.

1. Delete extra lines. Click the graph then drag the corners to make it smaller.
2. Type **Your Name** at the top of the report. To change from centered to left-aligned, click **Tools>Align left**.
3. Right click the ReportPad folder , then select **Print Report**.

d. Optional: Instead of printing the report from the folder, Right click the ReportPad folder then select **Move to Word Processor**.

1. You will be prompted for a name such as LinearRegress.
2. The contents will be transferred to your word processing program in rich text format then erased from ReportPad. Use your word processor to edit and print the document.

The report is shown on the next page.

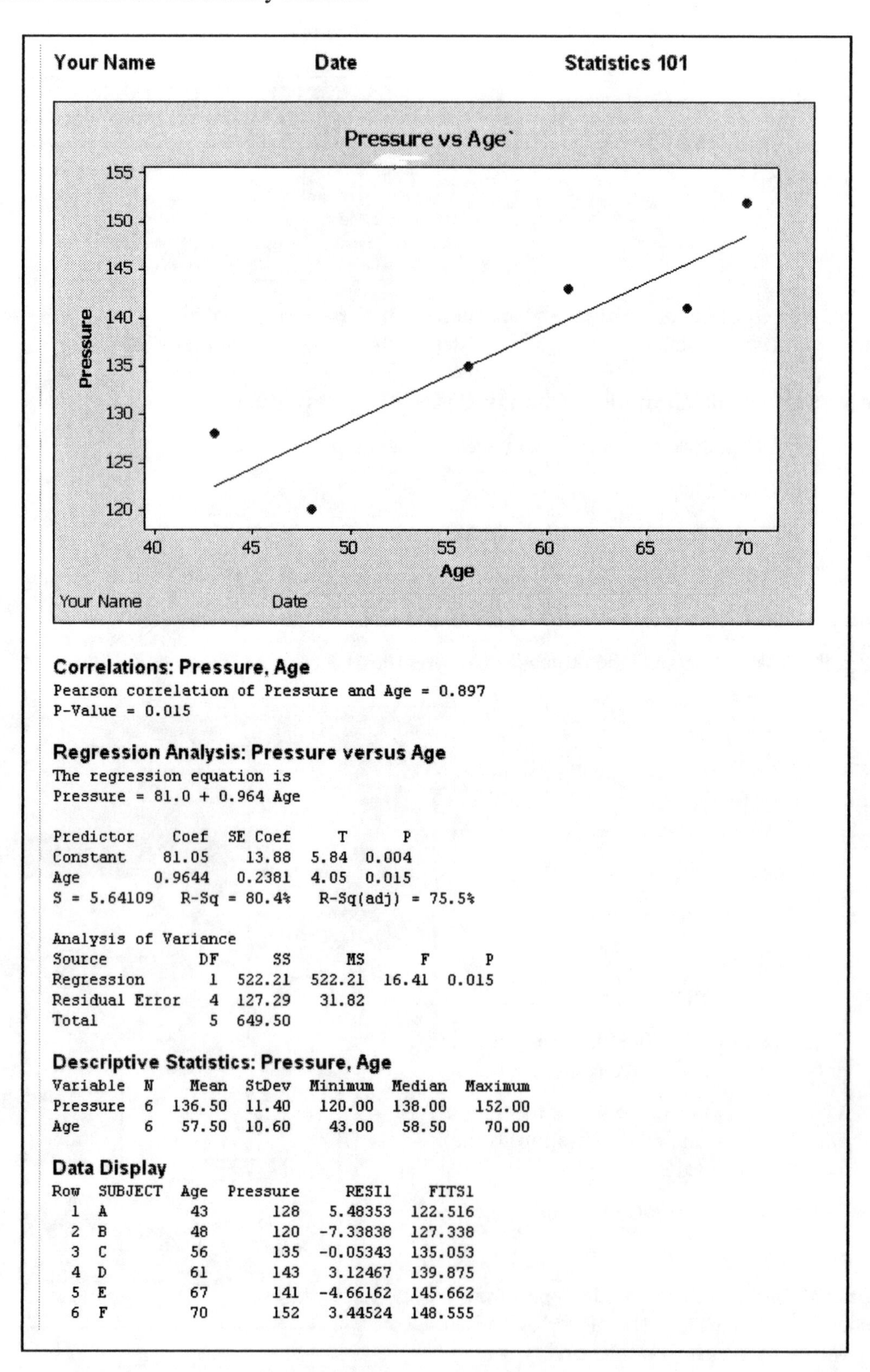

Correlations: Pressure, Age

```
Pearson correlation of Pressure and Age = 0.897
P-Value = 0.015
```

Regression Analysis: Pressure versus Age

```
The regression equation is
Pressure = 81.0 + 0.964 Age

Predictor     Coef  SE Coef     T      P
Constant     81.05    13.88  5.84  0.004
Age         0.9644   0.2381  4.05  0.015
S = 5.64109   R-Sq = 80.4%   R-Sq(adj) = 75.5%

Analysis of Variance
Source          DF      SS      MS      F      P
Regression       1  522.21  522.21  16.41  0.015
Residual Error   4  127.29   31.82
Total            5  649.50
```

Descriptive Statistics: Pressure, Age

```
Variable  N    Mean  StDev  Minimum  Median  Maximum
Pressure  6  136.50  11.40   120.00  138.00   152.00
Age       6   57.50  10.60    43.00   58.50    70.00
```

Data Display

```
Row  SUBJECT  Age  Pressure      RESI1     FITS1
  1  A         43       128    5.48353   122.516
  2  B         48       120   -7.33838   127.338
  3  C         56       135   -0.05343   135.053
  4  D         61       143    3.12467   139.875
  5  E         67       141   -4.66162   145.662
  6  F         70       152    3.44524   148.555
```

Step 11. Select **File>New>Minitab Project** if continuing or **File>Exit** to quit. Save the project as **Chapter10.MPJ.** Keep going. On the next page there is a worksheet to fill out for each simple linear regression analysis.

Completed Worksheet Here is the completed worksheet for the previous example. The location of the answers are circled and numbered in the report on the next page.

Simple Correlation and Regression Worksheet

Find the requested information in the computer output and write in the result on this worksheet.

1) r = +0.897 Calculate the correlation coefficient.

2) Describe the pattern you see in the graph. Does the correlation coefficient verify this pattern?

Strong, positive linear correlation. Yes since r is close to +1.

3) t = +4.05 What is the value of the test statistic?

4) Test the significance of the correlation coefficient. Show the steps of the hypothesis test.

Use α = 0 .05. H_0: $\rho = 0$ test statistic $t = 4.05$
H_1: $\rho \neq 0$ P-value $= .015$
Reject H_0, there is significant correlation.

5) Write out the equation of the regression line.

Pressure = 81.0 + .964 Age

6) Use the regression equation to predict a value of y for a given value of X.

(70 , 148.6) Plot this point on the graph. See scatterplot.

7) Determine the point that represents the mean of X and the mean of Y.

(57.5 , 136.5) Plot this point on the scatterplot. See scatterplot.

8) m = 0.9644 What is the value of the slope coefficient?

9) Interpret the slope for this data.

Pressure increases by .964 for every year in age.

10) 80.4% What is the value of the coefficient of determination?

11) 19.6% What is the value of the coefficient of non-determination?

12) 5.641 What is the value of the standard error of the estimate?

13) 649.50 What is the value of the total variation from the mean? (sum of squares total)

14) 127.29 What is the value of the sum of the squared residuals?

15) 522.21 What is the value of the sum of the squares regression?

16) .804018 Divide the sum of squares regression by the sum of squares total. $\frac{522.21}{649.50}$

$= R^2$

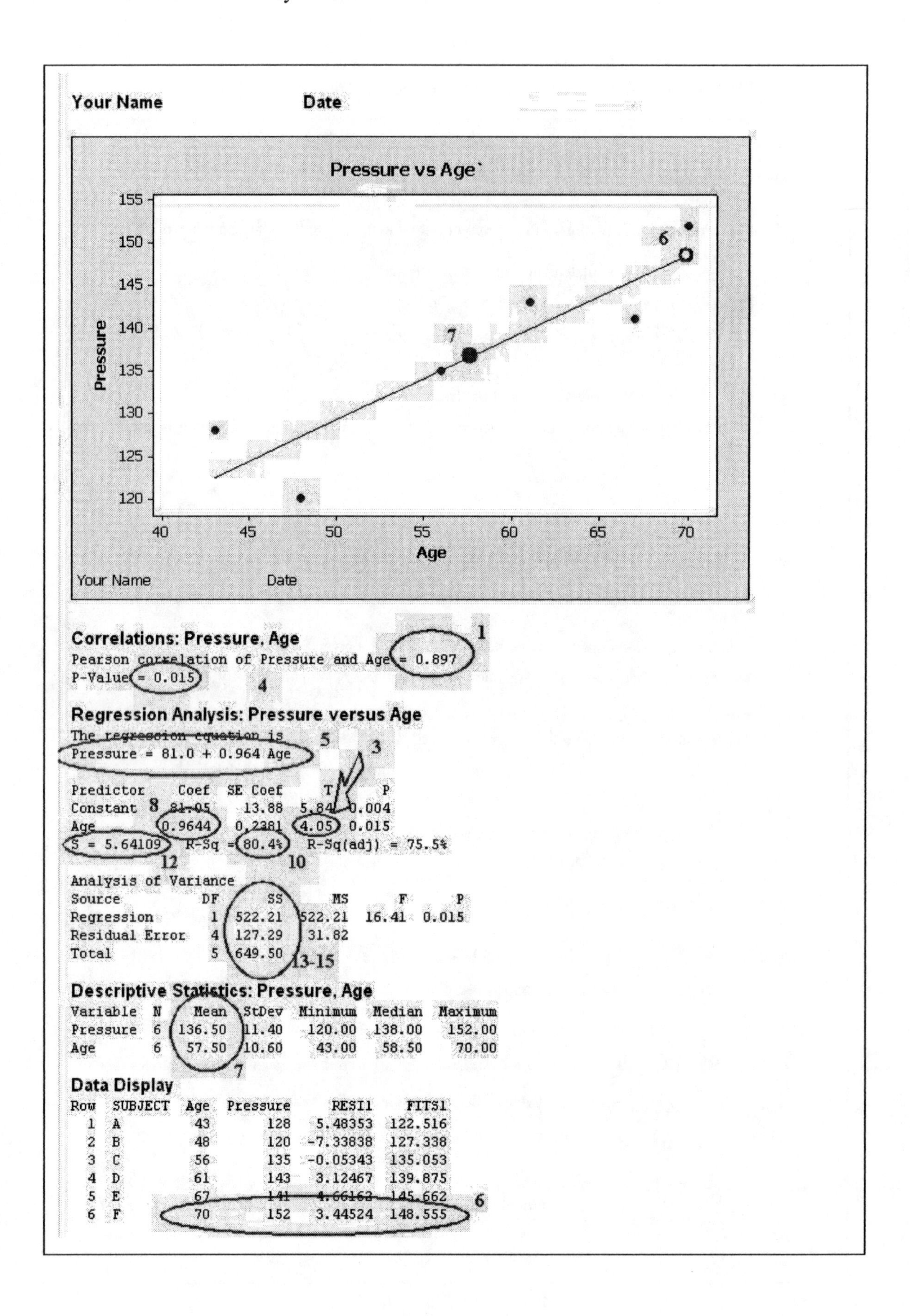

Your Name **Date**

Correlations: Pressure, Age

```
Pearson correlation of Pressure and Age = 0.897
P-Value = 0.015
```

Regression Analysis: Pressure versus Age

```
The regression equation is
Pressure = 81.0 + 0.964 Age

Predictor     Coef  SE Coef     T      P
Constant     81.05    13.88  5.84  0.004
Age         0.9644   0.2381  4.05  0.015

S = 5.64109   R-Sq = 80.4%   R-Sq(adj) = 75.5%

Analysis of Variance
Source          DF      SS      MS      F      P
Regression       1  522.21  522.21  16.41  0.015
Residual Error   4  127.29   31.82
Total            5  649.50
```

Descriptive Statistics: Pressure, Age

```
Variable  N    Mean  StDev  Minimum  Median  Maximum
Pressure  6  136.50  11.40   120.00  138.00   152.00
Age       6   57.50  10.60    43.00   58.50    70.00
```

Data Display

```
Row  SUBJECT  Age  Pressure      RESI1    FITS1
  1  A         43       128    5.48353  122.516
  2  B         48       120   -7.33838  127.338
  3  C         56       135   -0.05343  135.053
  4  D         61       143    3.12467  139.875
  5  E         67       141   -4.66162  145.662
  6  F         70       152    3.44524  148.555
```

Annotations: 1, 3, 4, 5, 6, 7, 8, 10, 12, 13-15

Simple Correlation and Regression Worksheet

Find the requested information in the computer output and write in the result on this worksheet.

1) r = ________ Calculate the correlation coefficient.

2) Describe the pattern you see in the graph. Does the correlation coefficient verify this pattern?

3) t = ________ What is the value of the test statistic?

4) Test the significance of the correlation coefficient. Show the steps of the hypothesis test.

   Use $\alpha = 0.05$.

5) Write out the equation of the regression line.

6) Use the regression equation to predict a value of Y for a given value of X.

   (_______ , _______) Plot this point on the graph.

7) Determine the point that represents the mean of X and the mean of Y.

   (_______ , _______) Plot this point on the scatterplot.

8) m = ________ What is the value of the slope coefficient?

9) Interpret the slope for this data.

10) _____________ What is the value of the coefficient of determination?

11) _____________ What is the value of the coefficient of nondetermination?

12) _____________ What is the value of the standard error of the estimate?

13) SST _________ What is the value of the total variation from the mean?

14) SSE _________ What is the value of the sum of the squared residuals?

15) SSR_________ What is the value of the sum of the squares regression?

16) _____________ Divide the sum of squares regression by the sum of squares total.

10-5 Data Analysis

From the Datank file, choose two variables that might be related, such as serum-cholesterol and weight. Select a random sample of thirty-five.

Step 1. Open the file using **File>Open Worksheet.**

Step 2. Select

Calc>Random Data>Sample From Columns.

- **a.** Type in **35** for the number of rows.
- **b.** Double click the two columns for C7 Serum-Chol and C6 Weight.
- **c.** Store the samples in Chol and Wt.
- **d.** Click **[OK].**

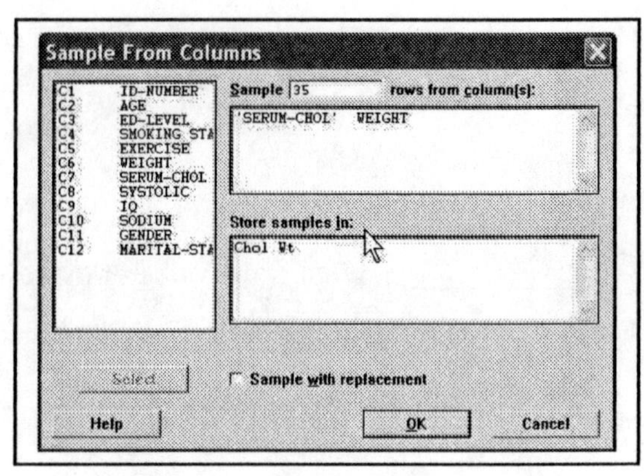

Thirty-five rows of data will be randomly selected and stored in the two new columns.
Your sample will be different from the example shown here. Chol is the Y variable.
Here is a quick summary of the commands you will need to use.

Step 3. Obtain a complete correlation and regression analysis using Chol, then Wt.

1. Make a scatterplot. **Graph>Scatterplot>Simple with Regression**
2. Calculate the correlation coefficient. **Stat>Basic Statistics>Correlation**
3. Obtain the regression equation. **Stat>Regression>Regression**
4. Obtain the descriptive statistics. **Stat>Basic Statistics>Display Descriptive Stats**
5. Display the data. **Data>Display Data.**
   Drag the mouse over Chol Wt Res1 Fits1 then click **[Select]** .
   Note: If the sample is large, don't display the data in the worksheet. Skip step 5.
6. Create and print a report. **Use the report pad to print a copy of the results**

In the end of this regression there may be a new section, Unusual Observations. MINITAB is telling us that there was one observation that was extreme. The letter R indicates this data point is quite a bit higher

```
Unusual Observations
Obs        Wt      Chol        Fit    SE Fit   Residual   St Resid
 19       156    164.00     215.60      3.78     -51.60     -2.43R
R denotes an observation with a large standardized residual
```

or lower than the regression line compared to the rest of the data. Its Y value is extreme. If there is an X instead of an R, that would indicate a data point that has undue influence in the regression because it's X value is extreme compared to most values of X. It is worth checking to make sure the data was entered correctly. This data value may have undue influence on the regression result. There can be more than one of these observations in the same data set. The output also tells you this data point is in row 19 of the worksheet. There is a significant relationship between cholesterol and weight because the P-value is .032. Cholesterol = 169 + 0.298 Weight.

Analyzing Trend in a Time Series

Use the data in exercise 15 in the review for chapter 2. Do a correlation and regression analysis, then forecast the minimum wage for 2010. Even though there are some special commands for time series data, use correlation and regression analysis.

Wage is the dependent variable, Y.
Year is the independent variable, X.

→	C1 Year	C2 MinWage
1	1960	1.00
2	1965	1.25
3	1970	1.60
4	1975	2.10
5	1980	3.10
6	1985	3.35
7	1990	3.80
8	1995	4.25
9	2000	5.15
10		

Min Wage Time Series

Step 1. Enter the data for the federal minimum hourly wage in the years as shown.

Step 2. Select **Stat>Regression>Fitted Line Plot.**

a. Double click C2 MinWage for the Response variable.

b. Year is the Predictor variable.

c. Click on **[Storage],** then check the options for Residuals and Fits.

d. Click **[OK]** twice.

This command is another way to obtain the scatterplot and regression results. It also has **[Options]** for storage and more advanced models such as quadratic regression.

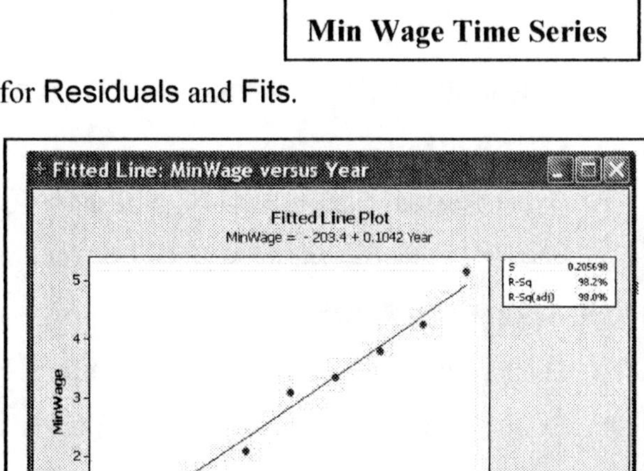

Fitted Line Plot : Minimum Wage

Step 3. Select **Stat>Basic Statistics>Correlation.**

a. Double click MinWage and then Year to select them for the variables.

b. Click **[OK].** r = +.990

The regression analysis will be displayed in the Session window and a Graph window will open with the scatterplot. The line of the regression equation will also be displayed in the Graph window. There is a strong increasing trend in this data.

```
Regression Analysis: MinWage versus Year
The regression equation is   MinWage = - 203.4 + 0.1042 Year
S = 0.205698   R-Sq = 98.2%   R-Sq(adj) = 98.0%
Analysis of Variance
Source       DF       SS        MS        F       P
Regression    1   16.2760   16.2760   384.67   0.000
Error         7    0.2962    0.0423
Total         8   16.5722
```

The regression equation is
Wage = - 203.4 + 0.1042 Year.

The slope is $ .1042. The minimum wage has increased about 10¢ a year over the past forty years.
To predict the minimum wage for 2005, substitute into the formula for Year.

Wage in 2010 ≈ -203.4 + .1042*2010 = $6.04

We would estimate that the minimum wage in 2010 will be $6.04.

Step 4. Print a report.

Step 5. Save the project as **Trend.MPJ**.

Step 6. Exit MINITAB or start a new project.

10-6 Multiple Regression (Two or More Independent Variables)

Section 6 Example 15: A nursing instructor wishes to see if a student's grade point average and age are related to the student's score on the state board examination. Score is the dependent variable, Y. Scatter diagrams cannot be made with three variables!

↓	C1-T	C2	C3	C4
	Student	GPA	Age	Score
1	A	3.2	22	550
2	B	2.7	27	570
3	C	2.5	24	525
4	D	3.4	28	670
5	E	2.2	23	490

Step 1. Enter the data for the example into three columns of MINITAB.
Name the columns **GPA, Age,** and **Score.**

Step 2. Select **Stat>Basic Statistics>Correlation.**

a. Double click on each variable starting with C4 Score, the Y variable.
b. Double click GPA then Age.
c. Click **[OK].**

In the Session window a matrix will show the correlation coefficient for each pair of variables.

Correlations: Score, GPA, Age

```
        Score    GPA
GPA     0.845
        0.072

Age     0.791  0.371
        0.111  0.539

Cell Contents: Pearson correlation
               P-Value
```

Step 3. Select **Stat>Regression>Regression.**

a. Double click on SCORE, the Response variable.
b. Double click each Predictor variable, GPA, then Age. There are two.

Step 4. Click on **[Storage],** then check the box for Residuals and the box for Fits.

Step 5. Click **[OK]** twice. The regression equation is: Score = - 44.8 + 87.6 GPA + 14.5 Age.

R-Sq(adj), ninety-six percent of the variation from the mean is explained by the correlation with GPA and Age.

To predict a score, use the age and grade point average of the individual.

Regression Analysis: Score versus GPA, Age

```
The regression equation is
Score = - 44.8 + 87.6 GPA + 14.5 Age

Predictor     Coef  SE Coef      T      P
Constant    -44.81    69.25  -0.65  0.584
GPA          87.64    15.24   5.75  0.029
Age         14.533    2.914   4.99  0.038
S = 14.0091   R-Sq = 97.9%   R-Sq(adj) = 95.7%


Analysis of Variance
Source          DF       SS      MS      F      P
Regression       2  18027.5  9013.7  45.93  0.021
Residual Error   2    392.5   196.3
Total            4  18420.0

Source  DF   Seq SS
GPA      1  13145.2
Age      1   4882.3
```

Step 6. Close MINITAB. Save this project as **MultiRegress.MPJ.**

Chapter 10 Textbook Problems Suitable for MINITAB

Page		Section	Exercises
Page	506	10.3	12 - 27
Page	516	10.4	12 - 35
Page	530	10.5	15 - 22
Page	536	10.6	none
Page	541	Review	1 - 12
Page	543	Data Analysis	1 - 3
Page	544	Data Project	1, 2

Many of the problems in this chapter use common data.

For example, the data for problem 12 is used in section 3 and again in section 4.

10.3 Page 506	10.4 Page 516	10.5 Page 530
12	12	
13	13	15, 19
14	14	16, 20
15	15	17, 21
16	16	18, 22
17	17	
18	18	
19	19	
20	20	
21	21	
22	22	
23	23	
24	24	
25	25	
26	26	
27	27	
EC	28 - 30	

Chapter 10 Endnotes:

Chapter 11 Other Chi-Square Tests

11-1 Introduction

The chi-square distribution can be used to test a hypothesis about a standard deviation or variance as we learned in Chapter 8. In this chapter we learn three more ways the chi-square distribution is used.

1--Test a hypothesis about how well a sample fits a particular distribution.
2--Test a hypothesis about a relationship between two qualitative variables.
3--Test a hypothesis about three or more proportions.

11-2 Goodness of Fit

Section 2 Example 1: A market analyst wished to see whether consumers have a preference among five flavors of a new fruit soda. A sample of 100 people provided the following data in the form of a frequency distribution. If there is no preference, equal proportions would prefer each flavor.

Calculate the Test Statistic for Goodness of Fit; Uniform distribution

C1	C2	C3	C4
Flavor	O	Px	
Cherry	32	.2	
Strawberry	28	.2	
Orange	16	.2	
Lime	14	.2	
Grape	10	.2	

There is no menu command for this test. You will type in the first three columns, then use the calculator and formulas to obtain the chi-square test statistic. These are the steps required to do the test statistic.

Step 1. Enter the **Flavors** into **C1** and observed counts into **C2**. Name the columns **Flavor** and **O.**

Step 2. Mentally calculate the proportion that would be in each category, $\frac{1}{k}=\frac{1}{5}=.2$ then enter that value in each row of **C3 Px.** Your table should now match the one above.

In the next steps we will calculate n, k and df then the expected counts, chi-square, and the P-value. Here goes!

Step 3. Select **Calc>Calculator**.

a. Type **K1** for the storage variable then type the formula **SUM('O')** and click **[OK].**

b. Edit the last dialog box.

c. Type **K2** for the variable, **N('O')** for the formula, and click **[OK].**

The N in this formula must be a capital letter, N not n.

d. Edit the last dialog box.
Type **K3** for the variable, **K2 -1** for the formula, and click **[OK].**

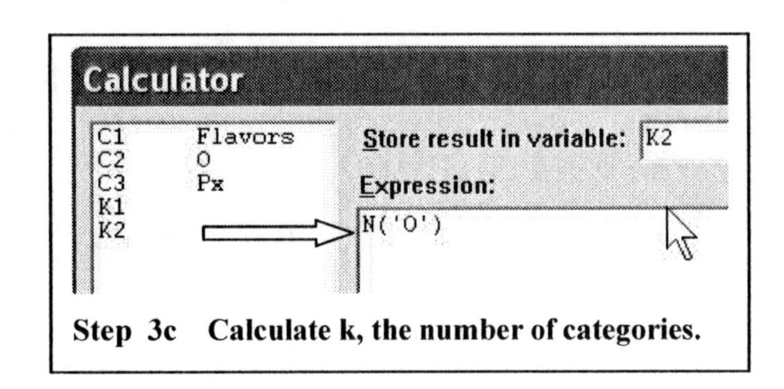

Step 3c Calculate k, the number of categories.

Step 4. Click the Project Manager Window icon.

Step 5. In the constants folder of the Project Manager window rename the constants, n, k, and df. Right click each constant then click Rename and type the name of the variable.

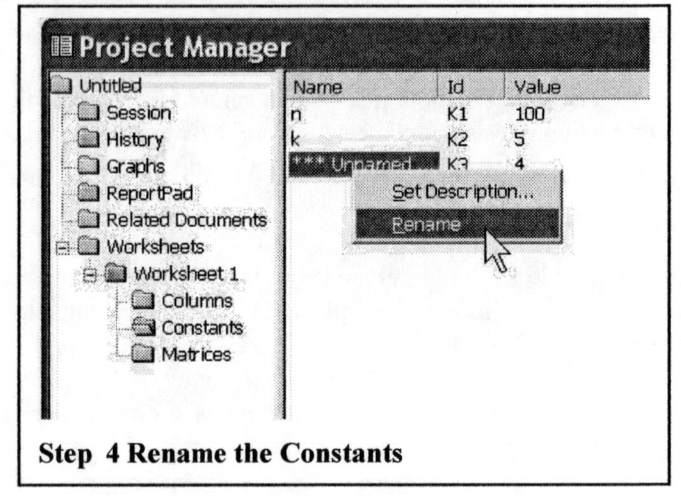

Step 4 Rename the Constants

Step 6. Calculate the expected counts in column 4. Select **Calc>Calculator.**

a. Type **E** for storage.
b. In the Expression: box type the formula **n*Px,** then click **[OK].**

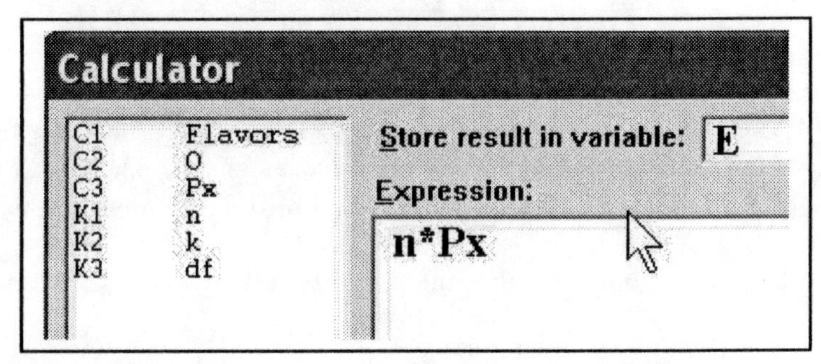

Calculate χ^2

Step 7. Calculate the chi-square test statistic using the formula. Select **Calc>Calculator.**

a. Type **K4** in the box for Store result in variable:.
b. In the Expression: box enter the formula, then click **[OK].**

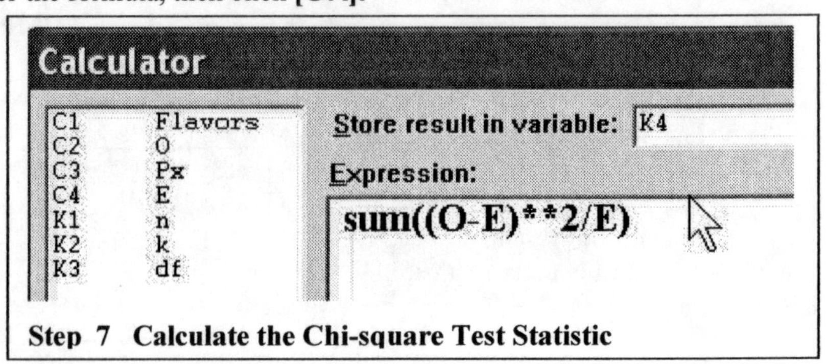

Step 7 Calculate the Chi-square Test Statistic

The chi-square test statistic, 18, will be stored in K4, a constant.
The P-value is not displayed. It must also be calculated.

Calculate the P-value for χ^2

This test is always right-tailed.

Step 8. Select

Calc>Calculator>Probability Distributions>Chi-Square.

Probability density
Cumulative probability
Noncentrality parameter: 0.0
Inverse cumulative probability
Noncentrality parameter: 0.0
Degrees of freedom: df
Input column:
Optional storage:
Input constant: K4
Optional storage: k5
OK

a. Click the option for Cumulative probability:.

Press <Tab>.

b. Type **K3** or **df**.

c. Type **K4** for the Input constant: and **K5** for Optional storage:.

d. Click **[OK]**.

Determine the probability in the right tail.

e. Select **Calc>Calculator,** then type **K6** for storage and **1– k5** for the formula.

f. Click **[OK]** once more.

g. In the Constants folder, rename K4 and K6 **chi-square** and **P-value.**

Step 9. Select **Data>Display Data**.

a. Drag the mouse over the columns and constants to highlight them.

b. Click **[Select]** then **[OK]**.

Data Display

```
n               100.000
k               5.00000
df              4.00000
chi-square      18.0000
K5              0.998766
P-value         0.00123410

Row  Flavors      O   Px    E
  1  Cherry      32   0.2  20
  2  Strawberry  28   0.2  20
  3  Orange      16   0.2  20
  4  Lime        14   0.2  20
  5  Grape       10   0.2  20
```

The results will be displayed in the Session window.

Step 10. The P-value is 0.00123410 rounded to 0.001.

Step 11. Reject the null hypothesis.

The proportion of each flavor is not 20 percent.

The complete hypothesis test is shown.

H_0: Consumers show no preference for flavors. (the population distribution is uniform)

H_1: Consumers show a preference. (the distribution is not uniform)

α = .05

From MINITAB:

Test statistic $\chi^2 = 18$ df = k-1= 4

P-value = .001

Conclusion: Reject H_0

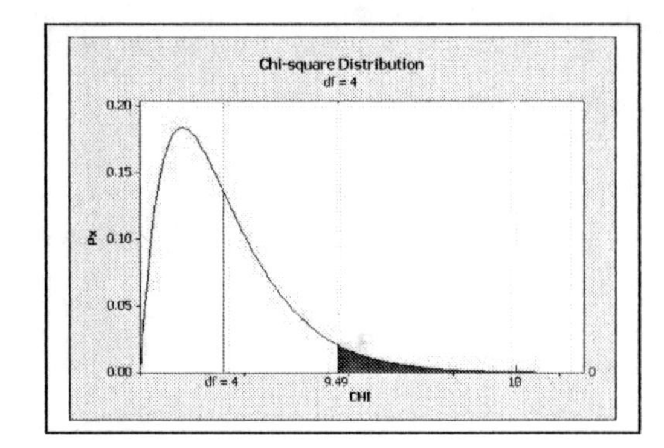

Summarize: There is a difference in the proportion of consumers who prefer each flavor. The test does not tell us which are different, only that they are not the same. The graph is not part of the output but is included for reference.

Steps 3 through 9 have been stored in goodness.MTB, a macro file for future use. The next example will use the same process after the worksheet has been set up. The first three columns must contain the labels, the observed frequencies and the probability for each category.

Calculate the Test Statistic for Goodness of Fit; Non-uniform Distribution

Section 2 Example 3: The advisor of an ecology club at a large university believes that the club consists of 10 percent freshmen, 20 percent sophomores, 40 percent juniors, and 30 percent seniors. The membership this year consisted of 14, 19, 51 and 16 respectively. At $\alpha = .10$, is the distribution of students the same as hypothesized?

Step 1. Select File>New>Minitab Worksheet.

- **a.** Type the labels for each category into C1. Name the column **Status.**
- **b.** Enter the observed frequencies in C2. Name the column **O.**
- **c.** Type the percentages in decimal form into C3. Name the column **Px.**

The worksheet with data entered is shown.

↓	C1-T	C2	C3
	Status	O	Px
1	Fr	14	0.1
2	So	19	0.2
3	Jr	51	0.4
4	Sr	16	0.3
5			

Step 1 Data entry

Calculate the Chi-square test statistic

Complete steps 3 through 9 in the previous example OR

Step 2. Execute the macro file. Select **File>Other Files> Run an exec...,** then choose goodness.MTB.
The test statistic is 11.2

Step 3. The P-value is .011.

Step 4. Reject the null hypothesis.

Step 5. The proportions are significantly different.

Data Display

```
n               100.000
k               4.00000
df              3.00000
chi-square      11.2083
P-value         0.0106511

Row   Status    O    Px    E
  1   Fr       14   0.1   10
  2   So       19   0.2   20
  3   Jr       51   0.4   40
  4   Sr       16   0.3   30
```

Step 6. Select **File>Save Project** or click the disk icon to save the project as **Chapter11.MPJ.**

11-3 Tests Using Contingency Tables

The test of independence of variables is used to determine whether two characteristics are related. The test of homogeneity of proportions determines if three or more proportions are equal. H_o: $p_1 = p_2 = p_3 = \ldots p_k$. Both tests use the chi-square distribution and a contingency table.

Example 11-5: A sociologist wishes to see whether the number of years of college a person has completed is related to his/her place of residence. A sample of 88 people is selected and a contingency table is completed as follows. Each cell contains an observed frequency excluding totals. At $\alpha = .05$ can the sociologist conclude that a person's location is independent of degree?

	C1	C2	C3
Location	**NoCollege**	**Bachelor**	**Graduate**
Urban	**15**	**12**	**8**
Suburban	**8**	**15**	**9**
Rural	**6**	**8**	**7**

Step 1. State the hypotheses. This step is not done with MINITAB.

H_o: Place of residence is unrelated to number of years of college completed.

H_1: Place of residence is dependent on the number of years of college completed.

Calculate the Chi-square Test Statistic from a Contingency Table

Step 2. Select **File>New>Minitab Worksheet** to open a fresh worksheet.

a. Enter the three columns of data with the observed frequencies. They are highlighted in the table above. Row labels and totals are NOT entered. Name the columns **NoCollege, Bachelor,** and **Graduate.**

b. Select **Stat>Tables>Chi-square test. (Table in Worksheet).**

c. Drag the mouse over the three columns to highlight them and click **[Select],** then **[OK].**

The Session window will contain the table with row totals, column totals, and expected counts. At the end of the table the calculation of chi-square for each cell is shown and the test statistic.

d. The test statistic is 3.006.

Step 3. The P-value is .557.

Step 4. Can't reject H_o.

Step 5. There isn't enough evidence in the sample to conclude there is a relationship between location of residence and years of education.

Chi-Square Test: NoCollege, Bachelor Degree, Graduate Degree

```
Expected counts are printed below observed counts
Chi-Square contributions are printed below
expected counts
                         Bachelor  Graduate
           NoCollege       Degree    Degree  Total
    1             15           12         8     35
               11.53        13.92      9.55
               1.041        0.265     0.250

    2              8           15         9     32
               10.55        12.73      8.73
               0.614        0.406     0.009

    3              6            8         7     21
                6.92         8.35      5.73
               0.122        0.015     0.283

Total             29           35        24     88
Chi-Sq = 3.006, DF = 4, P-Value = 0.557
```

This MINITAB command does exactly the same process required to calculate the statistic by hand. The rows and columns are totaled. The expected counts are calculated with a formula:

$$E = \frac{row\ total * column\ total}{Total}$$

Then the chi-square test statistic is calculated using this formula:

$$\chi^2 = \sum \frac{(O-E)^2}{E}$$

The columns must contain the observed frequencies. If the columns contain a file of data, a different command is required.

Calculate the Chi-square Test Statistic from Data

Test the claim that smoking is independent of gender, that is, women are just as likely to smoke as men. Use $\alpha = .05$. The Databank file contains a column of data for gender with an F for female and an M for male. Another column contains the smoking status using three levels of smoking:

0 doesn't smoke

1 smokes less than one pack a day

2 smokes one pack or more a day.

Step 1. State the hypotheses.

Ho: Smoking status is independent of gender.

H_1: Smoking is dependent on gender.

Calculate the test statistic and P-value. It is possible to make the contingency table and calculate the chi-square test statistic all with the same command.

Step 2. Calculate the test statistic and P-value.

a. Select **File>Open Worksheet.** Find the filename in the list and double click it.

b. Select **Stat>Tables>Cross Tabulation.**

1. For rows: double click the column for 'SMOKING STATUS.'

2. Press <Tab>, then For columns: double click the column for GENDER.

3. Click **[Chi-square].**

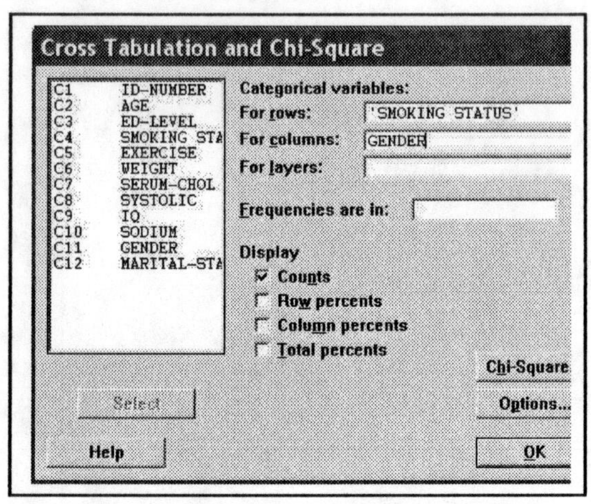

4. Check the buttons for **Chi-Square analysis** and for **Expected cell counts.**

This is the step that makes MINITAB calculate the test statistic and P-value. The observed counts will be determined and displayed above the expected counts.

Cross Tabulation - Chi-Square

Display
- [x] Chi-Square analysis
- [x] Expected cell counts
- [ ] Raw residuals
- [ ] Standardized residuals
- [ ] Adjusted residuals
- [ ] Each cell's contribution to the Chi-Square statistic

Help | OK | Cancel

5. Click **[OK]** twice.

Tabulated statistics: SMOKING STATUS, GENDER

```
Rows: SMOKING STATUS   Columns: GENDER
             F       M      All

0           25      22       47
         23.50   23.50    47.00

1           18      19       37
         18.50   18.50    37.00

2            7       9       16
          8.00    8.00    16.00

All         50      50      100
         50.00   50.00   100.00

Cell Contents:      Count
                    Expected count
Pearson Chi-Square = 0.469, DF = 2, P-Value = 0.791
Likelihood Ratio Chi-Square = 0.469, DF = 2, P-Value = 0.791
```

c. The chi-square test statistic is .469.

Step 3. The P-value is.791.

Step 4. Do not reject the null hypothesis.

Step 5. Summary. There is not enough evidence to conclude that smoking level is dependent on gender. Men and women are just as likely to smoke or not smoke. If the null hypothesis is rejected, then row or column percents would be calculated (see chapter 4). These conditional probabilities would help explain who is more likely to smoke or not smoke.

Step 6. Select **File>Save Project** to save the project as **Chapter11.MPJ.**

Step 7. Select **File>New>Minitab Worksheet,** then continue.

Another Example: Section 3 Exercise 14

An instructor wishes to see if the way people obtain their current event news is dependent on their educational background. A survey of 400 high school and college graduates yielded the following contingency table. At $\alpha = .05$, test the claim.

	C1	C2	C3
	Television	**Newspapers**	**Other sources**
High School	159	90	51
College	27	42	31

Show the five steps of the hypothesis test.

Step 1. State the hypotheses:

H_o: Source of news is independent of education.

H_1: Source of news is related to level of education.

Step 2. Calculate the test statistic.

- **a.** Enter the three columns of observed frequencies into the new worksheet. Do not enter row labels or totals, only the counts.
- **b.** Select **Stat>Tables>Chi-Square Test (Table in Worksheet).** This is the correct menu selection since the observed counts are entered in the worksheet.
- **c.** Double click each column or drag the mouse over the names of the three columns in the list to highlight the three columns, then click **[Select]** and **[OK].**
- **d.** The chi-square test statistic is 21.347.

Step 3. The P-value is .000.

Step 4. Reject the null hypothesis.

Chi-Square Test: Television, Newspaper, Other

```
Expected counts are printed below observed counts
Chi-Square contributions are printed below expected
counts

            Television  Newspaper  Other  Total
    1              159         90     51    300
                139.50      99.00  61.50
                 2.726      0.818  1.793

    2               27         42     31    100
                 46.50      33.00  20.50
                 8.177      2.455  5.378

Total              186        132     82    400

Chi-Sq = 21.347, DF = 2, P-Value = 0.000
```

Step 5. There is enough evidence in the sample to conclude that source of news is related to education. The Session window is shown. How are the two variables related?

Step 6. Highlight this table then select **File>Print Worksheet.** The printer dialog box should have chosen the option for selection . When you click [OK] to print, just this table should be printed.

To describe the relationship, use conditional probabilities, either column or row percents. The dependent variable is the source of news. The independent variable is education level. Calculate percents across the (row) independent variable. Do these by hand or select **Tools>Windows Calculator** . Write them onto your printed copy along with your name. In this example divide each count was divided by the row total. Compare the percentages down each column.

Step 7. Write the results with pen or pencil on the printed copy.

	Television	**Newspapers**	**Other sources**
High School	159/300 = 53%	90/300 = 30%	51/300 = 17%
College	27/100 = 27%	42/100 = 42%	31/100 =31%
	186/400 = 47%	132/400 = 33%	82/400 = 21%

In this sample, 53 percent of the high school graduates get their news from television compared to 27 percent of the college graduates. Forty-seven percent of the sample of 400 get their news from the television. Thirty percent of the high school graduates get their information from newspapers compared to 42 percent of the college graduates. There are 17 percent of the high school graduates who get their information from other sources and 31 percent of the college graduates use other sources. Overall, 21 percent get their news from other sources.

Step 8. Save the project as **Chapter11.MPJ**. Continue.

Test for Homogeneity of Proportions

This test looks very similar to the test of independence. In both cases, the table will contain observed counts.

Section 3 Example 7: A researcher selected a sample of 150 seniors from each of three area high schools and asked each senior, "Do you drive to school in a car owned by either you or your parents?" At a significance level of .05, test the claim that the proportion of students who drive their own or their parent's car is the same at all three schools.

	C1	**C2**	**C3**	
Response	**School 1**	**School 2**	**School 3**	**Total**
Yes	18	22	16	56
No	32	28	34	94
	50	50	50	150

Step 1. State the hypotheses.

H_o: $p_1 = p_2 = p_3$

H_1: $p_1 \neq p_2 \neq p_3$ (At least one proportion is different.)

Step 2. Calculate the test statistic. Use **File>New>Minitab Worksheet** to open a fresh worksheet.

a. Enter the observed counts for the three columns. Do not include the columns or rows of totals. Do not enter the **Response** column.

b. Select **Stat>Tables>Chi-Square Test.**

c. Drag the mouse over the names of the three columns in the list to highlight the three columns, then click **[Select].**

d. Click **[OK].**

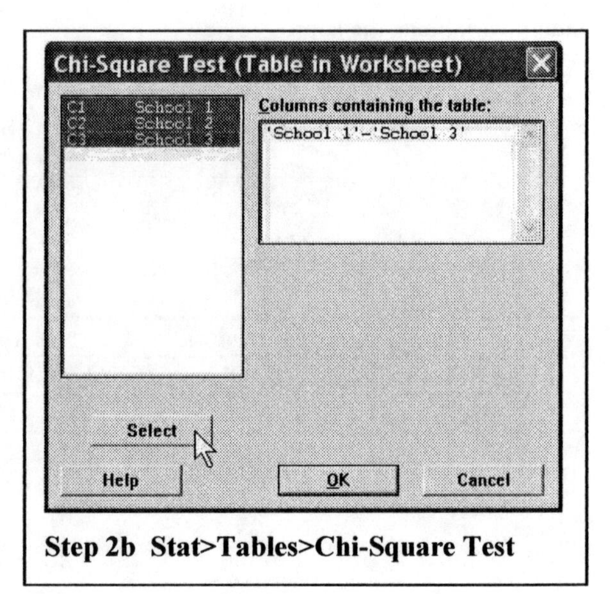

Step 2b Stat>Tables>Chi-Square Test

e. The chi-square test statistic is 1.596.

Step 3. The P-value is .450.

Step 4. Do not reject the null hypothesis.

Step 5. There is not enough evidence in the sample to conclude that the proportion of all students who drive their own or their parent's car to school is different between the three schools. The Session window is shown.

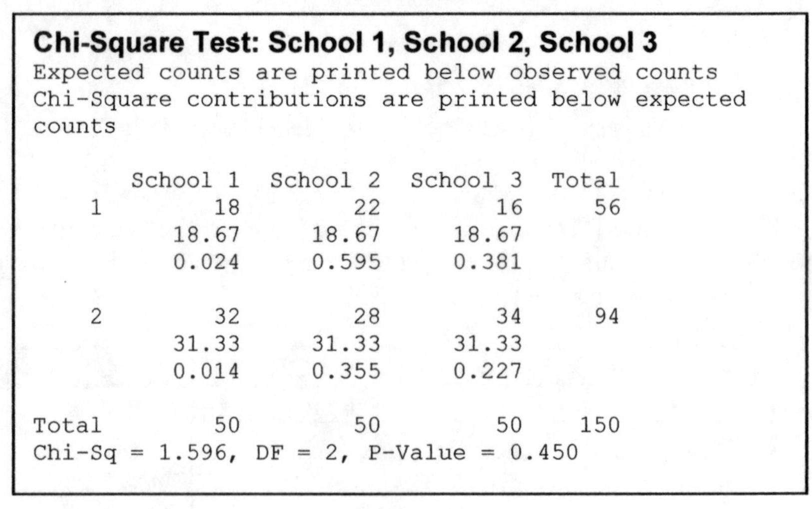

Chi-Square Test: School 1, School 2, School 3

```
Expected counts are printed below observed counts
Chi-Square contributions are printed below expected
counts

         School 1  School 2  School 3  Total
    1          18        22        16     56
            18.67     18.67     18.67
            0.024     0.595     0.381

    2          32        28        34     94
            31.33     31.33     31.33
            0.014     0.355     0.227

Total          50        50        50    150
Chi-Sq = 1.596, DF = 2, P-Value = 0.450
```

Step 6. Save the project. Print a report then Exit Minitab.

Chapter 11 Textbook Problems Suitable for MINITAB

Page		Section	Exercises
Page	556	11.2	5 - 17
Page	568	11.3	8 - 31
Page	577	Review	1 - 10
Page	579	Data Analysis	1 -5
Page	581	Data Project	1 - 3

Chapter 11: Endnotes

Chapter 12 Analysis of Variance

12-1 Introduction

Analysis of variance is used to compare three or more means. Ho: $\mu_1=\mu_2=\mu_3=..\mu_k$. A two-sample test is used when comparing two means. If the null hypothesis is rejected, there is evidence that at least one of the means is different. The F-test determines if there is a difference. The Tukey test is one test used to determine which pair(s) of means are different.

12-2 One-Way Analysis of Variance or ANOVA

Section 2 Example 1: A researcher wishes to try three different techniques to lower the blood pressure of individuals diagnosed with high blood pressure. The subjects are randomly assigned to three groups. The first group takes medication. The second group exercises and the third group diets. After four weeks, the reduction in each person's blood pressure is measured. At $\alpha = .05$, test the claim that there is no difference in the mean amount of reduction. The worksheet data is shown.

Step 1. State the hypotheses.

a. H_o: $\mu_1=\mu_2=\mu_3$

b. H_1: $\mu_1=\mu_2=\mu_3$

C1	C2	C3
Medication	Exercise	Diet
10	6	5
12	8	9
9	3	12
15	0	8
13	2	4

Calculate the Test Statistic

Step 2. Enter the data into 3 columns of a worksheet.

This data is "unstacked" since the quantitative measurement, the change in blood pressure, is separated into three columns.

a. Select **Stat>ANOVA>One-way (Unstacked).**

b. Drag the mouse over the three columns in the list then click on the **[Select]** button.

The **Responses** box shows the sequence Medication *through* Diet. A hyphen between column names indicates a sequence rather than a list. It is the same as the list Medication Exercise Diet. A space or a comma can be used to separate the items in the list if necessary.

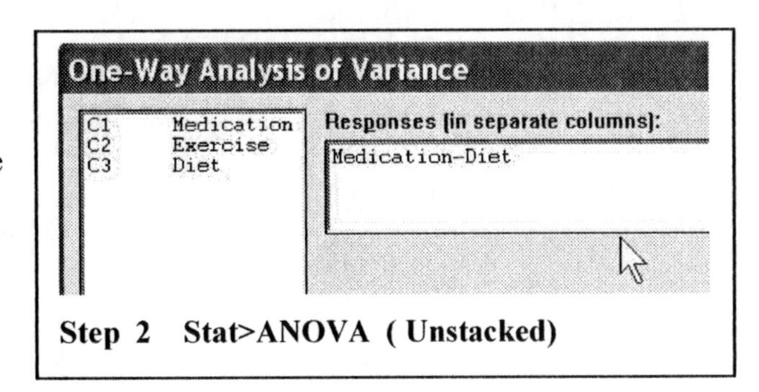

Step 2 Stat>ANOVA (Unstacked)

c. Click **[OK]**.

d. The test statistic, F, is equal to 9.17.

Step 3. The P-value is .004.

Step 4. Reject the null hypothesis.

Step 5. There is a difference in the means. The ANOVA table with the test statistic and P-value are displayed in the session window. The means for each group are also shown. The mean change with medication is 11.8 milligrams of Mercury. The mean difference for Diet is 7.6.

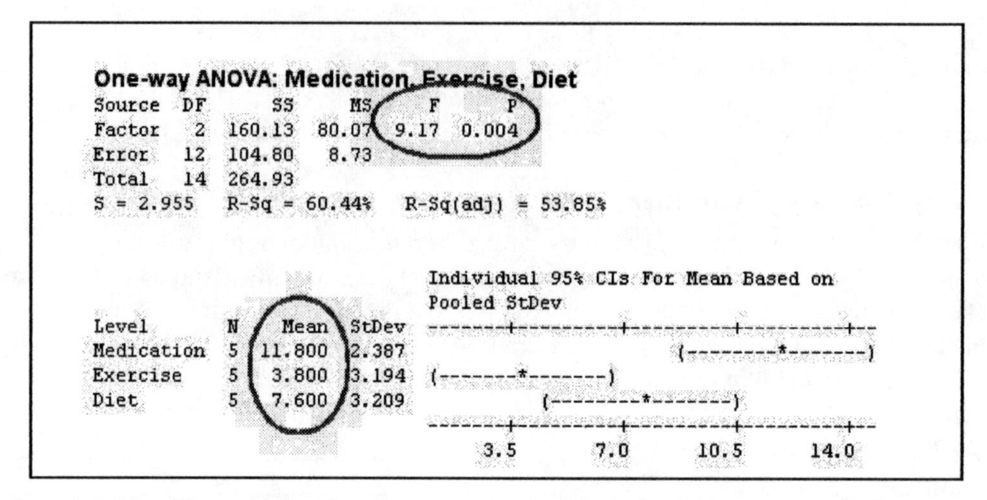

One-way ANOVA: Medication, Exercise, Diet

Source	DF	SS	MS	F	P
Factor	2	160.13	80.07	9.17	0.004
Error	12	104.80	8.73		
Total	14	264.93			

S = 2.955   R-Sq = 60.44%   R-Sq(adj) = 53.85%

```
                                Individual 95% CIs For Mean Based on
                                Pooled StDev
Level       N    Mean  StDev  -------+---------+---------+---------+--
Medication  5  11.800  2.387                          (--------*-------)
Exercise    5   3.800  3.194  (-------*--------)
Diet        5   7.600  3.209              (--------*-------)
                              -------+---------+---------+---------+--
                                   3.5       7.0      10.5      14.0
```

One procedure that identifies which pairs of means are significantly different is the Tukey test.

12-3 The Tukey Test

Section 3 Example 4 continues.

Calculate the Test Statistic for Stacked Data including the Tukey Test

Step 6. Click edit last dialog icon or select **Stat>ANOVA>Oneway (Unstacked).**

The settings in the dialog box should be the same.

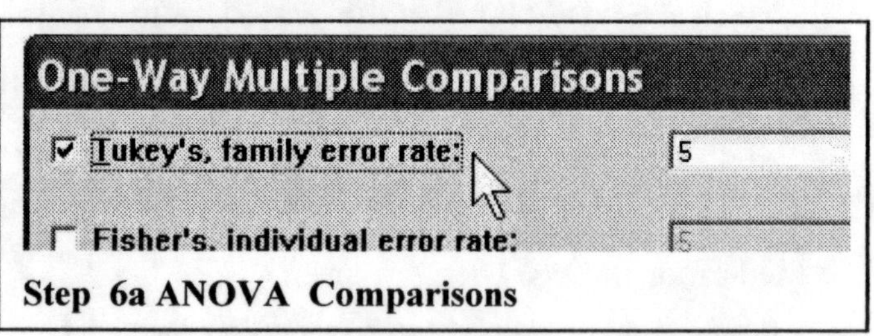

Step 6a ANOVA Comparisons

a. Click **[Comparisons].**
   1. Check the box for **Tukey.**
   2. Click **[OK]**.

b. Click [**Graphs**].
   Check the box for **Boxplots of data.**

c. Click **[OK]** twice.

A graph window will open showing the boxplots for the three groups. Lines connect the mean change for each group. There aren't outliers in any of the three data groups.

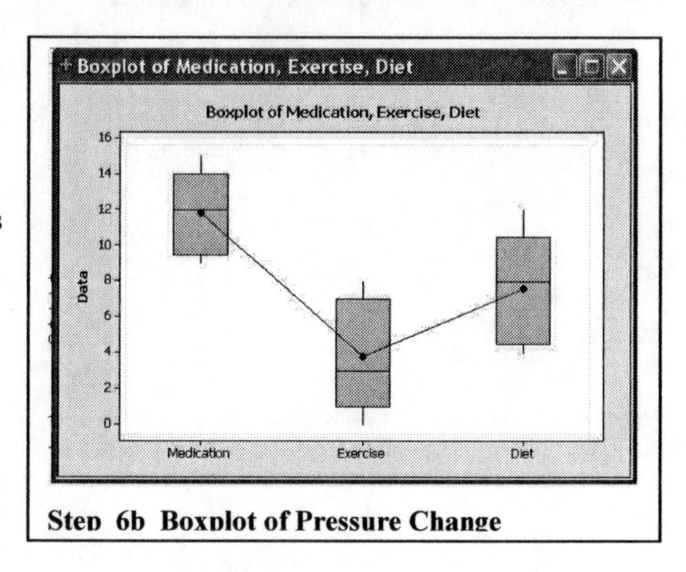

Step 6b Boxplot of Pressure Change

The Boxplot graphically shows that the data for diet and exercise "overlap" but Exercise and medication do not. The session window will contain a repeat of the ANOVA table from the previous section followed by the Tukey results. All possible pairs of treatments and the difference in the mean blood pressure reduction are displayed. There appears to be a difference in the mean reduction between medication and exercise but no significant difference in the reduction between diet and exercise or between diet and medication.

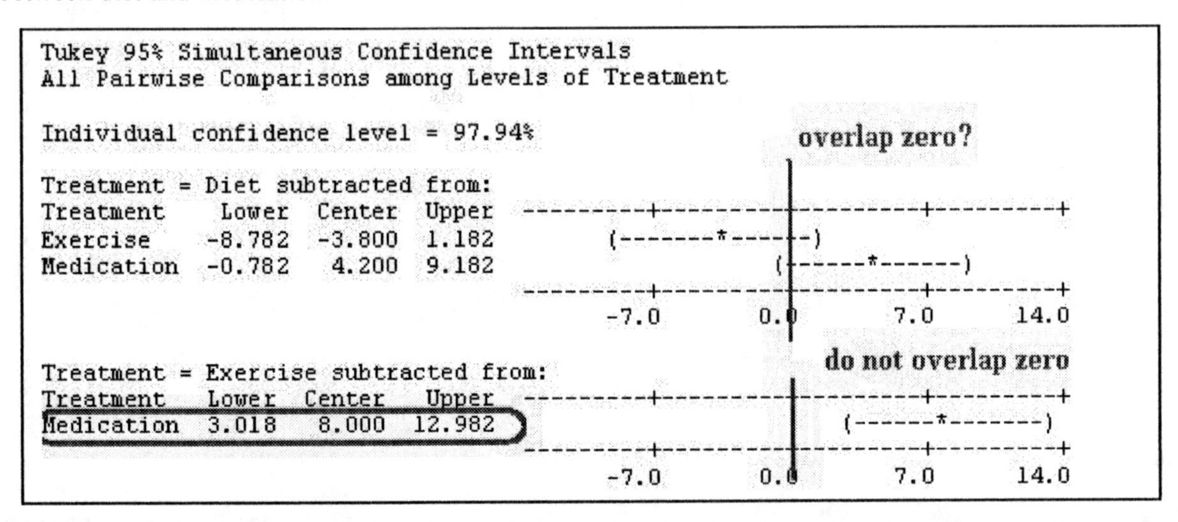

```
Tukey 95% Simultaneous Confidence Intervals
All Pairwise Comparisons among Levels of Treatment

Individual confidence level = 97.94%

Treatment = Diet subtracted from:
Treatment     Lower  Center  Upper  ---------+---------+---------+---------+
Exercise     -8.782  -3.800  1.182      (-------*-----)
Medication   -0.782   4.200  9.182               (------*------)
                                    ---------+---------+---------+---------+
                                          -7.0       0.0       7.0      14.0

Treatment = Exercise subtracted from:
Treatment    Lower  Center   Upper  ---------+---------+---------+---------+
Medication   3.018   8.000  12.982                   (------*-------)
                                    ---------+---------+---------+---------+
                                          -7.0       0.0       7.0      14.0
```


The difference in the mean reduction due to Exercise and Medication is 3 to 13 mmHg as highlogjted om the session window.

Step 7. Click the disk icon. Save the project as Chapter12.MPJ.

12-4 Two-Way Analysis of Variance

Example 12-5: A researcher wishes to see whether the type of gasoline used and the type of automobile driven have any effect on gasoline consumption. Two types of gasoline and two types of vehicles will be used in each group. There will be two automobiles in each of the 4 "blocks." The highlighted data, 26.7 is the miles per gallon for one two-wheel drive vehicle burning regular gasoline.

Gas	Two-wheel Drive	Four-wheel Drive
Regular	26.7 25.2	28.6 29.3
High Octane	32.3 32.8	26.1 24.2

A two-way analysis of variance will be used. The mean amount of gasoline consumption in miles per gallon (MPG) for each group will be compared.

C1 MPG	C2 GasCode	C3 TypeCode
26.7	1	1
25.2	1	1
32.3	2	1
32.8	2	1
28.6	1	2
29.3	1	2
26.1	2	2
24.2	2	2

Step 1 Data entry

Step 1. Select **File>New>Minitab Worksheet**.

Enter the data into three columns of a worksheet.

The data for this analysis has to be "stacked", one column for the quantitative variable and one column for each qualitative variable.

a. Type all of the gas mileage data in a single column named MPG.

b. A second column should contain codes identifying the Gasoline group, 1 for regular or 2 for high-octane. Categorical data could be used, i.e., the names of the types such as Regular and High-Octane.

c. The third column will contain codes identifying the type of automobile (1 for Two-Wheel or 2 for Four-wheel drive). All eight rows of data are shown.

In order for the Two-way ANOVA to work, the data must be entered in this format!

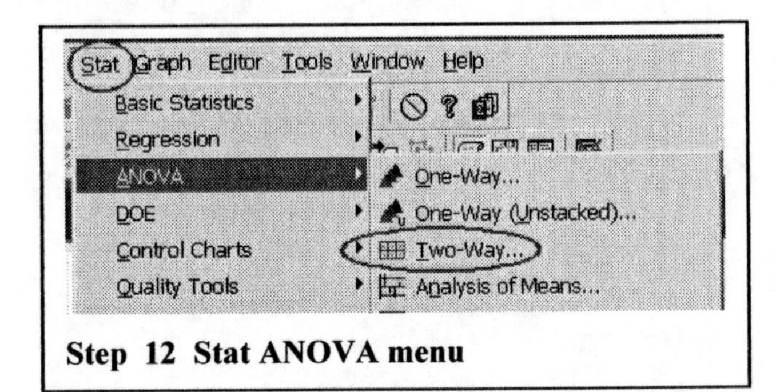

Step 12 Stat ANOVA menu

Step 2. Select **Stat>ANOVA>Two-way.**

a. Double click MPG in the list box for **Response**.

b. Double click GasCode for the **Row factor**.

c. Double click TypeCode for the **Column factor**.

d. Check the boxes for **display means**.

e. Click **[OK]**.

The completed dialog box is shown.
The session window displays the results.
Step 3. Find the P-values.

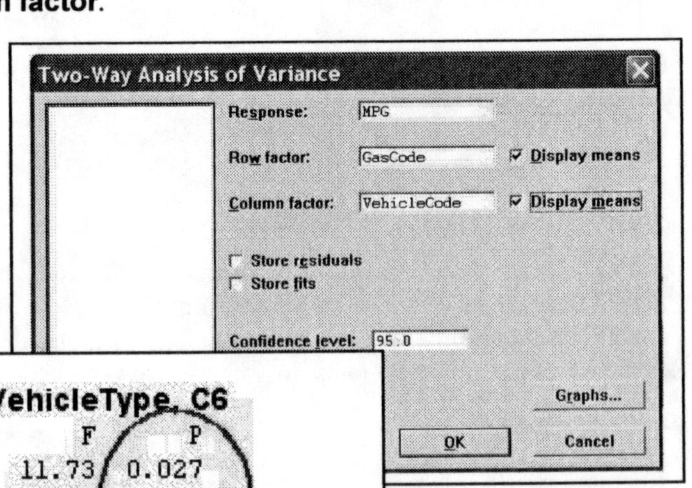

Two-way ANOVA: MPG versus VehicleType, C6

Source	DF	SS	MS	F	P
VehicleType	1	9.68	9.680	11.73	0.027
C6	1	3.92	3.920	4.75	0.095
Interaction	1	54.08	54.080	65.55	0.001
Error	4	3.30	0.825		
Total	7	70.98			

S = 0.9083   R-Sq = 95.35%   R-Sq(adj) = 91.86%

a. There is no significant difference in the mean MPG by gasoline type since the P-value is .095.

b. There is a significant relationship between the type of vehicle and gas consumption at .027. A difference in the mean MPG by either independent variable is called a main effect.

c. The interaction between type of vehicle and type of gas is statistically significant at .001. In other words, it appears that the MPG is affected by the type of vehicle and by the *interaction* of vehicle and gasoline types. This interaction should not be ignored.

Plot Interactions

Step 4. Select **Stat>ANOVA>Interactions Plot.**

a. Double click MPG for the **response variable** and "GasCode" and "TypeCode" for the **factors**.

b. Click **[OK]**.

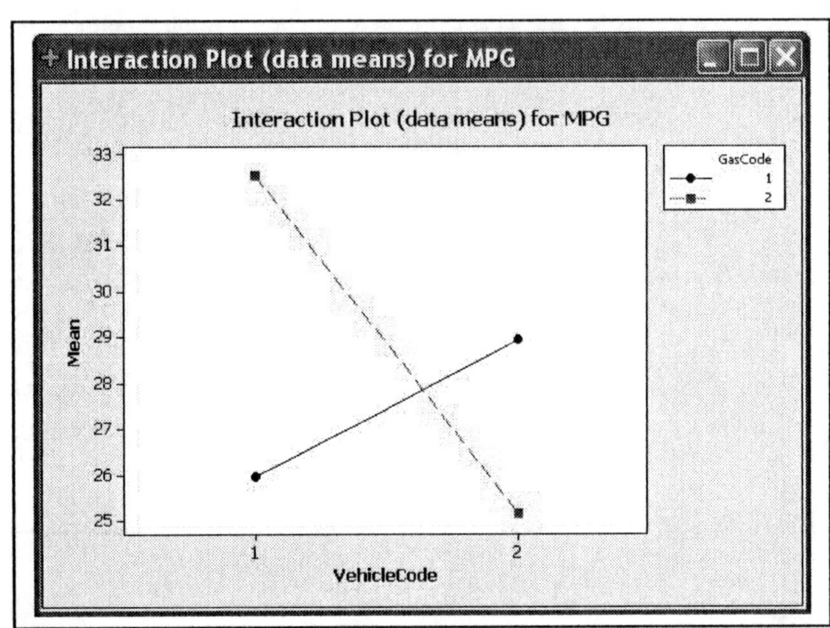

When the lines cross, there is interaction between the main effects, gasoline type and vehicle type. If the lines are parallel there is no interaction effect. If the lines neither intersect nor are parallel there may be some ordinal interaction but the main effects can be interpreted independently of each other.

To interpret the results, a graph is drawn by first calculating the mean for each main effect, graphing the mean of each group on a plot and interpreting the results. A cross-tabulation is a good way to calculate the means for each group from the raw data.

The codes for gasoline and vehicle type were entered as numeric codes. They could have been entered as text columns with the description for each type. This would make the output easier to read. Initially, they could have been entered this way. These instructions tell us how to change the existing codes to text.

Change the Codes from Numbers to Text

Step 1. Select **DATA>Code>Numeric to Text**. Press <Tab> .

a. Double click GasCode for the **from column.** Press <Tab> after each entry to move to the next box.

b. Type in **Gas** for the name of the new column, the Into column.

c. Type in the **1** for the first original value then **Regular** for the New code.

d. Type in the **2** for the second row of original values and then **HighOctane** for the new code.

e. Click **[OK]**.

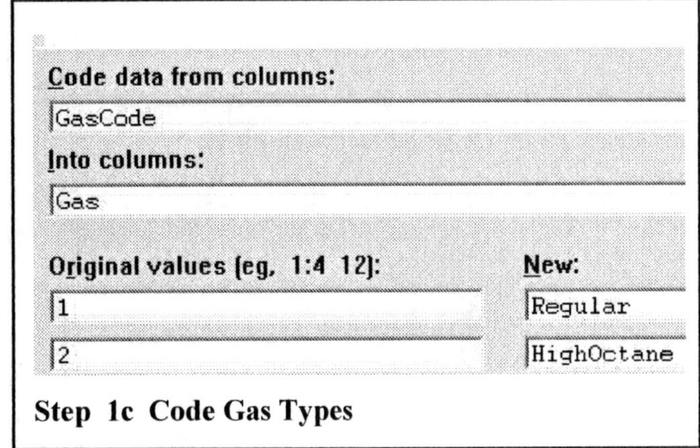

Step 1c Code Gas Types

A new column of text is created with gas categories.

Step 2. Click the **Edit Last Dialog** icon and repeat for the type of vehicle.

a. Enter **TypeCode** for the name in the from column.

b. Type in **Vehicle** for the name of the Into column.

c. Change the two codes to **TwoWheel** and **FourWheel**.

d. Click **[OK]**.

The worksheet will now contain 5 columns as shown.

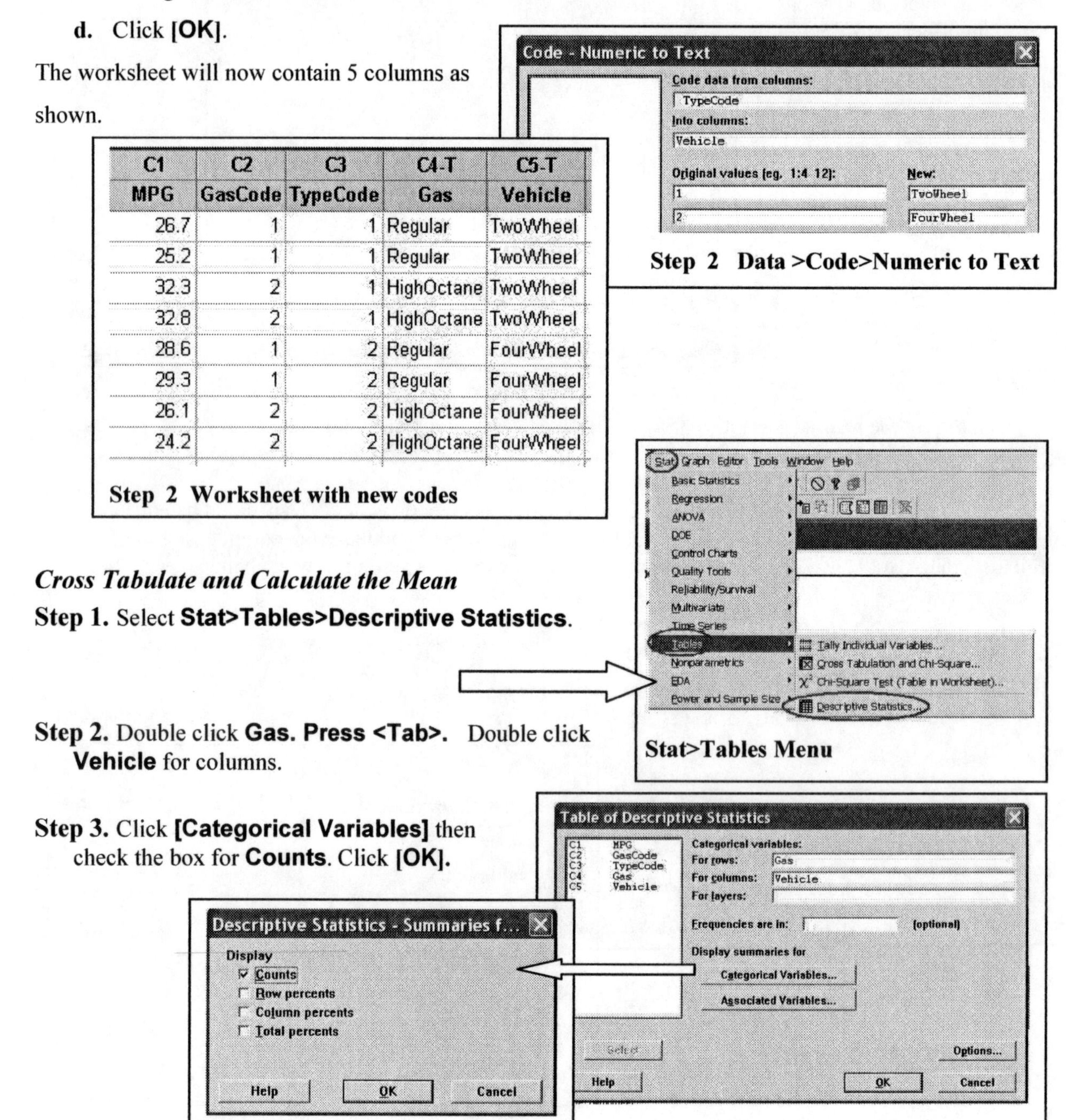

C1 MPG	C2 GasCode	C3 TypeCode	C4-T Gas	C5-T Vehicle
26.7	1	1	Regular	TwoWheel
25.2	1	1	Regular	TwoWheel
32.3	2	1	HighOctane	TwoWheel
32.8	2	1	HighOctane	TwoWheel
28.6	1	2	Regular	FourWheel
29.3	1	2	Regular	FourWheel
26.1	2	2	HighOctane	FourWheel
24.2	2	2	HighOctane	FourWheel

Step 2 Worksheet with new codes

Step 2 Data >Code>Numeric to Text

Cross Tabulate and Calculate the Mean

Step 1. Select **Stat>Tables>Descriptive Statistics**.

Stat>Tables Menu

Step 2. Double click **Gas. Press <Tab>.** Double click **Vehicle** for columns.

Step 3. Click **[Categorical Variables]** then check the box for **Counts**. Click **[OK]**.

Step 4. Click **[Associated Variables].**

a. Double click **MPG**.

This should be the column of data that you want to average. The crosstabulation will separate the data into the groups by gas and vehicle types and present the table in the session window.

b. Check the option for **Means**.

c. Click **[OK]** twice.

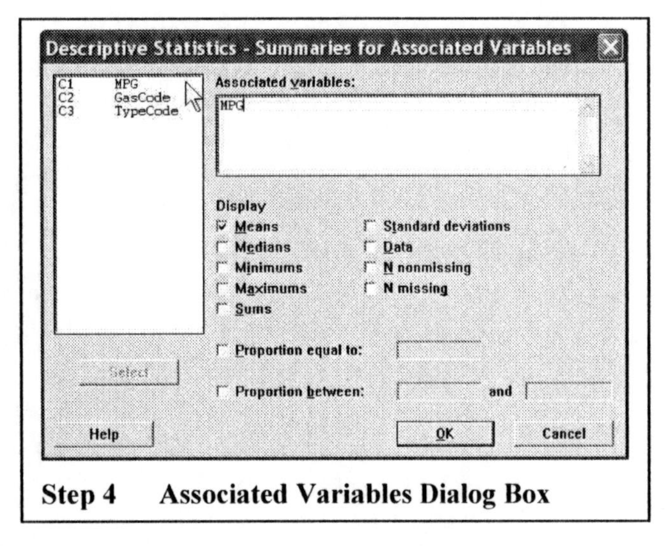

Step 4 Associated Variables Dialog Box

The result in the session window is a familiar table. The mean MPG for the four-wheel drive vehicles using high octane gasoline is 25.15 miles per gallon. The mean MPG for two-wheel drive vehicles using high octane gasoline is 32.550. Overall, the vehicles using high octane fuel averaged 28.850 miles per gallon.

Tabulated statistics: Gas, Vehicle

```
Rows: Gas   Columns: Vehicle

              FourWheel  TwoWheel    All

High Octane       25.15     32.55  28.85
                      2         2      4

Regular           28.95     25.95  27.45
                      2         2      4

All               27.05     29.25  28.15
                      4         4      8

Cell Contents:  MPG  :  Mean
                        Count
```

Step 5. Use **File>Exit** to close the program or **File>New>Minitab Project** to continue. When prompted, save the project as Chapter12.MPJ.

12-5 Data Analysis

A researcher hypothesizes there is a relationship between serum cholesterol and education level. The analysis would require a one-way analysis of variance because the dependent variable is quantitative and the independent variable is qualitative with 3 levels. The data collected is the Databank.mtw file. The data is already in stacked form since all of the measurements for cholesterol are in one column and the codes with education level are in a second column.

To do the analysis, we will recode the data so there are labels instead of numeric codes, an optional step but nice for the output. Then a one-way ANOVA will be done on the stacked data. The hypothesis test will be interpreted along with the Tukey test. Here goes!

Step 1. Open the Databank.MTW worksheet.

Step 2. To code the data select **Data>Code>Numeric to text**.

a. Double click 'Ed-Level' in the from column.

b. Type, **Education** for the Into column.

c. Type in a category for each original code as shown then click **[OK]**.

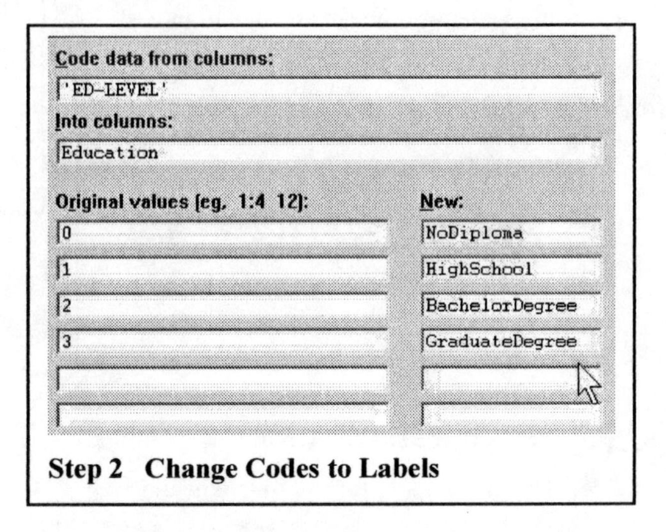

Step 2 Change Codes to Labels

Calculate the Test Statistic for Stacked Data including the Tukey Test.

Step 3. Select **Stat>ANOVA>Oneway**...

d. The Response variable is the quantitative variable, 'SERUM-CHOL.' Only numeric variables will be in the list.

e. Click in the box for Factor then select Education, the categorical variable.

f. Click [**Comparisons**] then select Tukey and click **[OK]** twice.

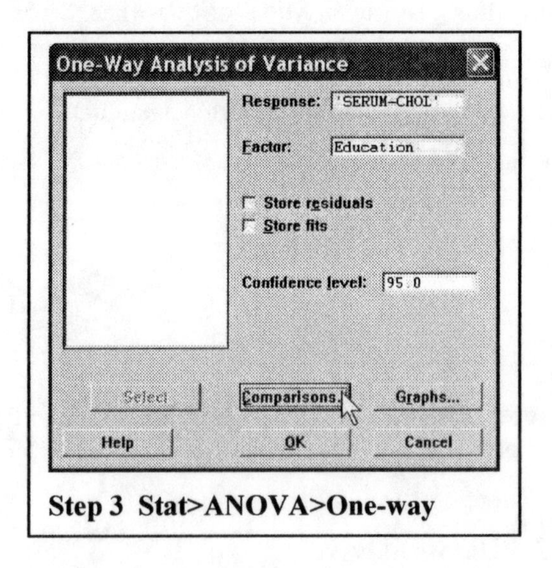

Step 3 Stat>ANOVA>One-way

Step 4. At a significance level of 0.05, the null hypothesis would be rejected. There is a significant relationship between cholesterol and education level.

According to the Tukey test, there is a significant difference in the mean cholesterol for those with a Bachelor's degree and those with No diploma. The mean for those with a Bachelor's degree is 207.44 and for those with no diploma the mean is 229.62, a difference of 22.18. There is no significant difference between any other pair.

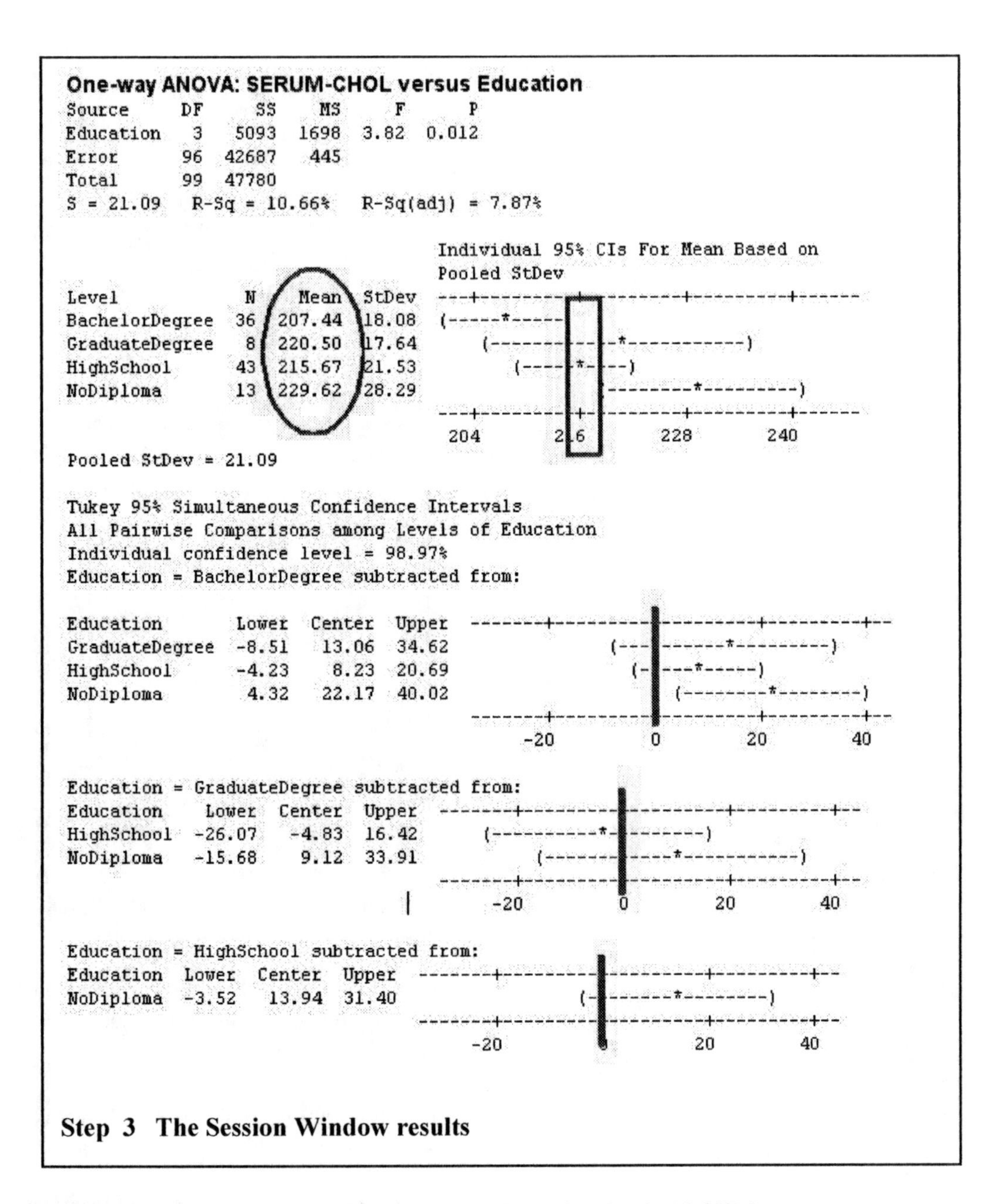

One-way ANOVA: SERUM-CHOL versus Education

```
Source      DF     SS    MS     F      P
Education    3   5093  1698  3.82  0.012
Error       96  42687   445
Total       99  47780
S = 21.09   R-Sq = 10.66%   R-Sq(adj) = 7.87%

                                    Individual 95% CIs For Mean Based on
                                    Pooled StDev
Level             N    Mean  StDev  ---+---------+---------+---------+------
BachelorDegree   36  207.44  18.08  (-----*-----)
GraduateDegree    8  220.50  17.64      (-------------*-----------)
HighSchool       43  215.67  21.53         (----*----)
NoDiploma        13  229.62  28.29             (---------*---------)
                                    ---+---------+---------+---------+------
                                     204       216       228       240
Pooled StDev = 21.09

Tukey 95% Simultaneous Confidence Intervals
All Pairwise Comparisons among Levels of Education
Individual confidence level = 98.97%
Education = BachelorDegree subtracted from:

Education        Lower  Center  Upper  -------+---------+---------+---------+--
GraduateDegree   -8.51   13.06  34.62              (----------*---------)
HighSchool       -4.23    8.23  20.69                (----*-----)
NoDiploma         4.32   22.17  40.02                   (--------*--------)
                                       -------+---------+---------+---------+--
                                            -20         0        20        40

Education = GraduateDegree subtracted from:
Education     Lower  Center  Upper  -------+---------+---------+---------+--
HighSchool   -26.07   -4.83  16.42     (----------*--------)
NoDiploma    -15.68    9.12  33.91          (-------------*-----------)
                                    -------+---------+---------+---------+--
                                         -20         0        20        40

Education = HighSchool subtracted from:
Education   Lower  Center  Upper  -------+---------+---------+---------+--
NoDiploma   -3.52   13.94  31.40              (--------*--------)
                                  -------+---------+---------+---------+--
                                       -20         0        20        40
```

Step 3 The Session Window results

Step 5. Close the program. Save the project as **DataAnalysis12.MPJ**.

Change the Order of a Qualitative Variable

The list of education levels are in alphabetical order. That is the default for text data. It is possible to change this order. Here is how.

Step 1. Select **Stat>Tables>Tally Individual Variables..**.

a. Double click C13-T Education.

b. Click **[OK]**.

The frequency distribution is created in the session window. The labels are in alphabetical order.

Tally for Discrete Variables:

Education	Count
BachelorDegree	36
GraduateDegree	8
HighSchool	43
NoDiploma	13
N=	100

Step 2. In the worksheet, click any cell in the text column, **Education**. There will be a block cursor and the entire cell will be highlighted. It doesn't matter which cell you click.

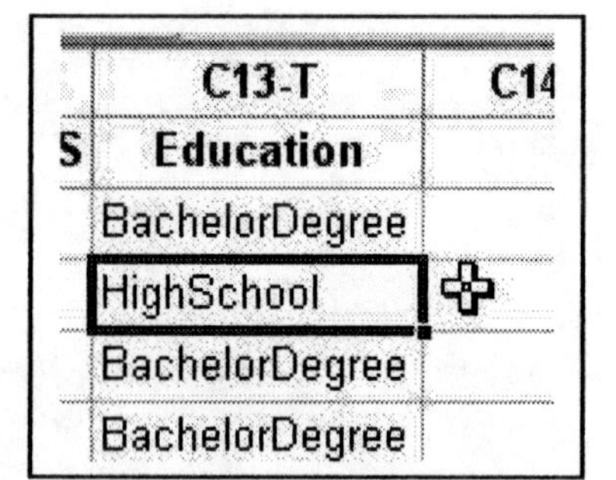

a. Right click. Select **Column >Value order** in the pop-up menus.

Step 2a Column>Value Order

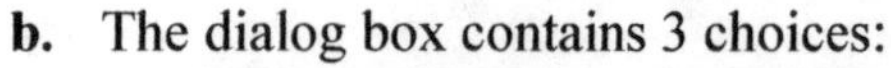

b. The dialog box contains 3 choices:
1. Alphabetical order, the default.
2. Order of appearance in the column.
3. Check the option for **User Defined.**

In the Define an order list, each code is listed in alphabetical order.

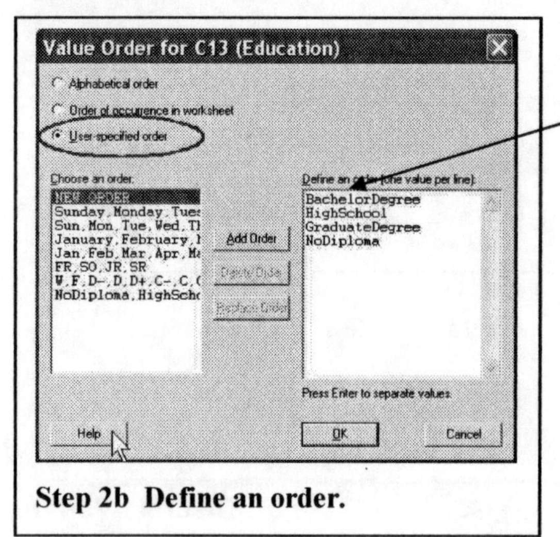

Step 2b Define an order.

c. Move the cursor before the B in Bachelor and press <Enter>. This should open up a blank line.

d. Highlight the line with NoDiploma. Right click and select Cut.

e. Move the mouse pointer to the blank line at the top of the list. Right click then select Paste. One move done.

f. Continue to cut and paste them until they are in the order of education. Delete any blank lines at the end of the list.

g. Click [Add order] then **[OK]** to complete the process.

The list should look like this when finished.

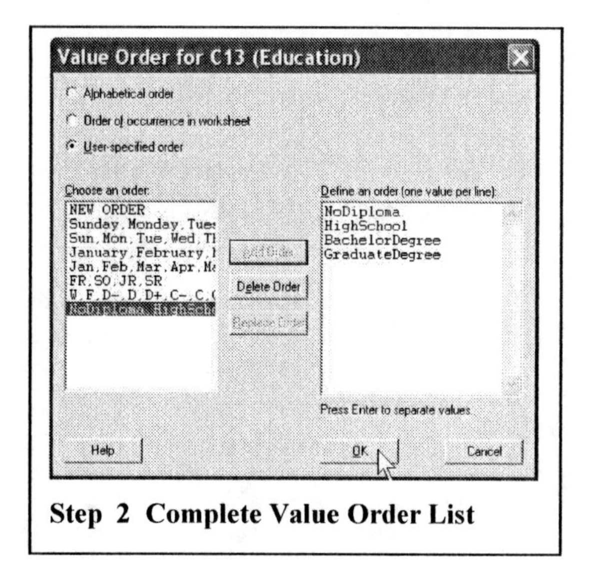

Step 2 Complete Value Order List

Step 3. To see that this has worked, select **Stat>Tables>Tally Individual Variables.**

a. Check **Counts**, **Percents** and **cumulative percents**.

b. Click **[OK]**.

Step 4. Select **Stat>Quality Tools>Pareto Chart**.

a. Double click Education for Chart Defects in.

b. Click **[Options]** and type an appropriate title such as Education Level of Employees.

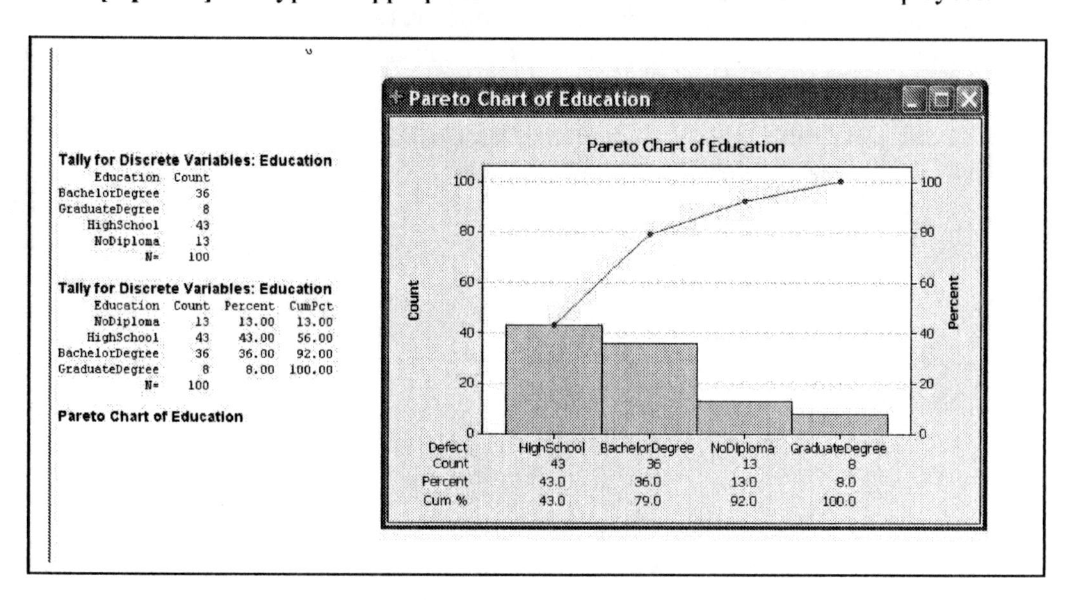

Tally for Discrete Variables: Education

Education	Count
BachelorDegree	36
GraduateDegree	8
HighSchool	43
NoDiploma	13
N=	100

Tally for Discrete Variables: Education

Education	Count	Percent	CumPct
NoDiploma	13	13.00	13.00
HighSchool	43	43.00	56.00
BachelorDegree	36	36.00	92.00
GraduateDegree	8	8.00	100.00
N=	100		

Pareto Chart of Education

c. Click **[OK]** twice. The resulting table will show the frequency distribution for education before and after applying the value order. The Pareto Chart will show the bars by order of frequency.

Value Order Help

You may change this order any time during a session. The order is saved with the worksheet and the project. Use the [Help] button on the Value Order dialog box if you need the instructions. These are the instructions for

> "**Define an order:** Use this box to define your own order.
> See To order values in a text column by defining your own order."

To order values in a text column by defining your own order
main topic

1 In the Data window, click in at least one cell of the text column(s) you want to order.

2 Choose **Editor > Column > Value Order**.

3 Choose **User-specified order** if it is not already selected. The text values from the column(s) appear in the **Define an order** box. Change the order of those values, cutting and pasting as needed.

- To cut: highlight the value and press Control+X.
- To paste: position the cursor where you want to paste and press Control+V. The pasted value appears after the cursor.
- Separate values so there is one value per line: Press Enter to start each new line.

4 If you want to save the format for future use, click **Add order**. The order will be added to the **Choose an order** box.

5 Click **OK**.

Note If a column contains values which are not included in the ordering scheme, those values are processed alphabetically after the values which do appear in the ordering.

Step 5. Close the program and save the project as DataAnalysis12.MPJ.

Chapter 12 Textbook Problems Suitable for MINITAB

Page		Section	Exercises
Page	597	12.2 and 12.3	8 - 19
Page	608	12.4	10 - 15
Page	615	Review	1 – 9
Page	617	Data Analysis	1 - 4
Page	620	Data Project	

Chapter 12: Endnotes

Chapter 13 Nonparametric Statistics

13-1 Introduction

Hypothesis tests for means, variance and proportions require that the populations being sampled are normal. Nonparametric tests are an alternative when that condition is not known to be true or there is evidence to the contrary.

13-2 Ranking and Ranks

Some tests require the use of ranks. Here are several commands that may be useful. The data represents the number of snow cones sold each day at a convenience store. The data used in these examples are from Example 1: The number of snow cones sold at a convenience store for a sample of 20 days is recorded.

```
SnowCones
     18    43    40    16    22    30    29    32    37    36
     39    34    39    45    28    36    40    34    39    52
```

Copy a Column of Quantitative Data

Step 1. Click the **maximize** icon of the worksheet then type this data into C1 of a MINITAB worksheet.

a. Select **Data>Copy>Columns to columns**.
b. Double click SnowCones to select it for the Copy from column.
c. Click the drop down list then select In current worksheet in columns. Press <Tab>.
d. Type Cones in the dialog box to name the new column.
e. Uncheck the option to Name the columns containing the copied data.
f. Click **[OK]**. There should now be an identical set of data in C2.

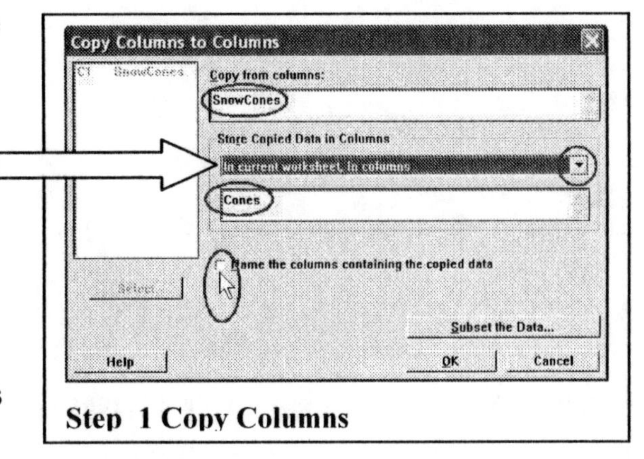

Step 1 Copy Columns

Sort a Column of Quantitative Data

Step 2. Select **Data>Sort.** In the sort dialog box enter the name of the new column in 3 places:

a. Sort column(s).
b. By column.
c. Click the option for Columns[s] of current worksheet then type in **Cones.**
d. Click **[OK].**

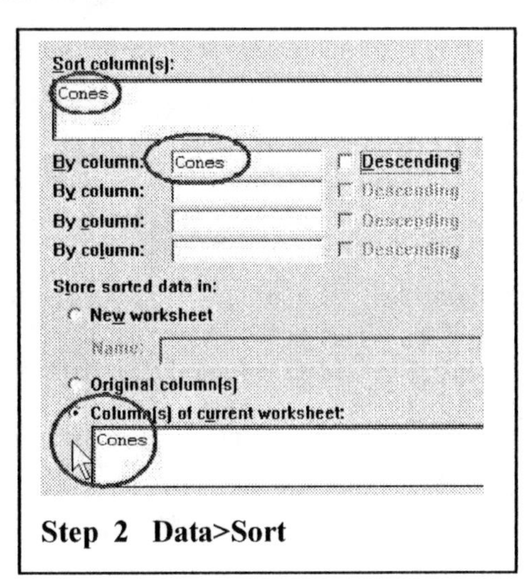

Step 2 Data>Sort

This will take the data in C2, sort it from smallest to largest and put the ordered list back into C2. If the box for Descending were checked, the data would be sorted from highest down to the lowest.

Calculate the Ranks for a Column of Quantitative Data

Step 3. Select **Data>Rank**.

a. Double click Cones for Rank data in:.

b. In the box for Store ranks in type in the name of a new column, **Ranks**.

c. Click **[OK]**.

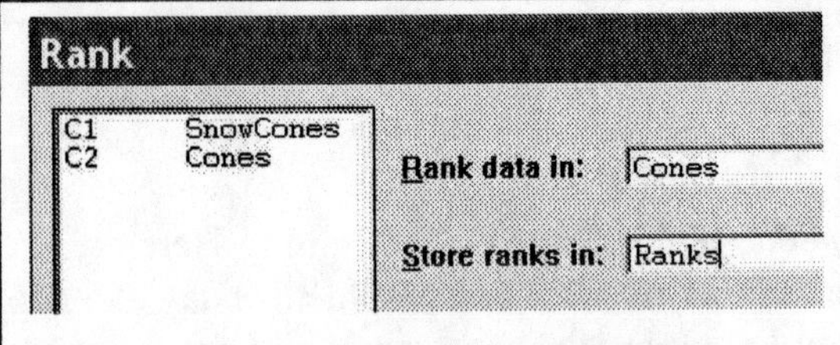

Step 3 Convert the data to Ranks

The worksheet with the three columns of data are shown.

↓	C1 SnowCones	C2 Cones	C3 Ranks
1	18	16	1.0
2	30	18	2.0
3	39	22	3.0
4	36	28	4.0
5	43	29	5.0
6	29	30	6.0
7	34	32	7.0
8	40	34	8.5
9	40	34	8.5
10	32	36	10.5
11	39	36	10.5
12	34	37	12.0
13	16	39	14.0
14	37	39	14.0
15	45	39	14.0
16	39	40	16.5
17	22	40	16.5
18	36	43	18.0
19	28	45	19.0
20	52	52	20.0

Step 3 Worksheet with Ranks

The ranks can be created without sorting the data first, however, the ranks are also in order if Cones are sorted. Ranks for duplicate values are averaged. In the list for Cones there are two values of 34. Their rank of 8th and 9th number are averaged to get the 8.5. If there were 3 duplicates like number 39, the three ranks of 13, 14 and 15 are averaged to get the 14.

This data will be used in the next section.

13-3 Sign Test

The sign test is the alternative to the one-sample t test when the sample size is small and the shape of the distribution is not known to be normal.

Example 13-1: Test the hypothesis that the median number of snow cones sold is 40. Use $\alpha = .05$.

Step 1. State the hypotheses.

H_0: Median = 40

H_1: Median $\neq$ 40

Calculate the Test Statistic and P-value for the Sign test

Step 2. The original data for Example 1 is already entered in column one of the worksheet.

- **a.** Select **Stat>Nonparametrics>1-Sample Sign Test**
- **b.** Double click the SnowCones column in the list box.
- **c.** Click on Test median. In the text box for the median value enter the hypothesized value, **40**.
- **d.** Click **[OK]**.

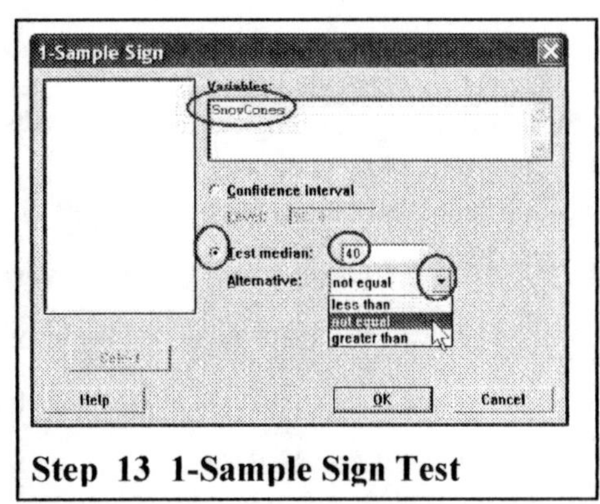

Step 13 1-Sample Sign Test

Step 3. The P-value is .0075.

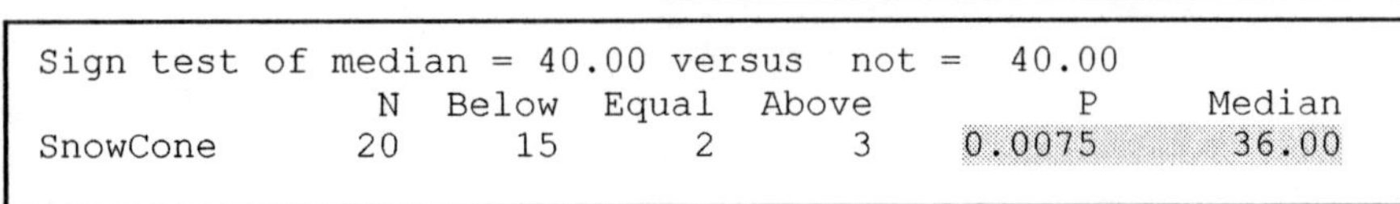

```
Sign test of median = 40.00 versus  not =  40.00
                   N  Below  Equal  Above          P    Median
SnowCone          20     15      2      3     0.0075     36.00
```

Step 4. Reject the null hypothesis.
Step 5. The median is probably not 40. The median of the sample is 36 snow cones.

Paired Sample Sign Test

To test an hypothesis about the difference in the means for two dependent samples from normal populations, the matched pairs-t test was used (Chapter 9). When the normal condition required for this test is uncertain, the paired sample sign test can be used instead. The procedure requires that we find the differences between all the pairs and then do the sign test on the differences.

C1	C2
Before	After
3	2
0	1
5	4
4	0
2	1
4	3
3	1
5	3
2	2
1	3

Example 13-3: A medical researcher believed the number of fear infections in swimmers can be reduced if the swimmers use earplugs. A sample of 10 people was selected and the number of infections before and after wearing ear plugs was recorded. Can the researcher conclude that wearing earplugs reduced the number of ear infections? $\alpha = .05$.

Step 1. Enter the data for Example 13-3 into a worksheet, only the Before and After columns are necessary. Calculate a column with the differences to begin the process.

Step 2. Select **Calc>Calculator.**

- **a.** Type **D** in the box for Store result in variable.
- **b.** Move to the Expression box then click on Before, the subtraction sign, and After. The completed entry is shown.

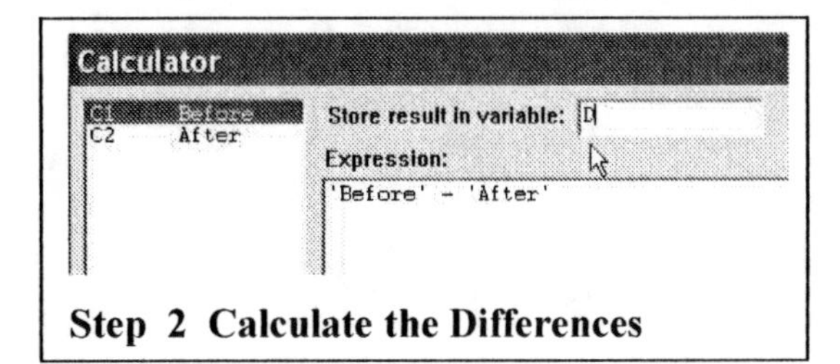

Step 2 Calculate the Differences

Step 3. Click **[OK]**.

MINITAB will calculate the differences and store them in the first available column with the name "D."

↓	C1-T	C2	C3	C4
	Swimmer	Before	After	D
1	A	3	2	1
2	B	0	1	-1
3	C	5	4	1
4	D	4	0	4
5	E	2	1	1
6	F	4	3	1
7	G	3	1	2
8	H	5	3	2
9	I	2	2	0
10	J	1	3	-2

Use the instructions for The Sign Test on the differences, D, with an hypothesized value of zero.

Step 4. Select **Stat>Noparametrics>1 Sample Sign.**

- **a.** Type the column name **D** for the variable.
- **b.** The hypothesized value of the median will be **0**.
- **c.** Change the Alternative to greater than then click **[OK]**.

Sign Test for Median: D

```
Sign test of median =  0.00000 versus not = 0.00000
    N  Below  Equal  Above         P      Median
D  10      2      1      7    0.1797       1.000
```

Step 5. The P-value is .1797.

Step 6. Do not reject the null hypothesis. There is not enough evidence to conclude the number of ear infections were reduced.

13-4 Wilcoxon Rank Sum Test (also called Mann-Whitney)

In the previous section, the 1-sample sign test was used to test an hypothesis for two independent samples. If the samples are independent the 2-sample t-test would be used assuming the two populations are normal. If that is unlikely, the nonparametric alternative is the Wilcoxon rank-sum test.

Example 13-4: Two independent samples of recruits are selected and the time in minutes it takes each recruit to complete an obstacle course is recorded as shown. At $\alpha = .05$, is there a difference in the amount of time it takes recruits to complete the course?

C1 Army	15	18	16	17	13	22	24	17	19	21	26	28
C2 Marines	14	9	16	19	10	12	11	8	15	18	25	

Step 1. Enter the data for Example 13-4 into two columns of a worksheet. Name the columns "Army" and "Marines." Note that the sample sizes are different. These are not matched pairs. They are independent samples.

Step 2. Select **Stat>Nonparametric>Mann-Whitney.**

- **a.** Double click Army to select it for in the First Sample.
- **b.** Double click Marines for the Second Sample.

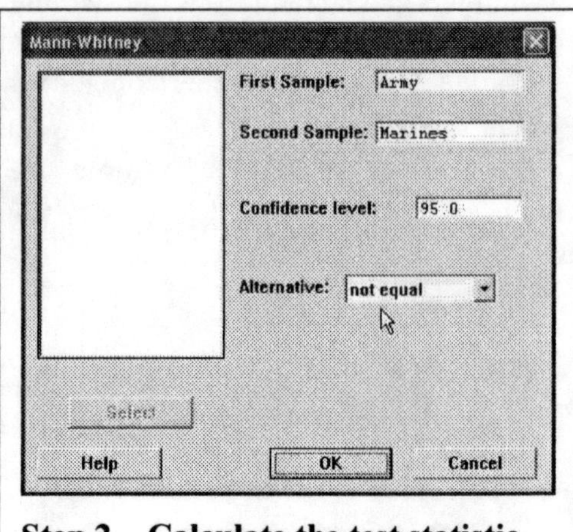

Step 2 Calculate the test statistic

Step 3. Click **[OK]**. The results are in the session window.

Mann-Whitney Test and CI: Army, Marines

```
            N  Median
Army       12  18.500
Marines    11  14.000

Point estimate for ETA1-ETA2 is 6.000
95.5 Percent CI for ETA1-ETA2 is (1.003,9.998)  W = 183.0
Test of ETA1 = ETA2 vs ETA1 not = ETA2 is significant at 0.0178
The test is significant at 0.0177 (adjusted for ties)
```

Step 4. The p-value for the test is .0177.

Step 5. Reject the null hypothesis. There is a significant difference in the times it takes the recruits to complete the course. In the sample the difference in the median values was 3.5 minutes.

MINITAB will calculate the test statistic from the raw data. To verify the calculations in the textbook for this example, stack the data into C3 Minutes with a second column (C4-T Group) to identify the branch of service.

Step 1. Select **Data>Stack>Columns**.

a. Select C1 Army and C2 Marines for variables.

b. Check option for Column of current worksheet then type in **Minutes** for the variable name.

c. Type **Group** for Store subscripts in.

d. Check the option for using subscripts.

e. Click **[OK]**.

Rank the stacked column, the number of minutes.

Step 2. Select **Data>Rank**.

a. Double click C3 Minutes.

b. Type in **Ranks** for storage.

c. Click **[OK]**.

C1 Army	C2 Marines	C3 Minutes	C4-T Group	C5 Ranks
15	14	15	Army	8.5
18	9	18	Army	14.5
16	16	16	Army	10.5
17	19	17	Army	12.5
13	10	13	Army	6.0
22	12	22	Army	19.0
24	11	24	Army	20.0
17	8	17	Army	12.5
19	15	19	Army	16.5
21	18	21	Army	18.0
26	25	26	Army	22.0
28		28	Army	23.0
		14	Marines	7.0
		9	Marines	2.0
		16	Marines	10.5
		19	Marines	16.5
		10	Marines	3.0
		12	Marines	5.0
		11	Marines	4.0
		8	Marines	1.0
		15	Marines	8.5
		18	Marines	14.5
		25	Marines	21.0

The data is shown here stacked and ranked. The last three columns contain the sorted data, a group identifier and the ranks for the data. MINITAB does these operations in the background in order to calculate the statistics for the Mann-Whitney test.

Step 3. Use **File>Save Project as ...** Save this project as Chapter13.MPJ.

Continue................

13-5 Wilcoxon Signed-Rank Test

The paired sample sign test was used to test the difference for two dependent samples when the populations are not assumed to be normal. It is a sign test on the differences that does not consider the magnitude of the differences. The Wilcoxon Signed Rank test is another alternative to the two sample matched pairs test.

Example 13-5: In a large department store, the owner wishes to see whether the number of shoplifting incidents per day will change if the number of uniformed security officers is doubled. A sample of seven days before security is increased and seven days after the increase shows the number of shoplifting incidents. $\alpha = .05$.

	C1	C2
Day	**Before**	**After**
Monday	7	5
Tuesday	2	3
Wednesday	3	4
Thursday	6	3
Friday	5	1
Saturday	8	6
Sunday	12	4

Test the median value for the differences of two dependent samples.

Step 1. Enter the data into two columns of a new worksheet. Name the columns Before and After.

Step 2. Calculate the differences using **Calc>Calculator**.

a. Type **D** in the box for Store result in variable.

b. In the expression box type **Before – After**.

c. Click **[OK]**.

Step 3. Select **Stat>Nonparametric>1-Sample Wilcoxon.**

a. Select **C3** D for the Variable.

b. Click on Test median. The value should be 0.

c. Click **[OK]**.

Wilcoxon Signed Rank Test: D

```
Test of median = 0.000000 versus median
not = 0.000000
           N
          for    Wilcoxon             Estimated
     N   Test   Statistic       P        Median
D    7      7        25.0   0.076         2.250
```

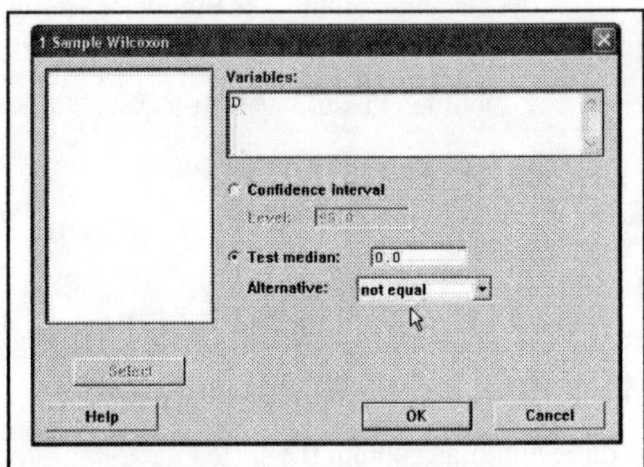

Step 3 Calculate the Test Statistic

Step 4. The P-value of the test is .076.

Step 5. Do not reject the null hypothesis. There is no significant difference in the number of shoplifting incidents.

13-6 The Kruskal-Wallis Test (also known as the H test)

Example 13-6: A researcher tests three different brands of breakfast drinks to see how many milli-equivalents of potassium per quart each contains. The following data was collected.

Step 1. State the hypotheses.
H_0: There is no difference in the amount of Potassium in three different brands of breakfast drinks.
H_1: There is a difference in the amount of Potassium in three different brands of breakfast drinks.

All of the numeric data must be in one column and the second column identifies the brand. The data for this test must be "stacked".

Step 2. Stack the data for example 13-6 into two columns of a worksheet.

a. First, enter all of the potassium amounts into one column.

b. Name this column **Potassium.**

c. Enter codes A, B or C for the brand into the next column.

d. Name this column **Brand.**

The worksheet is shown.

↓	C1	C2-T
	Potassium	Brand
1	4.7	A
2	3.2	A
3	5.1	A
4	5.2	A
5	5.0	A
6	5.3	B
7	6.4	B
8	7.3	B
9	6.8	B
10	7.2	B
11	6.3	C
12	8.2	C
13	6.2	C
14	7.1	C
15	6.6	C

Step 3. Select **Stat>Nonparametric>Kruskal-Wallis.**

The column with the brands in don't show until the cursor is in the Factor text box. The response variable must be quantitative and so only those show up in the list.

a. Double click C1 Potassium to select it for Response.

The cursor will jump to the Factor text box which can use the text column so it will appear in the list.

b. Double click C2 Brand for Factor.

c. Click **[OK]**.

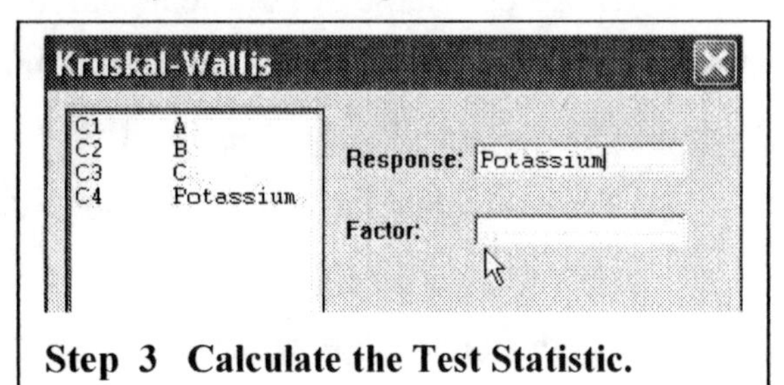

Step 3 Calculate the Test Statistic.

Kruskal-Wallis Test: Potassium versus Brand

```
Kruskak-Wallis Test on Potassium
Brand      N  Median  Ave Rank      Z
A          5   5.000       3.0  -3.06
B          5   6.800      10.6   1.59
C          5   6.600      10.4   1.47
Overall   15               8.0
H = 9.38  DF = 2  P = 0.009
```

The value H = 9.38 has a P-value of .009.

Step 4. Reject the null hypothesis.

Step 5. There is enough evidence in the sample to conclude there is a difference in the amount of Potassium.

Step 6. Click the **disk icon** to save the project and continue on.

Another example of the Kruskal-Wallis test follows with a more detailed analysis.

Exercise 7 Section 6: A meteorologist wishes to see if there is a difference in the number of deaths due to different types of severe weather. The raw data from the past six years are shown here. At $\alpha = .10$ is there a difference in the (median) number of deaths from the different weather conditions?

Lightning	Tornado	Flash Flood	Blizzard
39	30	46	54
41	39	55	43
73	39	45	39
74	53	109	35
67	50	62	56
68	32	30	48

Step 1. State the hypotheses.
Ho: There is no difference in the number of deaths due to severe weather.
H1: There is a difference in the number of deaths due to severe weather.

Enter the data into four columns of MINITAB or open the file: P-C13-S06-07.mtw. Stack the data.

Step 2. Select **Data>Stack>Columns**.

a. Drag your mouse over the four columns to highlight them then click **[Select]**.
b. Click the button for Column of current worksheet then enter the name of the new column that will contain all of the data, **Deaths.**
c. Type **Weather**, the name of the new column that will contain the group identifier, store subscripts in.
d. The button for Use variable names in subscript column should be checked.
e. Click **[OK]**.

Step 3. Calculate the test statistic for a one-way ANOVA with boxplots.

a. Select **Stat>ANOVA>One-way**.
b. The Response variable should be Deaths and the Factor should be Weather.
c. Click **[Graphs]** and check boxplots of data.
d. Click **[Comparisons]** and check Tukey's.
e. Click **[OK]**.

Step 4. Calculate the test statistic for the Kruskal-Wallis test.

a. Select **Stat>Nonparametrics>Kruskal-Wallis**.

b. Double click Deaths for Response and Weather for Factors.

c. Click [OK].

Step 5. Assemble your report.

a. Click on the Project Manager icon then the Session folder.

b. Right click on One way ANOVA then Append to Report.

c. Right click on Boxplots by Death and Weather then Append to Report.

d. Right click on Kruskal-Wallis Test then Append to Report.

e. Click on the Report pad folder. Add your personal information to the report. Drag the corners of the graphic containing the boxplot to shrink it a bit. Then Right click the Report folder and Print the Report.

See the next page.

MINITAB Project Report

Chapter 13 - 6

It appears from the boxplot that the distributions of deaths due to different weather conditions are not normal and are not similar in the amount of variation. The ANOVA also indicates that the variation in the four distributions is quite different.

Therefore, it would be more appropriate to use the Kruskal-Wallis test. The P-value of this test is .136, do not reject the null hypothesis. There is not enough evidence to conclude the median number of deaths if different depending on the weather condition.

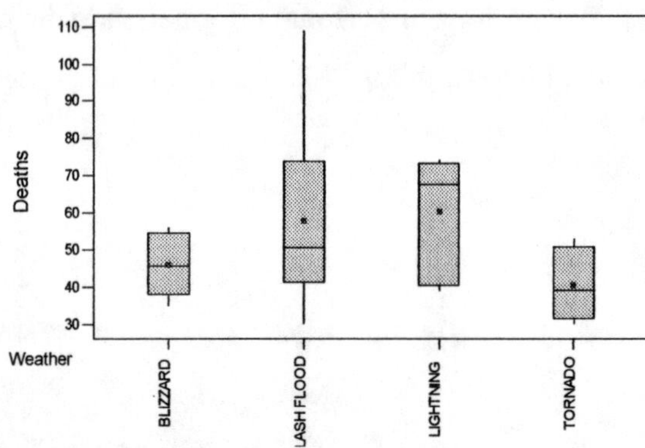

One-way ANOVA: Deaths versus Weather

```
Analysis of Variance for Deaths
Source     DF        SS        MS        F        P
Weather     3      1624       541     1.87    0.167
Error      20      5783       289
Total      23      7407
                                   Individual 95% CIs For Mean
                                   Based on Pooled StDev
Level       N      Mean     StDev  ---+---------+---------+---------+---
BLIZZARD    6     45.83      8.33      (---------*--------)
FLASH FL    6     57.83     27.29              (---------*--------)
LIGHTNIN    6     60.33     16.00                (--------*---------)
TORNADO     6     40.50      9.31  (---------*---------)
                                   ---+---------+---------+---------+---
Pooled StDev =    17.00               30        45        60        75
```

Kruskal-Wallis Test: Deaths versus Weather

```
Kruskal-Wallis Test on Deaths
Weather     N    Median   Ave Rank       Z
BLIZZARD    6     45.50       11.3   -0.50
FLASH FL    6     50.50       14.1    0.63
LIGHTNIN    6     67.50       16.9    1.77
TORNADO     6     39.00        7.8   -1.90
Overall    24                 12.5
H = 5.54  DF = 3  P = 0.136
H = 5.56  DF = 3  P = 0.135 (adjusted for ties)
```

13-7 Rank Correlation and the Runs Test for Randomness

Pearson's correlation coefficient to measure the association between two quantitative variables requires that the raw data be quantitative and normally distributed. If the level of measurement is not interval or higher and/or the distribution is non-normal, then the Spearman Rank correlation can be used to test the association. Convert the data to ranks and then calculate the correlation coefficient for the ranks.

Example 13-7. Two students were asked to rate eight different textbooks for a specific course on an ascending scale from 0 to 20 points. Test the hypothesis that there is a significant correlation between the two students' ratings. This ordinal level of data is shown.

Textbook	Student1	Student2
A	4	4
B	10	6
C	18	20
D	20	14
E	12	16
F	2	8
G	5	11
H	9	7

Step 1. Enter the ratings into 2 columns of a worksheet.

a. Name the columns **Student1** and **Student2**. Rank each columns (separate).

b. Select **Data>Rank**.

c. Double click **Student1** to select it for **Rank Data in.** Type the name of the new column, **Rank1**.

d. Click **[OK]**.

e. Click the **Edit last dialog icon** and change Student1 to **Student2** and Rank1 to **Rank2**.

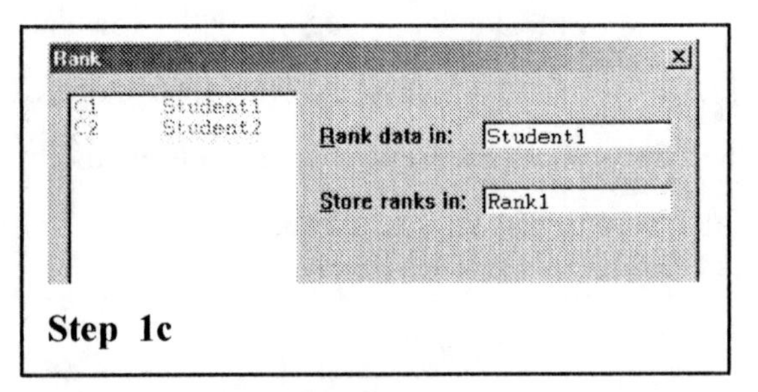

Step 1c

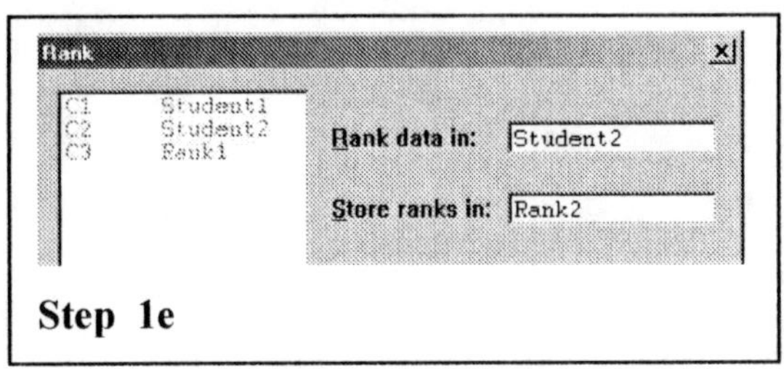

Step 1e

The completed worksheet is shown.
The ranks in the textbook are in descending order. The ranks here are in ascending order. Don't worry, the test statistic will be the same.

C1	C2	C3	C4
Student1	Student2	Rank1	Rank2
4	4	2	1
10	6	5	2
18	20	7	8
20	14	8	6
12	16	6	7
2	8	1	4
5	11	3	5
9	7	4	3

Calculate the correlation coefficient using ranks.

Step 2. Select **Stat>Basic Statistics>Correlation**.

- **a.** Double click each column of Ranks to select them for the Variables.
- **b.** Make sure the box for Display p-values is checked.
- **c.** Click **[OK]**.

In the session window, the correlation coefficient is .643. Because the ranks were correlated not the original data, this is the Spearman's rank correlation coefficient not a Pearson's correlation . The null hypothesis is not rejected since the P-value is .086. There is not enough evidence to conclude the ratings are different.

Correlations: Rank1, Rank2

```
Pearson correlation of Rank1 and Rank2 = 0.643   P-Value = 0.086
```

This procedure would also be used on quantitative data that is not normally distributed. Recall the example from Chapter 10 regarding the correlation of blood pressure and age. If we were doubtful that Age and/or Pressure have a normal distribution, then Spearman's correlation coefficient would be calculated to test for a relationship.

C2	C3	C4	C5
AGE(X)	PRESSURE(Y)	AgeRank	PressuRank
43	128	1	2
48	120	2	1
56	135	3	3
61	143	4	5
67	141	5	4
70	152	6	6

Example 10-1 Worksheet After Ranking Data

Each variable was ranked and the correlation done on the ranks.

Correlations: AgeRank, PressuRank

```
Pearson correlation of AgeRank and PressuRank = 0.886   P-Value = 0.019
```

The P-value is .019. The null hypothesis would be rejected. The Spearman's correlation would indicate there is a significant correlation between blood pressure and age.

Runs Test

Sequence is important! If the data were not presented in the same order they were selected, this test would not be appropriate. Example: The age of homes in a certain city were selected and recorded as shown. Can we conclude that the data resulted from a random selection?

Step 1. Sequence is important! Enter the data down C1 in the same order it was collected.

Do not sort it!

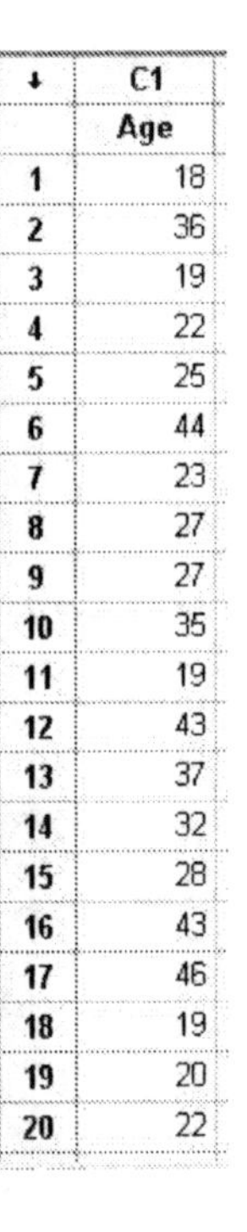

↓	C1
	Age
1	18
2	36
3	19
4	22
5	25
6	44
7	23
8	27
9	27
10	35
11	19
12	43
13	37
14	32
15	28
16	43
17	46
18	19
19	20
20	22

Step 2. Calculate the median and store it in a constant.

a. Select **Calc>Column Statistics**.

b. Check the option for Median.

c. Use C1 Age for the Input Variable.

d. Type the name of the constant, MedianAge in the Store result in text box.

e. Click **[OK]**.

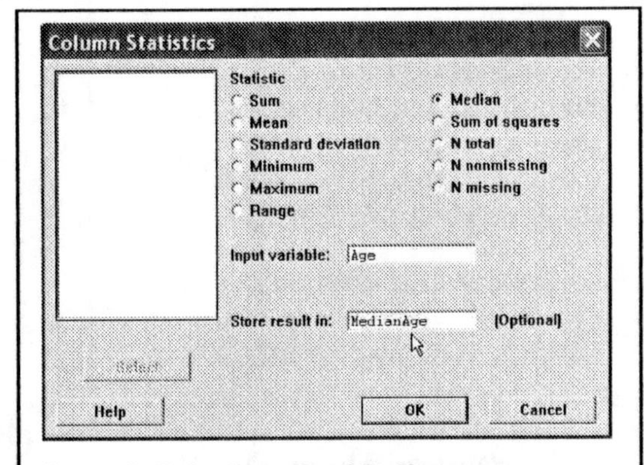

Step 2 Calculate the Median

Step 3. Select **Stat>Nonparametric>Runs Test**.

a. Select C1 Age as the variable.

b. Click the button for Above and below, then select MedianAge in the text box.

c. Click **[OK]**.

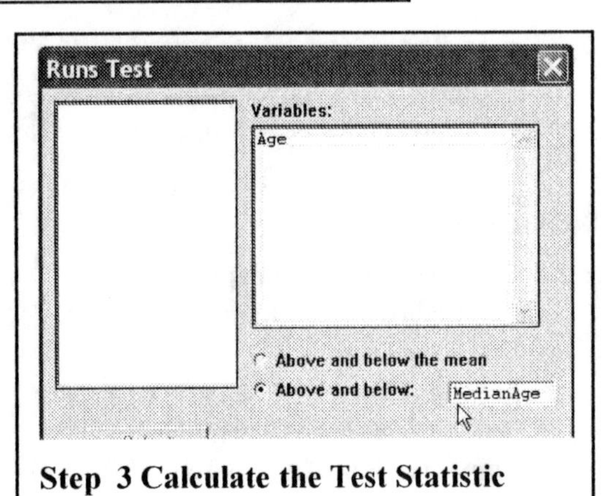

Step 3 Calculate the Test Statistic

The results will be displayed in the session window.

Runs Test: Age

```
Runs test for Age
Runs above and below K = 27

The observed number of runs = 9
The expected number of runs = 10.9
9 observations above K, 11 below
* N is small, so the following approximation may be invalid.
P-value = 0.378
```

Step 4. The P-value is .378.

Step 5. Do not reject the null hypothesis.
There is not enough evidence to conclude that the selection was not random.

13-8 Data Analysis

From the Databank file, select a sample of subjects and test to see if sodium levels are the same for smokers and non-smokers. Smoking has three levels. The measure for sodium is a continuous variable. If we can assume the data is normally distributed a one-way ANOVA would be used. Study a boxplot to help make that decision. The Kruskal-Wallis test is the alternative to a one-way ANOVA. Both techniques are applied.

Step 1. Use **File>Open Worksheet** then select Databank.MTP.
Select a random sample of 35 from the column of sodium and gender.

Step 2. Click **Calc>Random Data>Sample from Columns.**

- **a.** Enter 35 for the **Number of rows.**
- **b.** Press <Tab> to move to the dialog box for **column(s):** then double click Sodium and Smoking Status.
- **c.** Type the names of two new columns, **Sodium35** and **Smoke25** in the box for Store samples in.
- **d.** Check the box for **Sample with replacement**.
- **e.** Click **[OK]**.

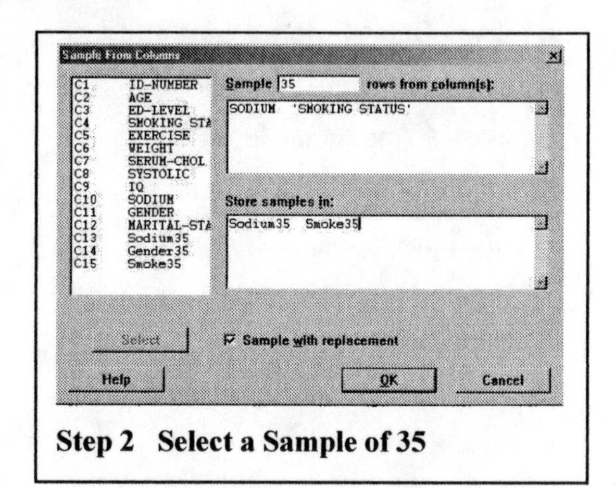

Step 2 Select a Sample of 35

Step 3. Calculate the test statistic for the Kruskal-Wallis test.

- **a.** Select **Stat>Nonparametrics**.
- **b.** The Response variable is Sodium35 and the Factor variable is Smoke35.
- **c.** Click **[OK]**.

Step 4. Use the Project Manager to append the boxplot and Kruskal-Wallis test to the report pad.

Step 5. Print the report pad then Close Minitab. Save DataAnalysis13.MPJ.

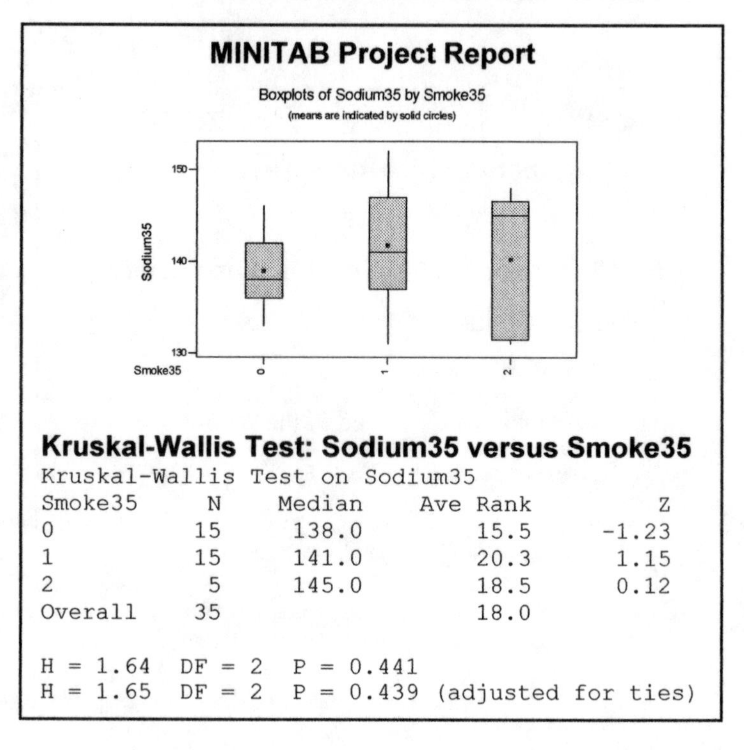

MINITAB Project Report

Kruskal-Wallis Test: Sodium35 versus Smoke35

```
Kruskal-Wallis Test on Sodium35
Smoke35       N     Median     Ave Rank         Z
0            15      138.0         15.5     -1.23
1            15      141.0         20.3      1.15
2             5      145.0         18.5      0.12
Overall      35                    18.0

H = 1.64  DF = 2  P = 0.441
H = 1.65  DF = 2  P = 0.439 (adjusted for ties)
```

Conclusion?
The boxplot indicates the sodium levels for those who smoke a pack or more a day is skewed right. The Kruskal-Wallis test would be appropriate.

The P-value for the test is .439. Can't reject the null hypothesis. There is not enough evidence in the sample to conclude the median sodium levels are different for the three smoking groups.

Chapter 13 Textbook Problems Suitable for MINITAB

Page	627	13.2	4 - 10
Page	632	13.3	5 - 20
Page	638	13.4	4 - 11
Page	644	13.5	9 - 13
Page	648	13.6	1 - 11
Page	658	13.7	5 - 23
Page	663	Review	1, 3 - 13
Page	662	Comparison of parametric and nonparametric tests.	
Page	665	Data Analysis	1 - 4
Page	667	Data Project	1, 2

Chapter 13: Endnotes

Chapter 14 Sampling and Simulation

14-1 Introduction:

Contrary to popular notion, a random sample is *not* haphazard, such as interviewing shoppers at a mall. Sampling plans are used to ensure every individual has an equal chance of being selected. Sampling techniques often use random numbers to select samples.

14-2 Common Sampling Techniques

Section 2 Example 1: Select a random sample of ten states. Four different sampling plans, ways of selecting the sample, will be demonstrated. Use MINITAB to randomly select the numbers. A list of all of the states is required.

Method 1: Select a ***simple random sample*** of ten integers from a uniform distribution.

Method 2: Select a ***simple random sample*** directly from the list of ID numbers, with replacement. Duplicates are allowed but the probability is the same for each subject.

Method 3: Select a ***simple random sample*** directly from the list of States, without replacement. There will be no duplicates.

Method 4: Select a ***systematic random sample*** from the list of states.

A numbered list of the population is required for the first four methods. Prepare the list in a worksheet.

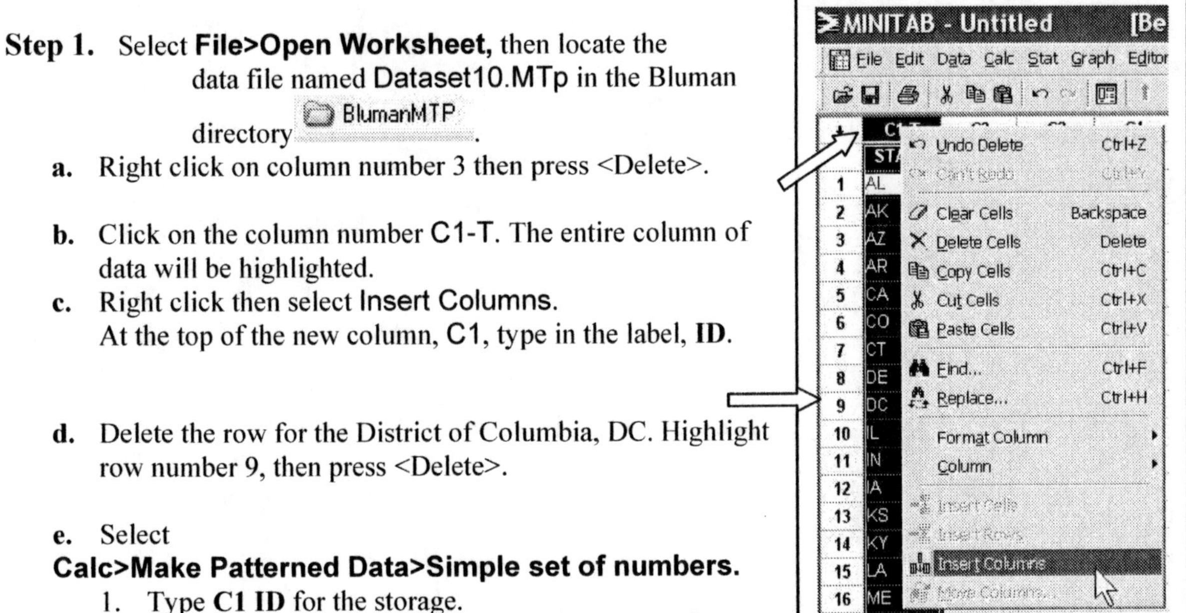

Step 1. Select **File>Open Worksheet,** then locate the data file named **Dataset10.MTp** in the Bluman directory BlumanMTP.

a. Right click on column number 3 then press <Delete>.

b. Click on the column number C1-T. The entire column of data will be highlighted.

c. Right click then select Insert Columns.
At the top of the new column, C1, type in the label, **ID**.

d. Delete the row for the District of Columbia, DC. Highlight row number 9, then press <Delete>.

e. Select
Calc>Make Patterned Data>Simple set of numbers.
1. Type **C1 ID** for the storage.
2. Type from 1 to 50 by 1 in the appropriate boxes.

f. Type in the names of the states for C3.

g. Name C2 **Abbr** and C3 **State**. The worksheet is ready.

Step 2. Select **File>Save Current Worksheet as...**then save the worksheet as **States.MTP**.
The list is shown on the next page. The database, a numbered list of all states, is required to use this method of sampling.

ID	Abbr	State
1	AL	Alabama
2	AK	Alaska
3	AZ	Arizona
4	AR	Arkansas
5	CA	California
6	CO	Colorado
7	CT	Connecticut
8	DE	Delaware
9	FL	Florida
10	GA	Georgia
11	HI	Hawaii
12	ID	Idaho
13	IL	Illinois
14	IN	Indiana
15	IO	Iowa
16	KS	Kansas
17	KY	Kentucky
18	LA	Louisiana
19	ME	Maine
20	MD	Maryland
21	MA	Massachusetts
22	MI	Michigan
23	MN	Minnesota
24	MS	Mississippi
25	MO	Missouri
26	MT	Montana
27	NE	Nebraska
28	NV	Nevada
29	NH	New Hampshire
30	NJ	New Jersey
31	NM	New Mexico
32	NY	New York
33	NC	North Carolina
34	ND	North Dakota
35	OH	Ohio
36	OK	Oklahoma
37	OR	Oregon
38	PA	Pennsylvania
39	RI	Rhode Island
40	SC	South Carolina
41	SD	South Dakota
42	TN	Tennessee
43	TX	Texas
44	UT	Utah
45	VT	Vermont
46	VA	Virginia
47	WA	Washington
48	WV	West Virginia
49	WI	Wisconsin
50	WY	Wyoming

Method 1: Simple Random Sample using Random Integers

Use MINITAB to create a list of random integers. These integers will be used to identify the ID number of the states included in the random sample. Every integer has the same probability of being selected and duplicates may be chosen.

Step 3. Select **Calc>Random Data>Integer.**

a. Type **10** for Generate rows of data.

b. Type the name of a column, **Random1** in the box for Store in column(s):.

c. Type **1** for Minimum value: and **50** for Maximum value: then click **[OK].**

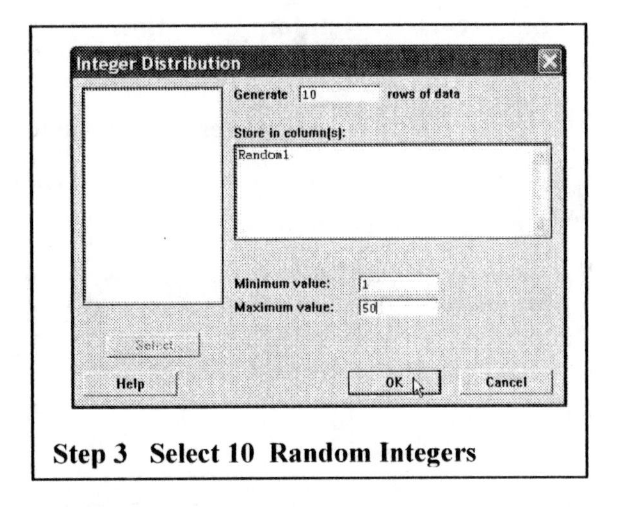

Step 3 Select 10 Random Integers

A sample of 10 integers between 1 and 50 will be displayed in column four of the worksheet. Expect your list to be different from the one shown. Because this command samples with replacement, an integer may be used more than once. Duplicates are possible. In this example, there are two twenty-sixes. If duplicates are not alright to use, ignore the duplicates. In that case, the sample size is nine.

Step 4. A list of the population is shown on the facing page. The highlighted states are the first sample. The states corresponding to these ID numbers were typed into C5.

C4	C5-T
Random1	**Sample1**
26	Montana
26	Montana
6	Colorado
17	Kentucky
20	Maryland
49	Wisconsin
23	Minnesota
44	Utah
5	California
28	Nevada

Method 1 Sample 1

Method 2: Simple Random Sample (without replacement)

To create a list of random numbers so there are no duplicates requires a different command. Instead of using the probability distribution, MINITAB will choose ten integers directly from the list of ID numbers.

Step 5. Select **Calc>Random Data>Sample From Columns.**

- **a.** Type **10** for the Sample rows from column(s): and **ID** for the name of the From column. This command gets the integers from the list in the worksheet.
- **b.** Type **Random2** as the name of the new column. Be sure to leave the option for Sample with replacement unchecked. Click **[OK].**

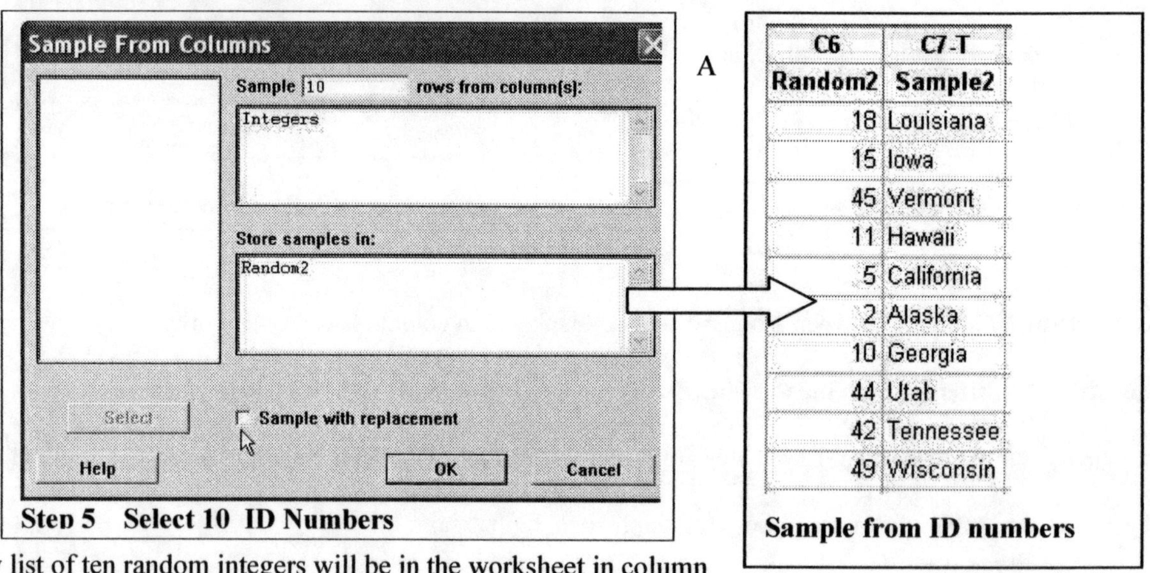

C6	C7-T
Random2	Sample2
18	Louisiana
15	Iowa
45	Vermont
11	Hawaii
5	California
2	Alaska
10	Georgia
44	Utah
42	Tennessee
49	Wisconsin

Step 5 Select 10 ID Numbers

Sample from ID numbers

A new list of ten random integers will be in the worksheet in column six.

There won't be duplicates in this list. There are ${}_{50}C_{10}$ = 10,272,278,170 samples of 10 that can be selected from a population of 50 in this way!

- **c.** Type the names of the states corresponding to the ID numbers into **C7**. Name the column **Sample2**.

Method 3: Simple Random Sample (without replacement)

Use the same command to select the sample directly from the list of states.

Step 6. Select **Calc>Random Data>Sample From Columns.**

- **a.** Sample rows from columns: should be left at 10.
- **b.** Press <Tab>, then double click C3 State to select it for From columns.
- **c.** Type **Sample3** in Store samples in:.
- **d.** Do not check the box for sample with replacement.
- **e.** Click **[OK].** A list of ten states is randomly selected.

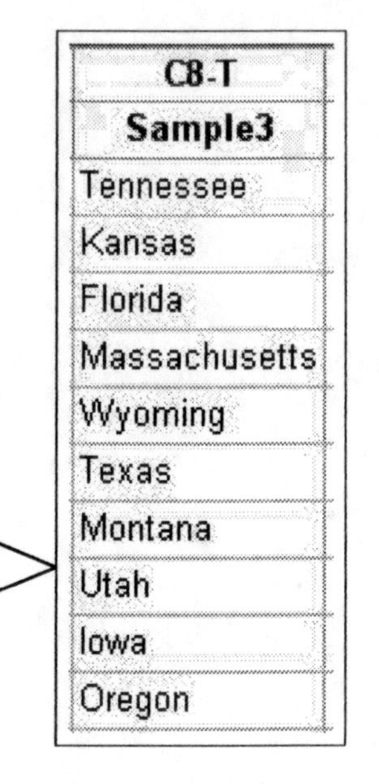

C8-T
Sample3
Tennessee
Kansas
Florida
Massachusetts
Wyoming
Texas
Montana
Utah
Iowa
Oregon

Method 4: Systematic Random Sample

Section 2 Example 2: Select a systematic sample of the fifty states. Every fifth state will be selected from the list using a randomly selected integer between 1 and 5 as the starting point.

Step 7. Use **Calc>Random Data>Integers** to select the first digit.

a. Select one random integer between 1 and 5. Store it in Random4.

b. Click **[OK].** In the first row of C9 is the selected digit 2. Yours may be different.

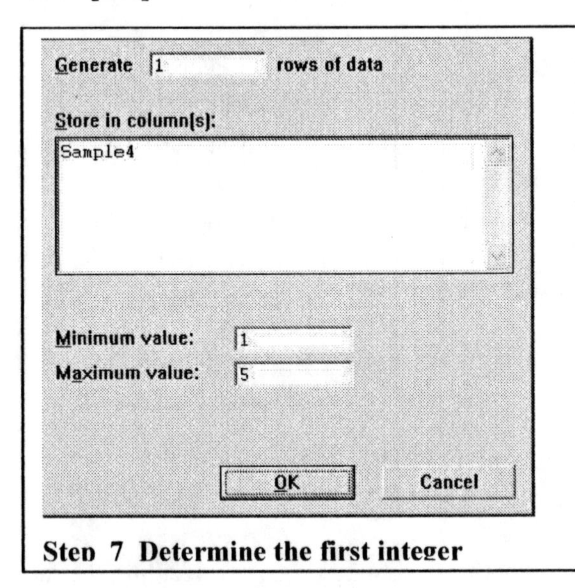

Step 7 Determine the first integer

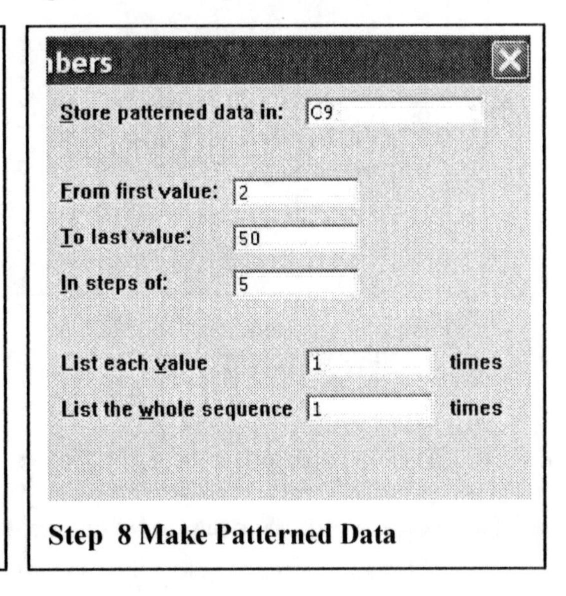

Step 8 Make Patterned Data

Create and store the list in Random4.

Step 8. Select **Calc>Make Patterned Data** and create the list beginning at this digit, 2 to 50 by increments of 5. The column contains the integers 2, 7, 12, 17, 22, 27, 32, 37, 42, 47.

This sequence of integers will be used in the next step. We could consult the database and find each state; however, there is a better way especially if the list is long. Create the sample. Here is how.

Step 9. Select **Data>Copy Columns to Columns.**

a. The cursor should be blinking in Copy from columns:. Double click C3 State.

b. Choose In current worksheet in columns from the drop-down list for Store Copied Data in Columns.

c. Click in the next box, then type **Sample4.**

d. Remove the check from the option to Name the columns.

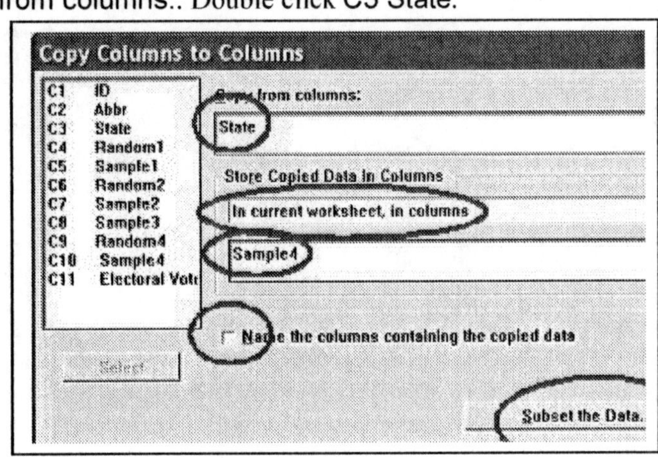

e. Click **[Subset the Data].**

f. Click the option for Row numbers: at the bottom of the dialog box and type in the sequence:

2, 7, 12, 17, 22, 27, 32, 37, 42, 47

g. Click **[OK]** twice.

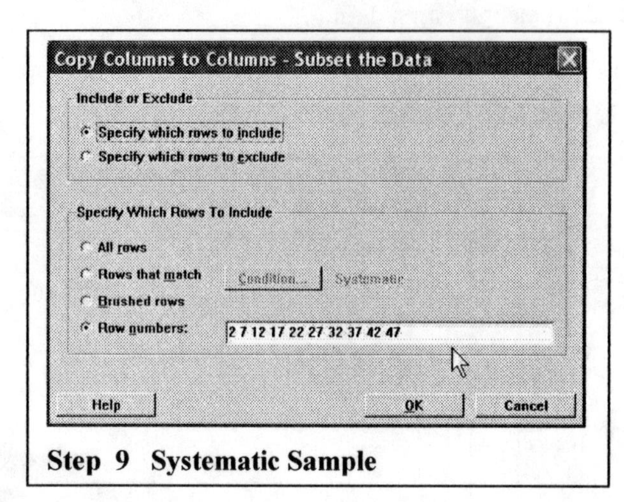

Step 9 Systematic Sample

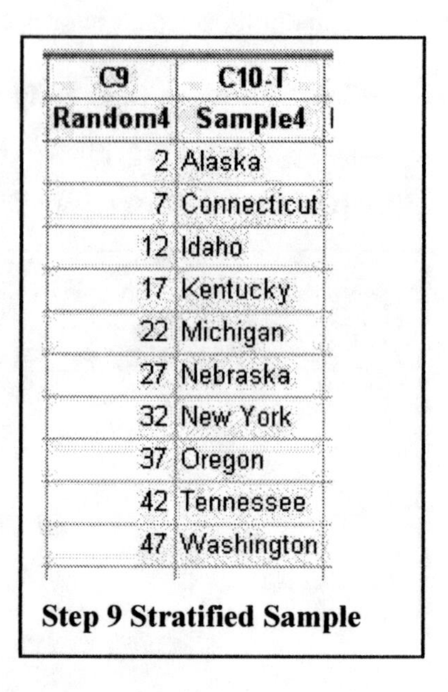

C9	C10-T
Random4	Sample4
2	Alaska
7	Connecticut
12	Idaho
17	Kentucky
22	Michigan
27	Nebraska
32	New York
37	Oregon
42	Tennessee
47	Washington

Step 9 Stratified Sample

Step 10. Click the disk icon to save the project as **Chapter14.MPJ.** Continue.

14-5 Simulation and Monte Carlo Techniques

MINITAB can be used to select the random numbers for a simulation. After deciding the distribution that will model the simulation, use an appropriate probability distribution to get random data.

Section 5 Example 6: A die is rolled until a 6 appears. Using simulation, find the average number of rolls needed. Try the experiment 20 times.
Plan: Random, uniform integers from 1 through 6 will model the roll of each die. In a worksheet, select 20 rows of 25. Each row will contain a trial of 25 dice rolls. Manually count how many trials were needed to obtain a 6. Summarize the result. Here goes!

Create a Sample of Random Digits to Simulate Rolling a Die

Set up a new worksheet. Make the column width smaller for viewing later.

Step 1. Select

Tools>Options >Data Window>General.

a. Change the Column width: to **4.**

b. There is an option here to turn off the warning about adding a new worksheet.

It is not checked.

c. Click **[OK].**

This step must be done BEFORE you open the new worksheet. It does not affect previously opened data. It stays in effect until you change it back!

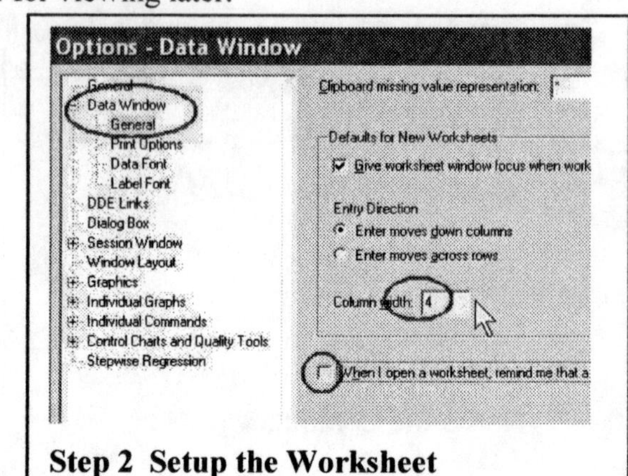

Step 2 Setup the Worksheet

Step 2. Select **File>New>Minitab Worksheet.**

Step 3. Select Calc>Random data>Integer. This is a uniform or "equally likely" distribution of integers.

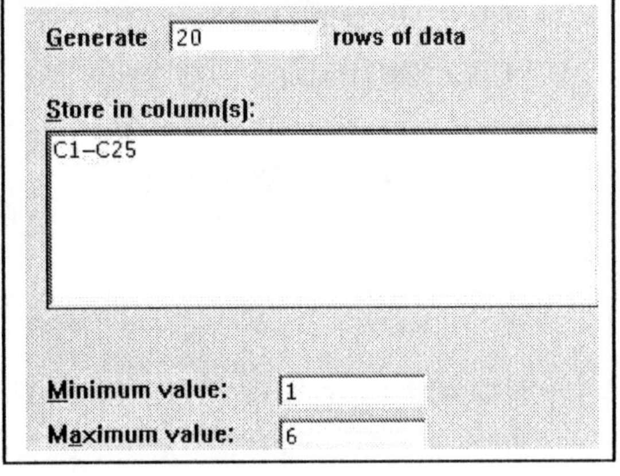

a. Type in **20** in Generate rows of data.

b. In the text box for Store in columns:, type **C1-C25.**

It is not likely to take more than 25 tosses to get a 6.

c. Type in the name of the column, **Die.**

d. From a Minimum value: of **1** to a Maximum value: of **6.**

e. Click **[OK].** Name C26 **n.**

The columns of data are generated. Your list will not be the same.

f. Click on the title bar of the worksheet then select **File>Print Setup.**

1. Change the orientation to landscape. Click **[OK].**
2. Select **File> Print Worksheet.**

Step 4. One row at a time, count how many trials it took to get a 6. Record this number in C26. Complete the count for your data. What will go in the next row for n? Hope you counted 8!

C1	C2	C3	C4	C5	C6	C7	C8	C9	C10	C11	C12	C13	C14	C15	C16	C17	C18	C19	C20	C21	C22	C23	C24	C25	C26
																									n
4	1	3	1	6	2	2	2	5	6	5	3	5	6	1	3	6	6	4	2	4	2	4	1	5	5
6	3	1	1	5	5	6	5	6	6	6	2	3	4	3	5	3	2	5	4	6	6	5	6	2	1
1	3	3	6	5	5	2	6	1	4	6	5	5	6	6	6	1	3	4	3	6	6	5	4	3	4
6	4	2	3	1	6	3	4	4	3	1	1	6	2	2	5	6	4	4	3	2	6	2	4	6	1
1	1	6	3	1	1	2	2	2	2	4	3	6	2	3	3	6	6	3	3	2	3	4	3	1	3
3	4	3	3	6	4	6	4	1	3	6	6	4	2	4	4	5	4	5	2	1	6	1	3	1	5
5	1	4	1	2	1	6	6	6	5	4	6	5	4	3	1	1	3	2	6	2	2	6	1	4	7
5	2	1	2	5	4	4	4	5	2	5	4	2	1	2	1	2	6	2	2	6	2	4	6	4	18
2	4	5	3	1	4	2	6	6	2	2	3	4	2	2	5	4	6	6	6	5	2	6	6	3	
4	3	5	1	5	4	5	5	4	4	2	5	1	6	3	5	1	2	6	2	2	1	1	5	1	
5	5	4	6	6	1	4	6	2	1	6	3	3	5	4	6	1	6	1	4	5	1	2	1	6	
6	4	1	6	1	4	3	6	4	5	2	1	2	1	2	2	3	3	6	6	2	3	6	5	4	
3	1	1	5	6	6	1	3	5	5	6	1	4	4	6	1	2	5	6	5	1	4	2	4	6	
5	5	6	5	3	6	1	2	6	6	1	2	5	4	6	6	5	5	6	1	1	3	4	5	1	
3	1	6	4	1	1	6	1	6	1	1	4	6	6	5	5	4	4	5	4	5	2	6	6	4	
4	4	3	3	2	5	5	3	3	2	4	2	6	5	6	4	4	6	5	4	4	1	6	4	3	
6	2	6	5	6	1	4	3	6	5	6	5	3	1	4	6	6	1	3	1	2	2	5	6	6	
5	5	6	2	3	3	3	4	2	6	6	2	3	1	2	3	4	6	6	6	6	1	5	6	2	
3	6	6	5	6	3	5	6	5	4	4	2	6	5	1	3	5	2	1	1	2	2	6	4	1	
1	6	6	6	5	3	4	6	1	4	4	5	2	5	4	3	6	6	4	1	5	1	4	6	5	

Summarize the results. Construct a frequency distribution for column 26 and calculate the descriptive statistics.

Step 5. Select **Stat>Tables>Tally Individual Variables.**

a. Select the C26 n for the Variable.

b. Check the options for Counts, Percents, and Cumulative percents.

c. Click **[OK].**

Tally for Discrete Variables: n

n	Count	Percent	CumPct
1	4	20.00	20.00
2	2	10.00	30.00
3	4	20.00	50.00
4	2	10.00	60.00
5	3	15.00	75.00
7	1	5.00	80.00
8	1	5.00	85.00
13	1	5.00	90.00
14	1	5.00	95.00
18	1	5.00	100.00
N=	20		

Sixty-five percent of the time it took 5 or less tosses to produce a 6. Occasionally, 10 percent of the time it took 7 or 8. There were 3 times out of 20 that it took 13or more rolls of the die to produce a 6. Eighteen was the highest.

Step 6. Select **Stat>Basic Statistics>Display Descriptive Statistics.**

- **a.** Select **C26 n** for the Variable.
- **b.** Click **[Graphs],** then select Boxplot.
- **c.** Click **[OK]** twice.

Descriptive Statistics: n

Variable	N*	Mean	StDev	Minimum	Q1	Median	Q3	Maximum
n	0	5.15	4.74	1.00	2.00	3.50	6.50	18.00

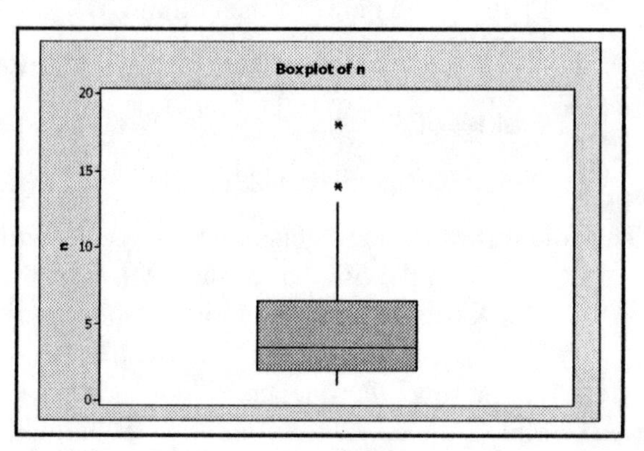

In the Session window the results for your table will be displayed similar to the summary for this data. In this sample, on average it took 5.15 trials before a 6 occurred. There were two outliers in this sample. It is not likely that 14 or 18 attempts would be needed to produce a 6.

Create a Sample of Bernoulli Outcomes

If the distribution needed has two possible outcomes, use a Bernoulli distribution. Bernoulli outcomes have two possibilities, success or failure. In MINITAB, Bernoulli outcomes will be a random sequence of zeros and ones. Zero is not a success. Specify the probability for getting a 1, a success.
Section 5 Exercise 21: A basketball player has a 60 percent success rate for shooting baskets. If she is awarded two free throws, find the probability that she will make one or both shots.

Step 1. Select **Calc>Random data>Bernoulli.**

- **a.** Type in **50** for Generate rows of data.
- **b.** Type **C1** and **C2** to Store in columns:.
- **c.** The probability of making each shot is 60 percent or .6.
- **d.** Enter .6 for the Probability of success:.
- **e.** Click **[OK].**

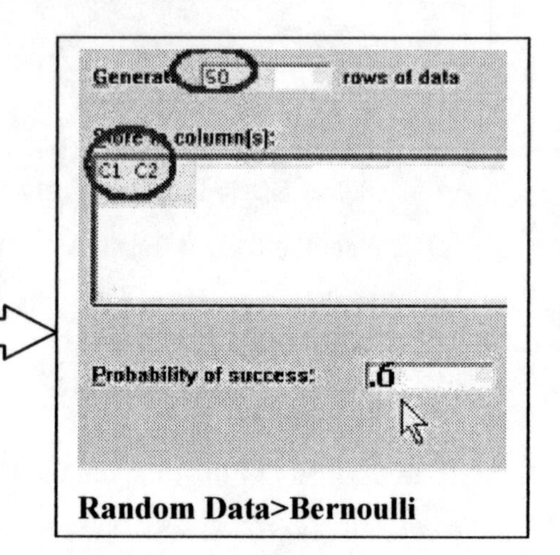

Random Data>Bernoulli

First add each row to get the number of baskets for the two tosses, a zero, one, or two are possible.

Step 2. Select **Calc>Row Statistics** then select C1 and C2 as the variables, and type **SUM** for Storage. Click **[OK].** The dialog box and the first nine rows of data are shown.

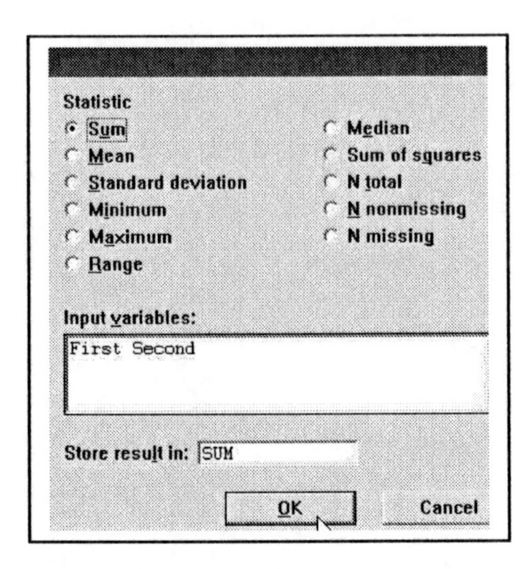

↓	C1	C2	C3
	First	Second	SUM
1	1	0	1
2	1	0	1
3	0	1	1
4	1	0	1
5	1	0	1
6	0	1	1
7	1	0	1
8	1	0	1
9	0	1	1

Summarize the data.

Step 3. Select **Stat>Tables>Tally Individual Variables.**

a. Double click C3 SUM to choose it for the variable.

b. Check the options for Counts, Percents and Cumulative percents.

c. Click **[OK].**

Tally for Discrete Variables: SUM

SUM	Count	Percent	CumPct
0	11	22.00	22.00
1	25	50.00	72.00
2	14	28.00	100.00
N=	50		

The probability of making both is 28 percent. About 50 percent of the time she will make one of the two. The probability of not making either is 22 percent. This is from the empirical probability of the simulation. The classical probability (using the multiplication rule of probability from in Chapter 4) of making both is .6*.6 = .36 or 36 percent.

Change the column width back to the default, **8.**

Step 4. Select **Tools>Options >Data Window>General.** Change the 4 back to an **8.**

Step 5. Save the project as **Chapter 14.MPJ .**

Step 6. Close MINITAB.

Chapter 14 Textbook Problems Suitable for MINITAB

Page		Section	Exercises
Page	681	14.2	11 - 19
Page	687	14.3	none
Page	693	14.4	8 – 21
Page	695	Review	1 - 18
Page	696	Data Analysis	1, 2, 4-5
Page	698	Data Project	1, 2

Chapter 14: Endnotes

Congratulations!

You have completed an introduction to MINITAB for Elementary Statistics.

APPENDIX: MACRO listings

Save each file as a text file.

FreqDist.mtb

```
Note   FreqDist.mtb
Note   A program to calculate the mean, variance and standards
Note   Deviation from a frequency distribution.
Note   Enter the midpoints in the first column and
Note   frequencies in the second column before running this macro.
Note   The columns will be named by MINITAB

Name C1 'X'
Name C2 'f'
Name C3 'fX'
Let 'fX' = 'f' * 'X'
Name C4 'fX2'
Let 'fX2' = 'f' * 'X'**2
Name K1 "n"
Sum 'f' 'n'.
Name K2 "SumX"
Sum 'fX' 'SumX'.
Name K3 "SumX2"
Sum 'fX2' 'SumX2'.
Let K4 = 'SumX' / 'n'
Let k5 = ('SumX2'-('SumX' **2/ 'n'))/('n'-1)
Let k6 = SQRT(K5)
Name K4 'Mean'
Name K5  'Variance'
Name K6 's'
Print K1-K6 C1-C4
```

WtMean.mtb

```
Note    WtMean.mtb
Note    This program will calculate the mean variance and standard
Note    deviation for a weighted mean or for a frequency distribution
Note    that is a population rather than a sample.
Name C1 'X'
Name C2 'f'
Name C3 'fX'
Let 'fX' = 'f' * 'X'
Name C4 'fX2'
Let 'fX2' = 'f' * 'X'**2
Name K1 "n"
Sum 'f' 'n'.
Name K2 "SumX"
Sum 'fX' 'SumX'.
Name K3 "SumX2"
Sum 'fX2' 'SumX2'.
Let K4 = 'SumX' / 'n'
Let k5 = ('SumX2'-('SumX' **2/ 'n'))/('n')
Let k6 = SQRT(K5)
Name K4 'Mean'
Name K5  'Variance'
Name K6 's'
```

```
Print  K1-K6 C1-C4
```

ChiFit.mtb

```
Note   ChiFit.mtb
Note   This program will calculate the chi-square statistic
Note   for goodness of fit.
Note   C1 = Labels
Note   C2 = Observed
Note   C3 = Px
Note   C4 = Expected
Note   To be valid,  all expected counts should be at least 5.
Let K1=SUM('Observed')/N('Observed')
Let K2=SUM(('Observed'-'Expected')**2/'Expected')
Let k5=Count('Observed')-1
Name k5 'df'
CDF k2 k3;
  ChiSquare K5.
Let K4=1-k3
Name k4 'Pvalue'
Print C1-C4 K1-K5
```

INDEX

G

H

I

J

K

L

M

T

U

V

W

Z